Zoo Studies

Living Collections, Their Animals and Visitors

Zoos and aquariums are culturally and historically important places where families enjoy their leisure time and scientists study exotic animals. Many contain buildings of great architectural merit. Some people consider zoos little more than animal prisons, while others believe they play an important role in conservation and education. Zoos have been the subject of a vast number of academic studies, whose results are scattered throughout the literature. This interdisciplinary volume brings together research on animal behaviour, visitor studies, zoo history, human–animal relationships, veterinary medicine, welfare, education, enclosure design, reproduction, legislation and zoo management conducted at around 200 institutions located throughout the world. The book is neither 'pro-' nor 'anti-' zoo and attempts to strike a balance between praising zoos for the good work they have done in the conservation of some species, while recognising that they face many challenges in making themselves relevant in the modern world.

Paul A. Rees was formerly a senior lecturer at the University of Salford, UK, where he taught zoo biology and wildlife law. He has taught a wide range of subjects, including biology, psychology and environmental science in colleges in the UK and Nigeria. He has a long-standing interest in large mammals, especially the ecology, behaviour and conservation of elephants and felids. His previous books include *An Introduction to Zoo Biology and Management* (2011), *Dictionary of Zoo Biology and Animal Management* (2013), *Studying Captive Animals* (2015), and *Elephants under Human Care* (2021).

Zoo Studies

Living Collections, Their Animals and Visitors

PAUL A. REES

CAMBRIDGE
UNIVERSITY PRESS

CAMBRIDGE
UNIVERSITY PRESS

Shaftesbury Road, Cambridge CB2 8EA, United Kingdom

One Liberty Plaza, 20th Floor, New York, NY 10006, USA

477 Williamstown Road, Port Melbourne, VIC 3207, Australia

314–321, 3rd Floor, Plot 3, Splendor Forum, Jasola District Centre, New Delhi – 110025, India

103 Penang Road, #05–06/07, Visioncrest Commercial, Singapore 238467

Cambridge University Press is part of Cambridge University Press & Assessment,
a department of the University of Cambridge.

We share the University's mission to contribute to society through the pursuit of
education, learning and research at the highest international levels of excellence.

www.cambridge.org
Information on this title: www.cambridge.org/9781108475068

DOI: 10.1017/9781108566049

First published 2023

Printed in the United Kingdom by TJ Books Limited, Padstow Cornwall

A catalogue record for this publication is available from the British Library.

Library of Congress Cataloging-in-Publication Data
Names: Rees, Paul A., author.
Title: Zoo studies : living collections, their animals and visitors / Paul A. Rees, University of Salford.
Description: Cambridge, United Kingdom ; New York : Cambridge University Press, [2023] | Includes
   bibliographical references and index.
Identifiers: LCCN 2022055775 (print) | LCCN 2022055776 (ebook) | ISBN 9781108475068 (hardback) |
   ISBN 9781108468725 (paperback) | ISBN 9781108566049 (epub)
Subjects: LCSH: Zoos. | Zoos–Educational aspects. | Zoos–Management. | Zoo animals. | Zoo visitors.
Classification: LCC QL76 .R445 2023 (print) | LCC QL76 (ebook) | DDC 590.73–dc23/eng/20230123
LC record available at https://lccn.loc.gov/2022055775
LC ebook record available at https://lccn.loc.gov/2022055776

ISBN 978-1-108-47506-8 Hardback
ISBN 978-1-108-46872-5 Paperback

For Katy, Clara and Elliot

Contents

Colour plates can be found between pages 238 and 239.

Preface

In this book I have attempted to discuss the various types of research done in, and on, zoos – hence the title *Zoo Studies* rather than *Zoo Biology* – based largely on the peer-reviewed academic literature. I have tried to cover as many different disciplines as possible, including animal behaviour, visitor studies, zoo history, human–animal relationships, zoo education, veterinary medicine, animal welfare, enclosure design, animal reproduction, zoo legislation, zoo management and many more.

This is not a book about the latest developments in zoo biology or, indeed, the latest developments in any particular area of zoo research. Although I have included many studies that have only recently been published, I have also referred to some very old studies, covering a period of around 200 years. I cannot claim that this is a comprehensive account of zoo-related research. However, I have tried to make it wide-ranging in the topics and zoological taxa it covers and in the geographical spread of the examples used.

In places I have drawn conclusions about the claims zoos make for their value to society and particularly their role in conservation. I have tried to be fair and to recognise the good work that many zoos do, and to 'call out' unfounded claims made with little or no evidence. The reader will have to decide whether or not I have struck a reasonable balance.

When I began the task of writing this book I thought I had a good grasp of the academic literature on zoos. I was wrong. I have a much better grasp of this now that the book is finished. I hope you enjoy reading and using it as much as I have enjoyed writing it.

Acknowledgements

I would like to express my gratitude to a number of people who have helped to make this book possible.

Without Dominic Lewis (formerly Senior Commissioning Editor for Life Sciences, Cambridge University Press) this book would not exist and I shall forever be grateful for his efforts and enthusiastic support during the commissioning process. Aleksandra Serocka (Editor, Life Sciences) has patiently guided me towards submission and very kindly accepted all of my excuses for not finishing the book sooner without question. I would also like to thank Jenny van der Meijden (Senior Content Manager) at Cambridge University Press, and Franklin Mathews Jebaraj (Senior Project Manager) of Straive Ltd., who efficiently managed the production of the book. This work was expertly copy-edited by Gary Smith, and I cannot thank him enough for his sterling efforts. Any errors that remain are entirely my responsibility.

Rachael Gerrie (Monitoring and Evaluation Officer) and Dr Tim Wright (Conservation Training Manager) generously provided me with the information about the research outputs of the Durrell Wildlife Conservation Trust used in Chapter 12.

Most of the figures have been produced from my own photographs, with the following exceptions. My daughter, Clara Clark, kindly provided the images used in Fig. 3.4. Stephen Fritz provided me with photographs of his elephant transport operations based in the United States that appear as Figs. 9.13b and 10.13b. Fig. 7.7 shows apparatus photographed by Kathryn Page, used by Helen Chambers (University of Salford) in a project on bear cognition. I am grateful to Kathryn for allowing me to reproduce this. Fig. 6.6 (Ota Benga) is reproduced courtesy of the United States Library of Congress, and the photograph of a quagga (Fig. 11.9) was made available by the Biodiversity Heritage Library (http://biodiversitylibrary.org/page/28201475/). My grandson, Elliot Clark, kindly assisted me with the typing of the index and I am most grateful for his help.

My friend Dr Alan Woodward drew Fig. 7.4, and this is reproduced here with permission from Elsevier, who also gave permission for the use of Fig. 8.6, an adapted version of Fig. 8.12 and Boxes 3.1 and 6.1, all of which were previously published in *Elephants under Human Care* (Rees, 2021). Fig. 9.3 was first published in Rees (2013) and is reproduced here with permission from Wiley-Blackwell.

Access to the booklet depicted in Fig. 2.4 was kindly given by Ian Trumble of Bolton Museum. Figs. 2.11 and 7.2 are images of material made available by Chetham's Library, Manchester. Hanif Gire kindly used his reprographic skills to assist in the preparation of Fig. 8.1 from material made available by the United States Patent Office (Serial No. 686,235). Prof. Robert Young (University of Salford) kindly drew my attention to several historical sources that were previously unknown to me.

Finally, I should like to confirm to my family that when they were talking to me about something other than this book, I really was listening.

Abbreviations

Sex ratios of animals living in zoos or listed in studbooks are commonly expressed as numbers using the following format:

1.4

males:females

or

2.7.1

males:females:unknown sex

AAZPA	American Association of Zoological Parks and Aquariums
ABM	animal-based measure (of welfare)
ACAP	Amphibian Conservation Action Plan
ACOPAZOA	Asociación Colombiana de Parques Zoológicos y Acuarios
AFT	Argyll Fisheries Trust
AI	artificial insemination
APHIS	Animal and Plant Health Inspection Service
ART	assisted reproductive technology
AZA	Association of Zoos and Aquariums
BIAZA	British and Irish Association of Zoos and Aquariums
BMI	body mass index
BSR	birth sex ratio
BTP	breeding and transfer plan
BUGS	Biodiversity Underpinning Global Survival*
CAZA	Canada's Accredited Zoos and Aquariums
CDC	Centers for Disease Control and Prevention
CEE	conventional environmental enrichment
CITES	Convention on International Trade in Endangered Species of Wild Fauna and Flora
CMR	cancer mortality risk
CZA	Central Zoo Authority
DEFRA	Department for the Environment, Food and Rural Affairs
DFTD	Devil facial tumour disease
EAAM	European Association for Aquatic Mammals
EAZA	European Association of Zoos and Aquaria
eDNA	environmental DNA

EEEV	eastern equine encephalitis virus
EEHV	elephant endotheliotropic herpesvirus
EEP	European Endangered Species Programme/EAZA Ex-situ Programme
ESA	Endangered Species Act
ESU	evolutionarily significant unit
EU	European Union
FAM	faecal androgen metabolite
FCM	faecal cortisol metabolite
FFI	Fauna & Flora International
FGM	faecal glucocorticoid metabolites
GAE	gross assimilation efficiency
GPS	Global Positioning System
GSMP	Global Species Management Plan
HEAT	Human-Elephant Alert Technologies
HR	home range
HVAC	heating, ventilation and air-conditioning
ICPA	Integrated Collection Planning and Assessment
IDA	*In Defense of Animals*
IR	immune reactive
ISB	International Studbook
ISIS	International Species Information System
IUCN	International Union for Conservation of Nature
IWRC	International Wildlife Rehabilitation Council
LAPS	Lexington Attachment to Pets Scale
MERS	Middle East respiratory syndrome
MPI	maintenance of proximity index
MVP	minimum viable population
NGO	non-governmental organisation
NWRA	National Wildlife Rehabilitators Association
PETA	People for the Ethical Treatment of Animals
PGC	Pennsylvania Game Commission
PIT	passive integrated transponder
PVA	population viability analysis
RCP	regional collection plan
RLI	Red List Index
RQI	relationship quality index
RS	rotating snake (illusion)
RSPB	Royal Society for the Protection of Birds
RZSS	Royal Zoological Society of Scotland
SARS	severe acute respiratory syndrome
SCNT	somatic cell nuclear transfer
s.d.	standard deviation
SEM	standard error of the mean

SMART	spatial monitoring and reporting tool
SNA	social network analysis
SSC	Species Survival Commission of the IUCN
SSP	Species Survival Plan®
TAG	Taxon Advisory Group
USDA	United States Department of Agriculture
USFWS	United States Fish and Wildlife Service
VORTEX	population viability analysis software
VR	virtual reality
WAZA	World Association of Zoos and Aquariums
WHO	World Health Organization
WildCRU	Wildlife Conservation Research Unit
WWF	Worldwide Fund for Nature/World Wildlife Fund
ZIMS	Zoological Management Information System
ZSI	Zoological Society of Ireland
ZSL	Zoological Society of London

1 Zoos and Research

1.1 Introduction

Zoo research has been something of a Cinderella subject until relatively recently. Most of the work that has been published on zoos has been biological in nature but few academics would describe themselves as zoo biologists.

A major problem for zoo researchers is that very often their work must be conducted in an ad hoc manner, in conditions that are beyond their control. Data are often 'snatched' in an opportunistic manner for all sorts of practical reasons: because of the weather, animal management and husbandry routines, changes to group composition, movements and deaths (or births) of animals, time constraints and a whole host of other events. Many studies of animals living in zoos have few subjects because zoos rarely keep large numbers of the same species (Table 1.1). However, there has been a trend towards multi-institutional studies in recent years. For example, Meehan et al. (2016) studied the welfare of elephants (*Loxodonta* and *Elephas*) in 68 North American zoos and Cronin et al. (2020) studied intragroup conflict among Japanese macaques (*Macaca fuscata*) in 10 zoos.

The purpose of this chapter is to review the studies that have attempted to examine the nature of, and trends in, research conducted on zoos and aquariums and the animals living in them.

1.2 The Advent of Dedicated Journals for Zoo Research

The Zoological Society of London (ZSL) has been publishing research since 1830, initially in the *Proceedings of the Zoological Society of London* and the *Transactions of the Zoological Society of London*, now the *Journal of Zoology* (Fig. 1.1). Many of the early papers published by ZSL reported the discovery of new species and matters of general zoological interest that did not relate directly to the Society's living collections. For example, Richard Owen published papers on the anatomy of the cheetah and giraffe, and the osteology of the orangutan (Owen, 1834; 1839a; 1839b) when he was Hunterian Professor in the Royal College of Surgeons, prior to his appointment as superintendent of the Natural History Department of the British Museum. Others papers, however, described animals received by the zoo from benefactors or collected by the zoo's staff on their many expeditions. In 1928, Joan

Table 1.1 Examples of studies conducted in zoos using five or fewer animals.

Authors	Study title	No. subjects
Elzanowski and Sergiel (2006)	Stereotypic behavior of a female Asiatic elephant (*Elephas maximus*) in a zoo	1
Gresswell and Goodman (2011)	Case study: training a chimpanzee (*Pan troglodytes*) to use a nebulizer to aid the treatment of airsacculitis	1
Zapico (1999)	First documentation of flehmen in a common hippopotamus (*Hippopotamus amphibius*)	1
Law and Tatner (1998)	Behaviour of a captive pair of clouded leopards (*Neofelis nebulosa*): introduction without injury	2
Xian et al. (2012)	Suckling behavior and its development in two Yangtze finless porpoise calves in captivity	2
Asa (2011)	Affiliative and aggressive behavior in a group of female Somali wild ass (*Equus africanus somalicus*)	3
Fischbacher and Schmid (2000)	Feeding enrichment and stereotypic behavior in spectacled bears	3
Franks et al. (2010)	The influence of feeding, enrichment, and seasonal context on the behavior of Pacific walruses (*Odobenus rosmarus divergens*)	4
Powell and Svoke (2008)	Novel environmental enrichment may provide a tool for rapid assessment of animal personality: a case study with giant pandas (*Ailuropoda melanoleuca*)	4
Dembiec et al. (2004)	The effects of transport stress on tiger physiology and behavior	5
Leighty et al. (2009)	GPS assessment of the use of exhibit space and resources by African elephants (*Loxodonta africana*)	5
Penfold et al. (2007)	Effect of progestins on serum hormones, semen production, and agonistic behavior in the gerenuk (*Litocranius walleri walleri*)	5

Procter, the Curator of Reptiles at ZSL, presented a paper on a Komodo dragon exhibited at one of its scientific meetings:

Procter, J. B. (1928). On a living Komodo Dragon *Varanus komodoensis* Ouwens, exhibited at the Scientific Meeting, October 23, 1928. *Proceedings of the Zoological Society of London*, 98, 1017–1019.

The first volume of the *International Zoo Yearbook* (IZYB) was published by the ZSL as a single bound volume in 1959 and thereafter more-or-less annually. This first volume contained, among other things, articles that focused on new developments in the keeping of great apes in captivity, along with accounts of new ape facilities at the zoos in Antwerp, Zurich, West Berlin, Tokyo and London. It also contained an eight-page 'International List of Animal Dealers'. This included one L. Gaillard of Buenos Aires, who specialised in supplying

Pumas, Jaguars, Maned Wolves, Wild Cats, Nandus and Wildfowl

and Heini Demmer of Vienna, who specialised in

Fig.1.1 The Zoological Society of London has been at the forefront of the publishing of zoological research since 1830. (A black and white version of this figure will appear in some formats. For the colour version, please refer to the plate section.)

Gorillas, Okapis, Pigmy Hippos, & other African Fauna; Kiang, Kulan, Siberian Tigers, Snow Leopards and various ruminants from Asia.

It is, of course, unthinkable now that an academic publication associated with zoos would publish such a list, although, at the time, it was perfectly legitimate for animal dealers to sell livestock to zoos, provided they complied with the relevant legislation in force at the time.

The IZYB contains reference sections on international studbooks for rare species and a list of the major zoos and aquariums of the world. In more recent volumes the main articles have focused on a particular theme, for example, reintroductions (Vol. 51, 2017), education (Vol. 50, 2016), reptiles (Vol. 49, 2015), freshwater fishes (Vol. 47, 2013) and bears and canids (Vol. 44, 2010).

The *Journal of Zoo and Wildlife Medicine* was originally published as the *Journal of Zoo Animal Medicine*. The first issue was published in 1970 by the American Association of Zoo Veterinarians, and included the following papers:

Krahwinkel, D. (1970). Primate anesthesiology. *The Journal of Zoo Animal Medicine*, 1(1), 4–9.

Fowler, M., and Gourley, I. (1970). Pyloric stenosis in a Bengal tiger (*Panthera tigris*). *The Journal of Zoo Animal Medicine*, 1(1), 12–16.

Russell, W., Herman, K. and Russell, W. (1970). Colibacillosis in captive wild animals. *The Journal of Zoo Animal Medicine*, 1(1), 17–21.

Fowler, M. and Mottram, W. (1970). Amputation of the tail in an Asian elephant. *The Journal of Zoo Animal Medicine*, 1(1), 22–25.

The first dedicated academic journal published as a periodical for general research on zoos and animals living in zoos was *Zoo Biology*. The first issue was published in 1982 and contained contributions from giants in their fields, including Frans de Waal, Hal Markowitz, Terry Maple (the founding editor) and Betsy Dresser. Along with colleagues, they reported on social behaviour in a chimpanzee (*Pan troglodytes*) colony, behavioural enrichment in Asian small-clawed river otters (*Aonyx cinereus*), computerised data collection and artificial insemination in Persian leopards (*Panthera pardus saxicolor*):

Maple, T. L. (1982). Toward a unified Zoo Biology. *Zoo Biology*, 1, 1–3.

Nieuwenhuijsen, K. and de Waal, F. B. M. (1982). Effects of spatial crowding on social behavior in a chimpanzee colony. *Zoo Biology*, 1, 5–28.

Foster-Turley, P. and Markowitz, H. (1982). A captive behavioral enrichment study with Asian small-clawed river otters (*Aonyx cinereus*). *Zoo Biology*, 1: 29–43.

Loskutoff, N. M., Ott, J. E. and Lasley, B. L. (1982). Urinary steroid evaluations to monitor ovarian function in exotic ungulates: I. Pregnanediol-3-glucuronide immunoreactivity in the okapi (*Okapia johnstoni*). *Zoo Biology*, 1, 45–53.

Dresser, B. L., Kramer, L., Reece, B. and Russell, P. T. (1982). Induction of ovulation and successful artificial insemination in a Persian leopard (*Panthera pardus saxicolor*). *Zoo Biology*, 1, 55–57.

Popp, J. W. (1982). Observations on the behavior of captive sitatunga (*Tragelaphus spekei*). *Zoo Biology*, 1, 59–63.

Gerth, J. M., Lewis, C. M., Stine, W. W. and Maple, T. L. (1982). Evaluation of two computerized data collection devices for research in zoos. *Zoo Biology*, 1, 65–70.

Recent interests in visitor studies, animal welfare and molecular biology as a conservation tool were reflected in the July/August 2017 issue of the journal, which carried articles about visitor engagement with a research demonstration on turtle cognition, the effects of visitors on ring-tailed lemur (*Lemur catta*) behaviour, measurement of stress in golden langurs (*Trachypithecus geei*) and the use of molecular tags in sexing birds of prey. The most recent issue of the journal (Vol. 41, March 2022) contains papers on the social organisation of Hamadryas baboons (*Papio hamadryas*), the attachment of zookeepers to the animals in their care, reproduction in giant pandas (*Ailuropoda melanoleuca*), population management in zoos and aquariums, nutrient analysis of ants used as food for pangolins (*Manis pentadactyla*), the dietary management of bears, coprophagy in gorillas (*Gorilla g. gorilla*) and the captive rearing of orphaned wild dogs (*Lycaon pictus*).

An attempt was made to establish a journal dedicated to aquarium research in 1997. Unfortunately, *Aquarium Sciences and Conservation* ceased publication in 2001 after just three volumes due to lack of contributors. Since 2013 two new open access journals dedicated to zoo research have been established. The first was an initiative of the European Association of Zoos and Aquaria (EAZA) – the *Journal of Zoo and Aquarium Research* – and its first issue was published in 2013. The second is the *Journal of Zoological and Botanical Gardens*, which was first published in

2020 and, to date, has published zoo-based studies almost exclusively, with very few papers of botanical interest.

1.3 What Do Zoo Researchers Research?

A number of papers have considered the nature of zoo research and how this has changed over time. Several of these have analysed the topics covered by articles in *Zoo Biology*. Care must be taken in interpreting these studies. It is undoubtedly the case that fewer zoo-based studies are published in *Zoo Biology* than are published elsewhere, and much of the emphasis of the journal is on work conducted in facilities accredited by the Association of Zoos and Aquariums (AZA). If dedicated zoo journals did not exist, zoo research would still be published in journals concerned with, for example, animal behaviour, reproduction and animal welfare, but analysing the topics studied by zoo researchers would be much more difficult. Furthermore, some zoo research is not about animals, and reports of this other work are dispersed among a very wide range of journals concerned with subjects such as visitor studies, cultural history, ethics and law. I am not aware of any review of zoo research that has attempted to gather together and analyse all of the outputs of those who have studied zoos in one form or another. What follows is a brief account of what we know about the nature of zoo research as seen through the eyes of zoo biologists.

1.3.1 Taxonomic Bias and Research Trends

In 1992 Devra Kleiman, working at the Department of Zoological Research at the Smithsonian's National Zoological Park in Washington, DC, reviewed the historical emphases on behavioural research in zoos and concluded that doing behavioural studies of excellence in zoos and aquariums had become more complicated than was previously the case (Kleiman, 1992). She attributed this to three factors. First, there had been significant changes in the aims and objectives of modern zoos, with an increasing focus on conservation. Second, there had been changes of focus in the science of animal behaviour. Third, there had been a tendency for trained animal behaviourists to take positions as curators and directors of zoos, leaving them no time to conduct research.

An analysis of the research subjects of 353 papers published in *Zoo Biology* between 1982 and 1992 found that 287 (81.3%) concerned non-human mammals (Hardy, 1996). Of these, 29.6% were studies of behaviour or behavioural ecology, a further 5.9% involved behavioural/environmental enrichment and 20.2% were studies of reproductive biology. Only 3.8% of all papers were concerned with genetics or population biology, and just 2.3% involved wildlife management. The remaining studies were concerned with nutrition and diet (3.5%), exhibit design and evaluation (1.2%), veterinary medicine (5.6%), captive management (24%) and morphology and development (5.6%).

An examination of 349 papers published in the same journal between 1996 and mid-2004 suggested a significant change in the direction of research carried out in zoos (Rees, 2005b). Reproductive studies replaced behaviour as the largest category (34%). This was followed by studies of nutrition, growth and development (19%) and behaviour and enrichment (17%). Only 2% of studies were concerned with ecology, field biology, conservation and reintroduction, but there was an increase in papers on taxonomy, genetics and population biology (10%).

An analysis of trends in the 395 research articles published in *Zoo Biology* in its first 15 years indicated a taxonomic bias towards mammals (73% of articles), with only 10% of articles on birds, 7% on reptiles and 7% on invertebrates (Wemmer et al., 1997). The predominant research areas were behaviour and reproduction, with, surprisingly, only a small number of papers on demography and genetics. Most papers were multi-authored and arose from institutions in the United States, with 26% of papers resulting from collaborative efforts between zoos and universities. The authors noted that almost one-third of papers were produced by the academic community and did not involve collaborations with zoos or aquariums. They concluded that there was a shortage of research-oriented staff in zoos and that the taxonomic bias discovered was in part the result of many authors publishing in taxon-specific journals.

Information from 991 articles published in *Zoo Biology* in its first 25 years (1982–2006) was evaluated by Anderson et al. (2008). They found that most articles were descriptive accounts that included inferential statistics and/or biological analyses, most concerned captive animals and they concentrated on mammalian behaviour and reproduction, especially in primates (35.5%). Carnivora were the second most popular subjects (23.4% of articles), followed by Artiodactyla (13.9%), Proboscidea (8.8%) and Perissodactyla (6.0%) (Fig. 1.2). The majority of first authors were affiliated with zoos or universities in the United States. Trends during this 25-year period indicated a significant increase in papers that were experimental and dealt with applied science, diet and nutrition, and a significant decrease in descriptive

Fig.1.2 A researcher studying white rhinoceros (*Ceratotherium simum*) in a British safari park.

papers, work concerning basic science, behaviour and population biology. The number of collaborative articles also increased with time.

A highly skewed distribution across 15 research subject categories was found in an analysis of 904 projects conducted in British and Irish zoos (Semple, 2002). Behavioural studies represented the largest category (40%), followed by studies of environmental enrichment (18%) and reproduction (8%). Studies of the genetics, ecology or conservation of a species were poorly represented, comprising just 5% of all projects.

The popularity of particular research areas is not necessarily reflected in the number of published studies, and discrepancies occur between the number of studies undertaken and the number published (Hardy, 1996). For example, behavioural and behavioural ecology studies made up 22.8% of 302 research projects carried out on mammals by zoo staff in 40 American zoos, but only 5.3% of studies published in the same period (Wiese et al., 1992). Only 19.5% of studies undertaken were concerned with reproductive physiology, but they accounted for almost 31% of all published studies. Natural history or fieldwork studies represented 23.1% of all published studies but only 16.6% of studies conducted.

Some taxa are very poorly represented in zoo research programmes. A survey of the research activity and conservation programmes of 52 North American zoo reptile and amphibian departments found that, of 164 technical papers produced between 1987 and 1997 by the 22 respondent institutions, 79% were conducted by just three institutions and only 16 field studies were reported (Card et al., 1998). Funding specifically for research was received by just one institution. The authors of the study concluded that zoo herpetology departments were not realising their potential for formalised research and conservation projects.

Lankard (2001) listed and categorised 957 publications produced in 1999–2000 by the member institutions of the AZA. Ecology/field conservation/reintroduction was the largest research category by far (27%). The second largest category was veterinary medicine/physiology (15%), followed by behaviour/ethology (9%). Studies of reproductive physiology/technology amounted to only 7% of the total, and nutrition accounted for just 3%. However, with regard to research output this is misleading because the documents examined encompassed a wide range of publications from status reports on individual taxa and recovery plans to papers on how to use a compass and how to make weather recordings. Publications ranged in quality from papers in peer-reviewed journals to technical handbooks, studbooks and items in newsletters.

In an investigation into the gaps in our knowledge of zoo animal management and welfare, Melfi (2009) identified three areas. First, research tended to focus on indicators of poor rather than good welfare. Second, animal husbandry and housing have been historically based on tradition rather than science. Third, a lack of species-specific baseline biological data for many species inhibits zoo research. Melfi noted that, at the time her work was published, studies of animal welfare in zoos had focused mostly on mammals, especially primates, large felids, bears and elephants.

The research contributions of institutions belonging to Canada's Accredited Zoos and Aquariums (CAZA) with respect to biodiversity conservation were analysed for the first time by Pyott and Schulte-Hostedde (2020). They found that CAZA members published most in the area of veterinary science, but there were publications in biodiversity conservation. The institution's age, research-oriented mission statements and financial assets were significant predictors of research productivity, but overall CAZA institutions published significantly less than members of the AZA based in the United States. This is not surprising as several AZA members have research institutes devoted to conservation.

Research output conducted between 2009 and 2018 has been systematically analysed by Rose et al. (2019). They examined 1,434 papers and the species holdings of zoos recorded in the IZYB (2009–2018). The authors concluded that zoo-themed research has been slowly filling research gaps for an increasing number of species. However, their data set confirmed the bias towards research on mammals, with Carnivora (154 papers) and Primates (294 papers) being represented in more papers than all those covering birds, reptiles, amphibians and fishes put together (204 papers). Rose et al. (2019) found just 17 papers published on zoo-applicable invertebrate research from 2011 to 2018, most of which were concerned broadly with welfare. During the time period examined, a steady increase in publications was detected only for birds. Rose et al. concluded that most publications lead to a specific advancement of knowledge, including the validation of research methodologies, and that zoo-themed research made a meaningful contribution to science. However, they noted that trends in species holdings were not reflected in publication trends.

Bajomi et al. (2010) examined 3,826 publications concerned with animal reintroductions and found that the literature was biased in favour of vertebrates (especially mammals and birds). They concluded that managers working with invertebrates and amphibians are less willing and/or less able to publish their work than those working with mammals and birds.

1.4   The Rise of Research as a Core Activity of Zoos

In Europe systematic scholarly work in menageries began in the mid-1600s and expanded during the Enlightenment (Baratay and Hardouin-Fugier, 2002). Early zoo research was focused on studies of anatomy, physiology and systematics.

By the end of the twentieth century there was evidence of an increased emphasis on research in American zoos. Stoinski et al. (1998) surveyed 173 North American zoos and aquariums and identified an increase in the role of research in AZA institutions in the previous decade, and a doubling of the number of researchers per institution since 1986. However, they found that the most common reasons for American zoos not conducting research were lack of funds, time and qualified personnel. More recently, the factors facilitating research in AZA-accredited zoos have been examined by analysing a questionnaire completed by 231 zoo professionals (Anderson et al., 2010). The majority of respondents conducted behavioural research on animals in a captive setting, held a curatorial position and had their salaries supported by the operating budget of their institution. Approximately 30% held doctorates, 19% held master's degrees, 34% held bachelor's degrees and 2% held other types of degree. The majority of respondents considered that they were part of a successful scientific programme and they judged that the two most important factors that contributed to this success were support from the chief executive officer and personnel dedicated to conducting scientific programmes.

A survey conducted by EAZA in 2005 found that only 25 (8.3%) of its 301 members had a research department and only about 33% had a research policy. Many EAZA members did not have a research budget and did not disseminate research findings in a publicly accessible format (Reid, 2007). EAZA subsequently launched a research strategy and action plan entitled *Developing the Research Potential of Zoos and Aquariums*.

In spite of this, most of the research conducted in EAZA zoos is produced by a small number of zoos. In the period 1998–2018 the 393 EAZA members contributed a total of 3,345 peer-reviewed papers to the scientific literature (Welden et al., 2020). During this time period more than two-thirds of EAZA members published, with the last decade of the period experiencing a three-fold increase. However, only seven institutions produced more than 100 papers each (representing 37% of the total). The top 10 publishing EAZA institutions produced a total of 1,458 papers: 43.6% of the total. The publication of this research led to an immediate response from staff at the ZSL, who claimed that their research had been under-represented by Welden et al. (2020) as much of the Society's work is published under the auspices of the Institute of Zoology, and individual staff are not associated with either of the ZSL's two zoos (Koldewey et al., 2020).

At least some of the recent interest in zoo research has been driven by changes to international and European Union law. Article 9 of the Convention on Biological Diversity 1992 requires parties – almost all of the countries in the world – to adopt measures for the *ex-situ* conservation of biodiversity. Regrettably the United States has chosen to remain outside the Convention.

Within the European Union, Article 9 has been implemented by the Zoos Directive (Council Directive 1999/22/EC of 29 March 1999), which requires zoos and aquariums to adopt a conservation role. One of the ways in which Member States may comply with the Directive is by undertaking research from which conservation benefits accrue. When I examined the nature of zoo research some five years after the adoption of the Directive in 1999 I concluded that most zoo research at that time had been concerned with animal behaviour, environmental enrichment, nutrition and reproduction, and was therefore largely irrelevant to *ex-situ* conservation (Rees, 2005b). I suggested that it was unlikely that zoos would increase their output of conservation-relevant research because most do not have adequate human or financial resources. The Directive did not make research a mandatory activity for zoos, it is merely one of the ways in which zoos may comply. They may also comply by engaging in training, information exchange or captive breeding. As most, if not all, zoos already engaged in at least one of these activities before the Directive was adopted, I argued that they could effectively comply by doing nothing new.

The leaders of modern progressive zoos want to engage with research. Many keepers are now well qualified at least in part because of the expansion of higher education courses aimed specifically at training them (see Section 3.14.1). In the United Kingdom, as the number of dedicated courses in universities has expanded there has, of necessity, been a concomitant expansion in specialist academic staff. More university staff interested in zoos has led to more postgraduate research, more collaboration with zoos, more publications and more dedicated journals.

1.5   The Grey Literature

Zoo professionals are in a unique position to gather data on many aspects of the husbandry of their animals. In many, if not most, cases their efforts result in the collection of data that are unsuitable for publication in academic journals because they are not collected in a systematic manner, the sample size is too small to yield statistically significant results, the information collected is anecdotal in nature, or for some other reason. This does not mean, however, that the information has no value, and many interesting articles have been published in the 'grey literature': documents published outside the normal academic channels. This may be within husbandry manuals or zoo reports, or in publications such as *International Zoo News*, *Ratel* (the journal of the Association of British and Irish Wild Animal Keepers (ABWAK)), *Connect* (the magazine of AZA) and *The Shape of Enrichment*. Articles in publications of this type are not peer-reviewed, but many are written by experienced keepers, zoo curators and directors, and others with knowledge of zoos and animals living in zoos, including academics.

The knowledge and expertise of zoo professionals should not be underestimated. It is not uncommon for those of us who study the behaviour of animals living in zoos to inform keepers of the results of our scientific studies only to be met with the response, 'Yes ... we know.' Indeed, if our scientific findings differed markedly from what experienced keepers know about their animals, we should perhaps question our methodology.

1.6   Zoo Research Is Not Just About Animals

Zoos have attracted interest from academics across a very wide range of disciplines and to the best of my knowledge there has been no attempt at a comprehensive survey of all of the zoo-related peer-reviewed work that has been published. This would be difficult because, although a small number of dedicated journals exist, most of which are affiliated to zoological organisations – *Zoo Biology* (AZA), *Journal of Zoo and Wildlife Medicine*, *Journal of Zoo and Aquarium Research* (EAZA), *International Zoo Yearbook* (ZSL) and *Journal of Zoological and Botanical Gardens* – a great deal of zoo research is published in journals that have a wider remit. My own work on zoos and animals living in zoos has appeared in *Zoo Biology*, the *Journal of Applied Animal*

Welfare Science, Applied Animal Behaviour Science and *International Zoo News*, and, less predictably, the *Journal of Zoology*, the *Journal of the Bombay Natural History Society*, the *African Journal of Ecology*, the *Journal of Thermal Biology*, *Oryx* and the *Journal of International Wildlife Law and Policy*.

1.7   Conclusion

Zoos feature in research from a very wide range of academic disciplines, from animal behaviour and veterinary science to history, sociology, law, ethics, architecture, visitor studies and tourism. The studies that have been made of zoo research have focused to a very large extent on the animals. One of the main purposes of this book is to introduce the reader to the many other aspects of zoos that have been studied, including those which have evaded detection by the reviews focused purely on animal biology.

2 Defining Zoos, Their Culture and Visitors

2.1 Introduction: What Is a Living Collection?

I have used the term 'living collection' in the title of this book, and this needs some explanation. A zoo, botanical garden, museum or art gallery often refers to the 'things' that it keeps as its 'collection'. In some cases the contents of this collection are inanimate objects; in others they are alive; and in some there are hybrid collections of living and non-living things.

Logically, a 'living collection' is a group of living things kept together for some specific purpose. Some organisations have used the term 'collection' in a restricted sense. The Royal Botanical Garden, Kew, defines a living collection as

a group of plants grown for a defined purpose, including for reference, research, conservation, education or ornamental display (Smith and Barley, 2019)

thereby excluding all animals and other living things. However, increasingly zoos are referring to their animals as living collections, and this is reflected in the titles of their staff: Head of Living Collections (Edinburgh Zoo); Curator of Living Collections (Twycross Zoo); Deputy Director of Living Collections (Oregon Zoo). Some zoos have living collections comprising both animals and plants.

The term 'zoo' is a contraction of 'zoological gardens', and its use has fallen out of favour in recent times due to its association with animals kept behind iron bars and in barren cages and enclosures. Some zoos have rebranded themselves as environmental parks (e.g. Paignton Zoo Environmental Park), Jersey Zoo has been renamed *Durrell* (after its founder, Gerald Durrell), Knowsley Safari Park (Merseyside, England) has become Knowsley Safari, and South Lakes Wild Animal Park (Cumbria, England) has become South Lakes Safari Zoo. Other institutions have retained the term 'zoo' in their title, for example, Chester Zoo, San Francisco Zoo, Edinburgh Zoo, Vienna Zoo (Tiergarten Schönbrunn) and Taronga Zoo (Sydney).

Broxbourne Zoo in Hertfordshire, England, was rebranded as Paradise Park and Woodland Zoo in 1986. The name was changed to Paradise Wildlife Park in the 1990s and then its owners announced in 2022 that it would become Hertfordshire Zoo from 2024 (Paradise Wildlife Park, 2022).

Defining the term 'zoo' is difficult because, although it has a general and reasonably well-understood meaning in general usage, its legal meaning varies between legal jurisdictions (Rees, 2011; 2017). In the European Union it is defined by the Zoos

Directive (Council Directive 1999/22/EC of 29 March 1999 relating to the keeping of wild animals in zoos), Article 2, as:

Definition

For the purpose of this Directive, 'zoos' means all permanent establishments where animals of wild species are kept for exhibition to the public for 7 or more days a year, with the exception of circuses, pet shops and establishments which Member States exempt from the requirements of this Directive on the grounds that they do not exhibit a significant number of animals or species to the public and that the exemption will not jeopardise the objectives of this Directive.

This definition encompasses traditional zoos, safari parks, birds of prey centres, aquariums, marine parks and similar institutions. I have used this broad definition to define living collections in this book, and, like the Directive, I have used the term 'zoo' as shorthand for such collections. In some cases, I have also referred to studies of other captive animal populations such as those held by research facilities because they contribute to our understanding of some aspects of the behaviour or physiology of animals living in zoos.

The range of facilities that have been included within the term 'zoo' have been discussed by Nekolný and Fialová (2018). They have highlighted the difference in the use of terms in different disciplines. For example, the terms 'zoological gardens' or 'zoological park' are often used in biological studies, whereas in tourism studies the terms 'animal-based attraction' or 'animal attraction' are used (Mason, 2007). Nekolný and Fialová have noted that small zoos should not be dismissed as members of the zoo community as even the largest and best zoos in the world began as small animal collections.

2.2 What Are Zoos For?

Modern zoos have struggled to communicate a clear purpose to the public, but this was not always the case. In his history of the environmental sciences, Bowler (1992) claims that the Zoological Society of London (ZSL) was founded (in 1826) to

act as a showcase for Britain's colonial possessions

and that,

The display of exotic animals in zoos was a public manifestation of the industrialized nations' ability to dominate the world.

The Society's founders included Sir Stamford Raffles (colonial administrator), the Marquis of Lansdowne (Member of Parliament), Lord Auckland (barrister and Governor-General of India), Sir Humphry Davy (a chemist and inventor), Sir Robert Peel (Prime Minister), Joseph Sabine (a lawyer, tax inspector and Arctic explorer) and Nicholas Aylward Vigors (a soldier, barrister and Member of Parliament). Although some of these men had political careers and an interest in the colonies, many of them had a deep interest in natural history, as had many of the other founders, and the

Society began publishing research within just a few years of its founding. The first female Fellow of the Society was elected in 1838 (Lady Raffles), and Sir Charles Darwin became a Fellow in 1837. The ZSL was granted a Royal Charter by King George IV in 1829 and the objects of the Society were described thus:

Whereas several of our loving subjects are desirous of forming a Society for the advancement of Zoology and Animal Physiology, and the introduction of new and curious subjects of the Animal Kingdom; (ZSL 1829)

The cultural role of London Zoo (now known as ZSL London Zoo) in Victorian society has been discussed by Jones (1997). He noted that when the zoo opened in 1829 access was restricted to Fellows of the Society and their friends, but by the mid-1830s 'the more respectable members of the middle classes were also granted access'. The emphasis in these early years was very much on science and research rather than providing a spectacle for the public, and the Society's aims were focused on furthering the science of comparative anatomy and introducing domesticated foreign animal breeds. The Society maintained a breeding facility at Kingston Farm for this latter purpose (Ito, 2014), and also a museum that began life in the Society's offices in central London and then moved to the old Carnivore House in 1843 so that it could be open to visitors (Mitchell, 1929). In 1855 the Society announced the museum was to be closed and its specimens were sold to the British Museum, colleges and other provincial museums.

The Society went on to focus its efforts on the public display of living exotic animals from foreign lands to the general public. Jones argues that there is a sense in which the Zoo was established in conscious opposition to the touring circuses and menageries that were touring the country during the Victorian era.

Over time the Society's purpose has evolved and in 2021 the ZSL stated its purpose as:

To inspire, inform and empower people to stop wild animals going extinct. (ZSL 2021a)

One of the more unusual historical papers published about zoos appeared in *The American Historical Review* and described the murder of an elephant mahout, Said Ali, at London Zoo in August 1928 in the context of the place of the zoo in the imperial history of the United Kingdom. The article also considered the other tragedies linked to the zoo's elephants that had been discussed in the press (Saha, 2016).

The Bristol and Clifton Zoological Society was established in 1835 to

promote the diffusion of useful knowledge by facilitating observation of the habits, form and structure of the animal kingdom, as well as affording rational amusement and recreation to the visitors of the neighbourhood. (Brown et al., 2011)

The Society opened its zoo in 1836: the Bristol and Clifton Zoological Gardens.

Acclimatisation societies were popular in the mid-nineteenth century as the colonial powers sought to introduce species from their homelands into their colonies to provide food and sport or simply for ornamental reasons. In some cases acclimatisation societies were associated with zoological societies, and as such had a hand in the creation of zoological gardens. For example, in Australia, the Zoological Society of Victoria was founded in 1857 and replaced by the Acclimatisation Society of Victoria in 1861. The Society managed the Melbourne Zoological Gardens, which was established in 1862. In 1872 the Society became the Zoological and Acclimatisation Society of Victoria, reflecting its broader interests.

A street survey of the general public ($n = 200$) determined that the main role of zoos was perceived to be conservation. However, the public's views of animals in zoos included the negative perceptions of them being bored and sad. Visitors to Edinburgh Zoo had a more positive perception of zoo animals, a greater awareness of environmental enrichment and were influenced by the visual messages received as they moved through the zoo (Reade and Waran, 1996).

Shchukina et al. (2019) studied the image of the St. Petersburg Oceanarium in the minds of the public and concluded that it is perceived as having multifunctional roles in leisure entertainment and science education. They also found that visits to the Oceanarium had aesthetic, relaxation, cognitive and sociocommunicative value.

Most zoos now claim to have a conservation function, and this is reflected in their mission statements. Those who work in or with zoos have been propounding their value in conservation for at least 55 years (Conway, 1967; Jarvis 1967; Schomberg, 1970), long before most zoos were claiming conservation as one of their major objectives.

2.3 Mission Statements

Many modern zoos and aquariums have ambitious mission statements that attempt to encapsulate their purpose, with some common themes but a transient existence, as they evolve over time (Table 2.1).

A study of the mission statements of 136 zoos in the United States accredited by the Association of Zoos and Aquariums (AZA) (Patrick et al., 2007) identified education and conservation as the predominant themes and concluded that

zoos are in a unique position to provide environmental education and conservation education to large numbers of people.

When the mission statements of zoos accredited by AZA from 2004 were compared with those of 2014 it was apparent that conservation education had become a prominent goal (Patrick and Caplow, 2018). The majority focused on 'inspiration' and 'connection' – difficult things to measure – and the authors of the study suggested that zoos should focus on skill-building and action-oriented language, which can be evaluated (see the mission statement of Zoo Atlanta in Table 2.1).

In an attempt to find a link between the mission statements and the strategic activities of institutions, data from 173 zoos and 38 aquariums accredited by AZA were analysed by Maynard et al. (2020). They concluded that mission statements emphasising conservation were not linked to more conservation practices; only the

Table 2.1 Selected zoo mission statements from 2021.

Zoo	Mission
Chester Zoo, England	Preventing extinction
Paignton Zoo, England	...to conserve species and their habitats, and to inspire and empower people to help in the fight to protect wildlife
Zoo Atlanta, United States	We save wildlife and their habitats through conservation, research, education and engaging experiences. Our efforts connect people to animals and inspire conservation action
St Louis Zoo, United States	To conserve animals and their habitats through animal management, research, recreation, and educational programs that encourage the support and enrich the experience of the public
Smithsonian's National Zoo & Conservation Biology Institute, United States	We save species by using cutting-edge science, sharing knowledge and providing inspirational experiences for our guests
Barcelona Zoo, Spain	...to contribute to the conservation of wildlife and biodiversity of this planet working in cooperation with other zoos, administrations, organizations and university and scientific centres
Royal Burgers' Zoo, Netherlands	...to protect and preserve (endangered) animal species and to let as many people as possible experience the wonders of nature. Together with nature, we want to inspire people, teach people, influence their behaviour where possible and allow them to enjoy everything that nature has to give.
Two Oceans Aquarium, South Africa	...to inspire action for the future well-being of our oceans

number of zoo personnel – a proxy for organisation size – predicted the amount of zoos' conservation grants to partner organisations.

Carr and Cohen (2015) examined the websites of 54 zoos spread throughout the world to assess the image that they portrayed to the general public. They concluded that, although they presented a strong conservation message, it lacked depth; they also placed a strong emphasis on entertainment. The authors suggested that zoos need to present their conservation credentials in more detail and ensure that these, and their education message, are not adversely affected by the entertainment message. They recognised, however, that attracting enough visitors is essential to ensure economic viability.

References to the educational role of zoos in their mission statements are explored further in Chapter 3.

2.4 Who Goes to the Zoo, and Why?

A number of studies have determined, unsurprisingly, that zoos provide important opportunities for families to spend their leisure time together (e.g. Turley, 2001;

Sickler and Fraser, 2009). A study of 359 visitors to Hamilton Zoo, New Zealand, concluded that zoos provided opportunities for family trips (Ryan and Saward, 2004), and a survey of visitors to three Malaysian zoos found that they visited mainly for recreation and perceived that animals were kept primarily to attract visitors. These visitors perceived the zoos as places for conservation, education, research and recreation, in this order of priority (Puan and Zakaria, 2007). However, a study of 84 American students aged 14–18 years attending the same American high school found that they did not mention conservation with respect to zoos unless specifically asked about it, did not mention the role of zoos in species survival and did not view zoos as a source of conservation information (Patrick, 2006).

Of the 3,052 visitors surveyed at Zagreb Zoo, Croatia, most came as a family group (72 per cent), and 72 per cent came with children. Others came as couples (17 per cent), friends (9 per cent) or alone (2 per cent)(Knežević et al., 2016). Holzer et al. (1998) collected data from 750 visitors to Cleveland Metroparks Zoo in the United States and found that the motivation for their visit was, in order of importance, family togetherness, enjoyment, novelty seeking, education and relaxation. Those visitors that put greater value on the educational benefits of a zoo visit said they had often visited zoos as a child. These people were also more likely to visit a variety of zoos and expressed greater commitment to zoos than those that had not often visited zoos as children.

Although the motivation for many people to visit zoos is for a family day out with children, this is not always the case. Most of the 707 weekend visitors who completed questionnaires at Attica Zoological Park, Athens, Greece were defined as middle-aged (although 78.6 per cent were between 18 and 40) and highly educated (47 per cent had a university education), and the motivation for their visits was entertainment (questionnaires, $n = 707$). However, almost 57 per cent visited without children (Karanikola et al., 2020). Knežević et al. (2016) reported that 38 per cent of visitors to Zagreb Zoo, Croatia, possessed graduate degrees, and 14 per cent possessed undergraduate degrees.

DeVault (2000) has argued that the concept of 'family' is something that is actively constructed during a family visit to a zoo. She has shown that parents coordinate the viewing behaviour of the family at animal exhibits, pointing out to children which particular features deserve their attention (Fig. 2.1). Much of this behaviour is performed in a public setting, making it visible to others outside the family, with the adults demonstrating that they are highlighting important things properly. As they travel around the zoo, parents move directly and slowly between the exhibits, while children sometimes race ahead and sometimes lag behind.

Clayton et al. (2011) found that viewing animals in a zoo was primarily a social activity and sometimes enabled discussion about a shared conception about the relationship between humans and animals. They have suggested that zoos enable the development of an environmental identity that encourages concern for animals. Their study surveyed and observed zoo visitors and found that environmental identity, sense of connection with animals and perceived similarity with animals were all correlated with general environmental concern and interest in conservation. They found no

Fig. 2.1 A visit to the zoo promotes family interactions, with parents coordinating the viewing of younger members. Western lowland gorillas (*Gorilla g. gorilla*). (A black and white version of this figure will appear in some formats. For the colour version, please refer to the plate section.)

significant difference between survey responses before and after visiting an exhibit. However, this varied between exhibits, and those that made more comparisons with humans tended to evoke higher ratings of support for helping animals.

Turley (2001) noted that zoos are perceived as a place to take children for recreation, and although this encourages family visits it also deters older visitors. This presents a challenge for zoos as the demographics of human populations change.

If educational messages from zoos are to be effective, it is important to determine whether or not particular groups are excluded from visiting zoos for economic or other reasons. West and Virgene (1990) found that Detroit minorities (defined as non-white) were not under-represented in visitation rates of Detroit residents to Detroit Zoo in 1986, despite lower car ownership rates among non-whites. They found that adults between the ages of 25 and 34 years were the age group most likely to visit the zoo – that is, the age group most likely to have young children. The zoo attracted 70 per cent of both white and minority Detroit families with children under 16 years of age. Although financial factors mitigated against non-whites visiting the zoo, this was compensated by a greater proportion of minority families with children being drawn to the zoo for family outings.

2.5   How Often Do Visitors Visit?

It is impossible to determine accurately the number of visitors that attend the world's zoos and aquariums each year. Any data that exist are likely to be comprised of annual attendance figures from individual zoos, where it is not possible to determine if an attendance of 100 relates to 100 different individuals or 50 individuals who each visited the zoo twice. Some zoos, at least in the past, were only able to estimate visitor numbers. For example, most visitors to Knowsley Safari Park drive around the park in their own cars. In the past, visitors paid per car rather than per person, and as a result the park attendance data were estimated and sometimes reported as an identical number in consecutive years in the *International Zoo Yearbook* (IZYB)!

Almost one-third of visitors to Attica Zoological Park in Athens visited at least once a year, 8.5 per cent at least once a month and 1.3 per cent once a week (Karanikola et al., 2020). In south-west Nigeria, 78.3 per cent of visitors to the University of Ibadan and Obafemi Awolowo University Zoos were repeat visitors (Adetol and Adedire, 2018). First-time visitors to the Bronx, Central Park and Prospect Park Zoos in New York represented 29, 45 and 38 per cent of all visitors respectively (Bruni et al., 2008), and on average 7 per cent of visitors to these zoos visited more than 10 times per year. Knežević et al. (2016) reported that 88 per cent of visitors to Zagreb Zoo, Croatia, were repeat visitors, and only 12 per cent were first-time visitors.

2.6   Visitor Preferences: Which Species Do Visitors Want to See?

When it comes to deciding which species to keep, zoos have a dilemma: many popular species are common and relatively large while many rare species are small and rather unattractive (or at least not well known). Zoo professionals refer to the 'ABC animals': A for aardvark, B for bear, Z for zebra and so on. Children want to see elephants and rhinoceroses that are expensive to keep and take up a lot of space; many conservationists want to save invertebrates, amphibians and rare fishes. Turley (1999) noted that traditional zoos must balance the demands of the paying visitor with those of operating a credible conservation and education-oriented organisation.

A survey of the general public on the Island of Jersey (Channel Islands) asked what animals they would like to see in a zoo (Carr, 2016a). Large mammals dominated the responses and desirable animal traits included whether they were endangered, active and displayed intelligence. It is perhaps worth noting that the animal collection at Jersey Zoo (*Durrell*) focuses on endangered species but has relatively few large mammals. The same author surveyed 444 visitors to *Durrell* immediately after their visit (Carr, 2016b). In common with the previous study, mammals were identified as the favourite animals. Birds and reptiles were the least favourites. Visitors favoured cute and entertaining animals over those that they perceived as boring, hard to see, scary or smelly.

A study of the mammalian species kept by Prague Zoo, in the Czech Republic, found that the most popular animals were carnivores and ungulates, and the least popular were smaller rodents and afrosoricids (golden moles, otter shrews and tenrecs). The main factors that led visitors to rank species as beautiful were complex fur patterns and body shapes, and animal beauty was associated with the respondents' willingness to protect the species (Landová et al., 2018).

The hypothesis that the perceived beauty of parrot species affects the size of zoo populations has been tested by asking respondents to evaluate pictures of different species (Frynta et al., 2010). The respondents preferred species that were large, long-tailed and colourful, and the study found positive associations between perceived beauty and the size of worldwide zoo populations. In contrast, the species' IUCN (International Union for Conservation of Nature) listings appeared insignificant. The authors of the study concluded that zoos appear to keep preferentially beautiful parrots and pay less attention to conservation status.

According to the results of an international study of 458 zoos, the proportion of threatened species kept by a zoo is not an important factor driving attendance, and visitors are more interested in seeing large numbers of large species, particularly mammals (Mooney et al., 2020). The study also found that keeping unusual animals was associated with greater attendance and that keeping many small animals may also be an effective strategy for attracting visitors. Visitors to three Malaysian zoos preferred to see attractive and active animals rather than threatened and healthy animals (Puan and Zakaria, 2007).

Today's children are tomorrow's adults, and their attitudes to conservation are heavily influenced by the media. Interest in local biodiversity is important to its conservation, but there is evidence that children are largely interested in exotic species. Ballouard et al. (2011) studied a sample of 7–11-year-old children from 10 schools in France to determine their level of knowledge of animals and the species they were most likely to protect. They did this by comparing the data collected from a questionnaire survey with an internet content analysis of the children's keyword searches in Google. They found that the children's knowledge and concern for the protection of animals was mainly limited to material found on the Internet about a small number of exotic and charismatic species, such as giant pandas and polar bears. The children's identification rate of local animals was poor and they were more likely to protect unseen exotic animal species than local species. Ballouard et al. reported similar results from studies of children in Italy, Serbia, Slovakia, Spain, Portugal, Morocco, Nepal and Turkey. Regardless of the country, children referred to a few iconic mammals. They suggested that this was the result of a strong uniform media influence and that children were disconnected from local biodiversity.

A questionnaire survey of 562 children and adolescents aged 9–15 years in rural and urban areas of southern Norway found that dogs, cats, horses and rabbits were their favourite 'species' and that they least liked worms, spiders, crows and bees (Bjerke et al., 1998). Girls were more pet-oriented while boys preferred wild animals. Urban respondents liked animals more than did rural respondents, especially large carnivores. Interest in wildlife decreased with age and few respondents expressed a desire to save ecologically important species such as ants, bees and ladybirds from extinction. In their study of 1,297 Slovakian children aged 10–15 years, Prokop and Tunnicliffe (2010) found that boys were more favourably inclined towards animals that might pose a threat than were girls, and that having pets at home was associated with more positive attitudes to, and more knowledge of, animals.

There is some evidence of cultural differences in the preferences of individuals for particular taxa. When Arab and Jewish visitors to the Tisch Family Zoological Gardens in Jerusalem were asked to choose three animals they had seen in the zoo and describe the message they were intended to convey they recorded cultural differences in their choices (Tishler et al., 2020). Apes, elephants and penguins were chosen more frequently by Jewish participants, while grazers, reptiles and animals in the children's zoo were selected by Arab participants. There were no significant differences in the frequencies with which large predators (Felidae), raptors, 'Israeli fauna', birds and 'Australian animals' were selected by members of the two communities.

Lions, tigers, bears and other large carnivores are always popular at the zoo. Humans appear to be predisposed to pay more attention to the presence of dangerous predatory species than non-predators. Yorzinski et al. (2014) recorded the eye movements of men and women as they detected images of dangerous animals (lions and snakes) among arrays of non-dangerous animals and as they detected non-dangerous species among an array of dangerous species. They found that participants were quicker to locate targets when they were dangerous animals than when they were non-dangerous animals. When the targets were non-dangerous, participants spent more time looking at the dangerous non-target animals and looked at a larger number of these dangerous distractors. Yorzinski et al. suggested that dangerous animals capture and maintain the attention of humans because historical predation has shaped some aspects of visual orienting and its neural architecture.

A study of the effects of keeping animals as pets on children's concepts of vertebrates and invertebrates showed very strong bias towards rearing vertebrates and a general ignorance of invertebrates (Prokop et al., 2008). The authors suggested that science activities with animals should focus more on rearing invertebrates to improve children's knowledge and attitudes towards them. It may be that, despite zoos' efforts to interest visitors in rare but small and little-known species, we are all predisposed to pay more attention to large and dangerous animals as a result of our evolutionary history and childhood experiences.

Invertebrates are not especially popular with zoo visitors, but some zoos have invested in expensive exhibits for invertebrate taxa. A study of London Zoo's BUGS (*Biodiversity Underpinning Global Survival*; Fig. 2.2) exhibit – which houses a range of small invertebrates – determined that it both entertains and fascinates visitors, thereby enabling them to value nature (Chalmin-Pui and Perkins, 2017). Nevertheless, the authors concluded that BUGS fell short of its experiential potential because it neither resonated with visitors' everyday lives nor did it enable them to contribute personally to conservation.

Hoff and Maple (1982) studied the avoidance of reptile exhibits in zoos and found that, at most ages, females more often refused to enter such exhibits than did males. If they did enter, females spent less time in the exhibits than males. Teenagers were the age group least likely to refuse to enter reptile exhibits and spent more time in these exhibits than visitors of other ages. Myers et al. (2004) measured the emotional response of zoo visitors to three animal taxa (snake, okapi and gorilla) and found that

Fig. 2.2 The BUGS (*Biodiversity Underpinning Global Survival*) exhibit at London Zoo houses a range of small invertebrates.

the most important variables influencing emotions were the kind of animal, the subject's emotionality and the subject's relation to the animal, but not demographic variables.

Mäekivi and Maran (2016) have argued that attitudes towards different animal species are affected by people's psychological dispositions, the biosemiotic conditions – communication and signification (the conveying of meaning) in living things – cultural connotations and symbolic meaning. They suggest that which animals are selected for display in zoos and the manner in which they are presented and interpreted to the public are affected by biosemiotics and culture. Belle Vue Zoological Gardens in Manchester, England opened a Monkey House in 1881 which was constructed in the style of an Indian temple. The giraffe enclosure at Detroit Zoo has a backdrop suggestive of past links of the species to the Ancient Egyptian civilisation (Fig. 2.3). Mäekivi and Maran claim that:

These semiotic dispositions are further used as motifs in staging, personifying or de-personifying animals in order to modify visitors' perceptions and attitudes.

The authors suggest that, on the one hand, zoos personify individual animals to encourage the public to care about them as individuals, and on the other hand they de-personify them in order to raise welfaristic attitudes to the species as a whole. As a result, zoos may simultaneously propagate welfaristic and conservational attitudes

Fig. 2.3 The giraffe enclosure at Detroit Zoo in Michigan (2011). Instead of mimicking a modern-day African savannah, this exhibit makes a link between giraffes and the culture of Ancient Egypt. (A black and white version of this figure will appear in some formats. For the colour version, please refer to the plate section.)

among their visitors. They note that this is easier for some species than for others. The same species may be revered by some cultures but vilified by others.

2.7 The Psychological Benefits of Zoo Visits

There has been increasing concern in recent years about the alienation of people, especially children, from nature by virtue of a lack of experience of outdoor activities due to increased time spent using computers, urban lifestyles and concerns about safety. This appears to be associated with an increase in attention deficit disorder, obesity and depression in young people. Although it is not recognised as a psychological disorder, Louv (2005) coined the term 'nature deficit disorder' for this phenomenon.

Many people feel drawn to wild places due to what the ecologist Edward Wilson called 'biophilia':

The innately emotional affiliation of human beings to other living organisms. (Wilson, 1993)

Urban populations have limited opportunities to interact with nature and zoos offer exposure to exotic species that few people can afford to experience in the wild. Zoos are important places where people can connect with nature, and there is some evidence that a zoo visit may be good for human health. Sakagami and Ohta (2010) studied 233 visitors to two zoos in Japan. They assessed their psychological state using a questionnaire, their blood pressure and pulse rate, and the distance walked during their visit (using pedometers). They found that both zoo visits caused a decrease in blood pressure in participants, they walked more than 6,000 steps during their visit and their quality of life sub-scale scores were improved after the visit. They concluded that zoo visits had the potential to improve both health and the quality of life.

A visit to a zoo may, however, have little more health benefit than a visit to a park. The health benefits to older people of visiting a park and visiting a zoo were compared by Akiyama et al. (2021). They found that the visitors' cortisol and/or blood pressure fell after visiting the zoo and the park, and mood improved during the zoo visit. The authors concluded that visiting the park had almost the same health benefits as visiting the zoo. In a second study using different volunteers, salivary oxytocin increased and cortisol decreased during zoo visits, and after the visits oxytocin levels were lower for park visits than for zoo visits.

2.8   Zoos, Culture and Social Identity

It is difficult to imagine how much interest there would have been in a small collection of exotic animals in a world where glossy magazines, television and the Internet did not exist, and where most people travelled just a few miles from their place of birth during their lives.

In Victorian England the enthusiasm for curiosities of all kinds led to the exhibition of all manner of exotic animals in static and mobile displays, often alongside stuffed animals and other items such as waxworks and automata. Sharples's Museum of Curiosities in Nature and Art was opened in 1835 in rooms above the Star Inn on Churchgate in Bolton, England. According to the museum's catalogue, the live specimens included monkeys, golden eagles, exotic parrots and cockatoos (Fig. 2.4).

Travelling menageries were common in England in Victorian times. Queen Victoria herself was visited at Windsor Castle in 1847 by Wombwell's Menagerie (*The Illustrated London News*, 1847a) (Fig. 2.5). *The Illustrated London News* reported that:

In the course of Wednesday ... the servants of the Menagerie set about washing and cleaning the animals, and their dens and cages, so as to introduce the collection to Royalty in as presentable a manner as the short interval would allow.

In addition to travelling menageries, other animal shows existed throughout the nineteenth century. Staged scenarios of confrontations between humans and animals occurred, ranging from lion acts to large-scale re-enactments of war using cannons,

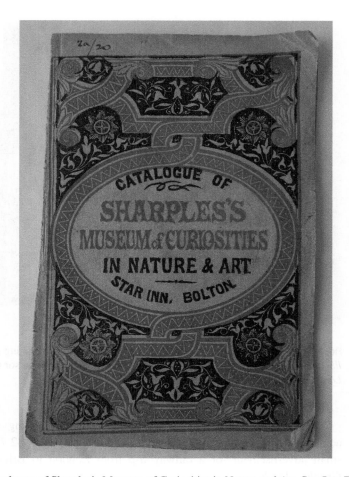

Fig. 2.4 Catalogue of Sharples's Museum of Curiosities in Nature and Art, Star Inn, Bolton, 1850 (courtesy of Bolton Museum).

gunpowder and trained horse actors (Tait, 2016). The historical development of the presentation of animals to the South Australian public by the Royal Zoological Society of South Australia has been discussed by Anderson (1995), from the menagerie-style caging of the late nineteenth century, through the Fairground Era (mid-1930s to the early 1960s) to the contemporary era of naturalistic enclosures.

Van Reybrouck (2005) contended that the cityscape is a domain for carving out social identity: a sense of self based on group membership. He examined zoological gardens and railway stations in nineteenth-century Europe – which developed around the same time and often in the same vicinity – and argued that both entities were instrumental in transforming cities into bourgeois spaces that provided gateways to the outside world. A visit to a zoo allowed people to experience faraway places – that most could not hope to visit – by being exposed to exotic animals and plants, sometimes set in naturalistic landscapes and associated with exotic buildings and other structures.

THE ROYAL VISIT TO WOMBWELL'S MENAGERIE

Fig. 2.5 Queen Victoria and other members of the royal family visiting the lions (source: *The Illustrated London News* (1847a). Wombwell's Menagerie at Windsor Castle. *The Illustrated London News*. 6 November 1847, pp. 297–298).

The development of railways was important in drawing visitors to zoos and aquariums from considerable distances. On 25 February 1882 *The Graphic* carried a small advertisement for a railway trip from London to Brighton Aquarium (*The Graphic*, 1882a):

THE GRAND AQUARIUM AT BRIGHTON. – EVERY SATURDAY. Cheap First Class Trains from Victoria at 10.55 and 11.50 a.m., and London Bridge at 9.30 a.m. and 12.0 noon, calling at Clapham Junction and Croydon. Day Return Fare – 1st Class, Half-a-guinea (including admission to the Aquarium and the Royal Pavilion Picture Gallery, Palace, and Grounds), available to return by any Train the same day, except Pullman Car Trains.

The distance from London to Brighton is approximately 50 miles. The fare was the equivalent of about £50 today.

2.8.1   Case Study: From Blackpool Circus to Blackpool Zoo

Zoos often have a curious history, frequently linked to the endeavours of a single animal enthusiast and sometimes to circuses with performing animals.

Blackpool is a large seaside town in Lancashire, on England's north-west coast. The Ordnance Survey map of Blackpool for 1893 (based on a survey made in 1891) shows a curious feature located on the promenade between the Royal Hotel and the market: a seal pond, a menagerie, an aviary and an aquarium. It was owned by Dr William Henry Cocker, the first mayor of Blackpool, who held this office between 1876 and 1879.

Dr Cocker was a surgeon, but in 1875 he gave up this profession to run Dr Cocker's Aquarium, Aviary and Menagerie (Live Blackpool, 2022; NFCA, 2022). It had existed on this site, in what had been the Prince of Wales Arcade and previously the mansion West Hey, since 1873. In 1890 the Blackpool Tower Company purchased the aquarium with the intention of building a replica of the Eiffel Tower on the site (Fig. 2.6). The now famous Blackpool Tower was built between 1891 and 1894, during which time Dr Cocker's attraction was kept open to the public. Eventually it became one of the tower's most popular attractions. In all, it was home to 57 species of fishes – saltwater and

Fig. 2.6 Blackpool Tower is a tourist attraction in Lancashire, England. It once housed a menagerie inside the building at its base. Some of the animals were used in the Blackpool Tower Circus, including Asian elephants. (A black and white version of this figure will appear in some formats. For the colour version, please refer to the plate section.)

freshwater – with the largest tank holding some 32,000 litres of salt water. Dr Cocker's aviary and menagerie were considered to be one of the finest collections of animals in the country and included lions, tigers and polar bears. When the tower was completed it incorporated a menagerie in the building at its base, as well as a circus.

The Blackpool Tower Zoo closed in 1969. In 1972 the new Blackpool Zoo opened on a site that was originally Blackpool Municipal Airport, just a few miles from the tower. The zoo was operated by the local council for many years, but is now owned by Parques Reunidos, an international operator of more than 40 leisure parks, zoos and aquariums.

2.8.2   Miniature Zoos: Zoo Models and Toys

William Britain & Sons (Britains Limited) of London began making lead soldiers in 1893 and then branched out into farm and zoo animals. The 1940 Britains Limited Zoological Series catalogue listed and described a wide range of painted hollow cast lead animals, keepers, enclosures and scenery (Table 2.2). These lead figures were later replaced with detailed plastic models, eventually ranging in size from an African elephant to a duck-billed platypus (Fig. 2.7). These models are now highly prized by collectors. The introduction to the 1940 catalogue states that the company had produced

a series of models taken wherever possible from life and reproduced with care as to detail, and which will give the 'collector' the opportunity of having his or her own Zoo at home and will, we hope, provide many long hours of pleasure and enjoyment, with perhaps some measure of education.

This hope is now, of course, mirrored in the mission statements of many major zoos. The history of the development of model zoo animals by Britains Limited has been described in detail by Cole (P. Cole, 2004) and more recently by Naish (2017).

The *Playmobil* toy zoo, although aimed at young children – of at least four years – contains a variety of foods for the animals and enrichment in the form of a suspended net containing foliage for the giraffes, a log for the elephants and a slide for the penguins (Fig. 2.8)! The animals are separated by low rails and transparent walls rather than the metal railings supplied with the Britains Limited zoo that were prevalent in the real zoos of the day.

2.9   Animal Personalities and Cultural Heritage

Many zoos have art located within their grounds. Some of it is thought provoking in nature – for example, representations of animals constructed from recycled materials – and some is commemorative – such as statues of founders, former staff or favourite animals. Those of animals are usually of long-dead favourites and sometimes of current residents. Occasionally they commemorate individuals that have died in tragic circumstances.

Table 2.2 Models listed in the Britains Limited Zoological Series catalogue 1940.

Model no.	Item	Model no.	Item	Model no.	Item
901	Indian elephant	925	Railings, straight section	949	Malay tapir
902	Kangaroo	926	Railings, curved section	950	Baby kangaroo or wallaby
903	Penguin	927	Standard post, two way, straight	951	Baby rhinoceros
904	Climbing monkey	928	Standard post, two way, right angle	952	Young Indian elephant
905	Hippopotamus	929	Standard post, three way	953	Bactrian camel with boy rider
906	Gorilla	930	Standard post, four way	954	Gorilla with pole
907	Zebra	931	Keepers	955	Wolf
908	Indian rhinoceros	932	Keeper seated astride	956	Walrus
909	Pelican (beak closed)	933	Eland bull	957	Red dear
910	Lion	934	Brown bear	958	Young crocodile
911	Lioness	935	American bison	959	Young hippopotamus
912	Adult giraffe	936	Bear cubs	960	Young rhinoceros
913	Pelican (open beak)	937	Giant tortoise	961	Young giraffe
914	Polar bear (seated)	938	Howdah for elephant	962	Lion cub
915	Chimpanzee	939	Boy or girl for howdah	963	Gazelle
916	King penguin	940	Baby hippopotamus	964	Sea lion
917	Nile crocodile	941	Tiger	965	Himalayan bear (sitting)
918	Bactrian camel	942	Wild boar	966	Polar bear (walking)
919	Coconut palm	943	Baby camel	967	Polar bear (standing)
920	Date palm	944	Baby elephant	968	Indian or water buffalo
921	Guenon monkey	945	Sable antelope	969	Giant panda
922	Ostrich	946	Stork	970	Baby pandas
923	Llama	947	Flamingo	971	The panda family
924	Gate with posts	948	Warthog		

Source: based on information made available by Brighton Toy and Model Museum, United Kingdom, www.brightontoymuseum.co.uk/index/Category:Britains_Zoo (accessed 1 August 2022).

Fig. 2.7 A selection of plastic models of zoo animals produced by William Britain & Sons (Britains Limited). The series included species ranging in size from a duck-billed platypus to an African elephant.

Fig. 2.8 Even toy animals are now provided with environmental enrichment. *Playmobil* toys are made by a German company and the current models (2022) include sets that can be combined to form a model city zoo. The orangutans are provided with an enrichment device that dispenses food.

Holtorf (2013) has discussed the zoo as a place of remembrance: a place where we are reminded of our childhood. We are also reminded of various past and present cultures, outstanding human and animal figures, the genetic heritage of evolution and human origins. This means that animals in zoos cannot be easily associated with wildlife in its natural setting alone. Hortorf argues that zoos are not just about animals; they are also about metaphorical places and memories that are socially conditioned, with undertones of colonialism and imperialism.

London Zoo has a statue of *Guy* the gorilla, who was a celebrity in the 1960s and 1970s and died in 1978; *Durrell* has a statue of *Jambo* the gorilla, who famously protected a small child who fell into his enclosure in 1986 (Fig. 2.9); and Cincinnati Zoo has a statue of *Harambe*, a male gorilla who was shot dead when zoo staff feared for the safety of a child who fell into his enclosure. A statue of a keeper with a bear in London Zoo commemorates *Winnie bear*, a black bear that was the inspiration for

Fig. 2.9 Statue of *Jambo* located near the Western lowland gorilla (*Gorilla g. gorilla*) exhibit at *Durrell* (Jersey Zoo) in the Channel Islands.

Winnie the Pooh, and Berlin Zoo has a statue of *Knut*, a young polar bear who drowned after falling into water in his enclosure as a result of a brain disorder.

Statues are sometime erected to honour the founders of zoos, their directors and other figures associated with zoos or conservation. Barcelona Zoo in Spain has erected a statue of St Francis of Assisi, the patron saint of ecology. Next to the statue is a small sign with a QR code that allows the visitor to access an audio recording in several languages that provides information about St Francis (Fig. 2.10).

The maps that zoos provide to their visitors are intended as wayfinding aids. However, they are also important historical documents. Maps from two American zoos (Philadelphia and Brookfield, Chicago) from between 1886 and 1949 were evaluated by Mary et al. (2008). They identified shifts in cartographic style, especially the representation of animals, from scientific 'plan maps' to tourist-oriented 'cartoon

Fig. 2.10 A statue of St Francis of Assisi in Barcelona Zoo. In 1979 Pope John Paul II proclaimed St Francis patron saint of ecology. Inset: A QR code is located near the statue that can be scanned by a smart phone to obtain more information via an audio recording.

maps'. They argue that historical zoo maps reveal past social norms and values concerned with zoos and their animals, and they help to reveal the cultural heritage of zoos (Fig. 2.11).

2.10   Zoo Landscapes and Culture

Axelsson and May (2008) have noted that landscape is playing an increasing role in the experience of zoo visitors and that what is considered the 'appropriate' landscape for any species is culturally variable, reflecting the cultural assumptions and aspirations of those who create them. They suggest that an African savannah in a zoo does not represent Africa as it is but Africa as we think it should be. Wild savannah animals do not live in isolation in a pristine grassland ecosystem, but alongside human communities, their buildings and their livestock.

Some modern exhibits attempt to simulate the coexistence of animals in a landscape controlled by people. The *Islands* exhibit at Chester Zoo, England, represents the habitats found in the tropical forests of six islands in South-East Asia – Sumatra, Papua, Sulawesi, Sumba, Bali and Panay – and includes an artificial waterway along which visitors explore the exhibit in boats, and a sub-tropical monsoon forest building (Fig. 2.12).

Shapland and Van Reybrouck (2008) noted that zoos are complex representations of the natural world and tell us much about cultural attitudes towards animals. While many zoos have been successful in protecting their architectural heritage, retaining historically important buildings presents a considerable challenge where space is in short supply and zoos seek to satisfy the demands of modern animal welfare standards and meet the expectations of the zoo-going public. Consequently, many structures of great artistic or architectural merit have been preserved but no longer house their original occupants. Shapland and Van Reybrouck discussed the complexities of reconciling natural and cultural heritage in relation to the removal of penguins from Lubetkin's Penguin Pool, built in London Zoo in 1934 (Fig. 2.13).

Bristol Zoo in England contains a number of historically and culturally important buildings and structures (Fig. 2.14). The domed Monkey Temple was constructed in 1928 to house a troop of around 100 rhesus macaques (*Macaca mulatta*). The design was cutting-edge at the time as the animals were retained by ditches and high walls, in contrast to the traditional bars used elsewhere. The structure is no longer suitable for housing animals, but it has been preserved as the centrepiece of an exhibit called *Smarty Plants* which illustrates how animals use and manipulate plants to survive. Shapland (2004) used the Monkey Temple as an archaeological case study to illustrate the way society's changing attitudes to nature have inspired different styles of zoo enclosure. The building was based on temples commonly found in Asia and was colonially inspired, based on the idea of exoticism and the separation of humans and animals. Shapland saw the integration of the temple into the *Smarty Plants* exhibit as a metaphor for the shift in the zoo's focus from the exotic to a modern ecological vision of the place of humans in nature.

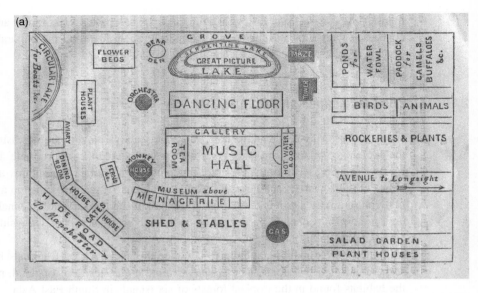

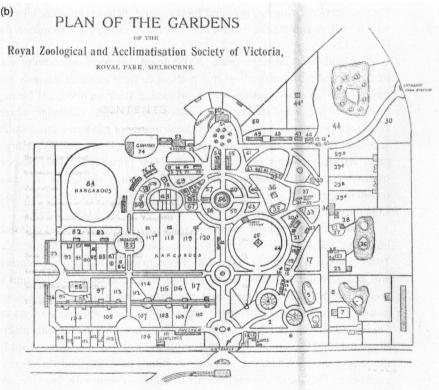

Fig. 2.11 (a) A plan of Belle Vue Zoological Gardens, Manchester, England as depicted in the Zoo Guide from 1872. (b) Plan of the Gardens of the Royal Zoological and Acclimatisation Society of Victoria, Melbourne, Australia as they appeared in 1913 (source: both images courtesy of Chetham's Library, Manchester, England).

Fig. 2.12 The *Islands* exhibit at Chester Zoo in Cheshire, England, represents the South-East Asian islands of Sumatra, Papua, Sulawesi, Sumba, Bali and Panay. Visitors approach the exhibits through a simulated village representing the environs of the people of the region before reaching a boat ride and exhibits containing orangutans, tigers, sun bears and many other species representative of South-East Asia. (a) Traditional boats; (b) local shop. (A black and white version of this figure will appear in some formats. For the colour version, please refer to the plate section.)

Fig. 2.13 The Penguin Pool was designed by Lubetkin and completed in 1934. It has not been used for penguins since 2004, but is protected because of its architectural importance as a Grade I listed building.

Victorian zoos retained many dangerous species behind vertical metal bars. This emphasised their dangerous nature while also being symbolic of humans' dominion over wild beasts. *Alfred* the gorilla arrived at Bristol Zoo in 1930. The original gorilla cage was constructed with vertical metal bars, but in 1956 the zoo opened a new gorilla house with curved bars. This shape softened the bars' appearance and made it easy for the animals to climb to the top of the cage, but they were bars nonetheless and, although the building itself is long gone, small sections of the bars are still used for decorative purposes in the zoo's gardens. Among other things, the zoo has also preserved a bear pole from the bear pit that was one of the original enclosures when the zoo opened in 1836.

Hall (2017) has argued that there is greater public awareness of the damage being done to the non-human components of the environment than at any time in the recent past, as exemplified by the reaction to killings of healthy animals by zoos, notably the death of a healthy giraffe called *Marius* and four lions at Copenhagen Zoo (see Chapter 6). The opportunities for people to gain satisfaction by supporting the conservation ethos provided by both zoos and archaeology have been discussed by Holtorf and Ortman (2008). They likened the arguments used for the need to protect endangered animals from extinction by keeping them in zoos to the increasing advocacy of *in-situ* preservation by archaeologists. The authors found similarities in the arguments

Fig. 2.14 Zoos are important places that remind adult visitors of their childhood. These photographs show structures of historical importance at Bristol Zoo, in England. (a) The last pole from the middle of the old bear pit. The pit itself was one of the original enclosures when the zoo was opened in 1836. (b) The Monkey Temple building was opened in 1928 and was home to around 100 rhesus macaques (*Macaca mulatta*) until the late 1980s. It is now a Grade II listed building. (c) These railings were once part of the bars of the gorilla house that was opened in 1956. (A black and white version of this figure will appear in some formats. For the colour version, please refer to the plate section.)

used in the conservation campaign concerning the protection of weathering rock carvings in Bohuslän in Sweden with those used to support rescue operations conducted by the zoo Nordens Ark (Nordic Ark) in the same area, including the benefits to future generations. Holtorf and Ortman considered the extent to which the rhetoric of conservation is an attempt to 'jump on the Green bandwagon' to gain public support and legitimacy, but concluded that humans have genuine desires to help save scarce resources.

2.11   When Do Visitors Go to the Zoo?

There have been few detailed analyses of the annual pattern of visitor attendance in zoos and aquariums. A study of the impact of weather and climate on visitor attendance at Chester Zoo – located in an area that experiences a temperate climate – found that daily visits increased with temperature up to 21°C but then decreased on hotter days (Aylen et al., 2014). Unsurprisingly, visitors avoided rainy days. Visitor attendance was largely influenced by the rhythm of the year and the pattern of public and school holidays, with 70 per cent of the variation in visit levels explained by a combination of the effects of time of year and weather. The authors considered the possible effects of climate change on visitation patterns and concluded that, although climate change was likely to redistribute visitors somewhat across the year, it was not likely to result in significant increases in spring and autumn due to the other factors that constrain visitation patterns. They concluded that zoos may need to adapt the physical environment as temperatures rise and rainfall decreases.

Hewer and Gough (2016a) found that the most influential weather variable affecting zoo visitation at Toronto Zoo, Canada – which experiences a continental climate – was temperature, followed by precipitation and then wind speed. Visitor numbers declined when temperatures exceeded 26°C in the shoulder season (between the peak and low seasons) and 28°C in the peak season. Even light rain caused visitor numbers to decline by almost 50 per cent. An analysis of the effect of extreme weather events on attendance at Toronto Zoo showed that they could have a significant effect on visitor numbers (Hewer, 2020). For example, the coolest 2 per cent of days in June (<15.4°C) caused a 43 per cent decrease in visitor numbers and in July and August temperatures >34.7°C resulted in a 53 per cent decrease. Similarly, visitors were found to avoid Phoenix Zoo, Arizona, on excessively hot days and Zoo Atlanta, Georgia, on excessively cold days (Perkins and Debbage, 2016). Other analyses of the effect of weather on attendance at Toronto Zoo and the implications of climate change have been presented by Hewer and Gough (2016b; 2016c).

The factors affecting worldwide attendance at zoos during a 40-year period have been reviewed by Davey (2007a) using data published in the IZYB. He identified a marked decline in visitor attendance in the 1960s and 1970s in most world regions, followed by increases in North American and British zoos since the 1980s. This study revealed positive relationships between zoo attendance figures and a country's population size and income. Unsurprisingly, Mooney et al. (2020) found that zoos close to large human populations have greater attendances than those that are not.

2.12 Zoo Economics

Zoos and aquariums are expensive to operate. The cost of keeping animals in captivity is not a new concern of those who manage them. In the last quarter of the nineteenth century, Mather (1879) published an article on public aquariums in which he discussed the problem of supplying 'a sufficient quantity of oxygen to completely consume all the feculent matter' in a large aquarium and a plan to reduce running expenses. Mather was the first director of the Cold Spring Harbor Fish Hatchery, which was established by New York State in 1883 to breed fish to stock lakes, ponds and streams.

Ward et al. (1998) noted that, unsurprisingly, larger animals are more expensive to maintain in zoos than are smaller animals. They suggested that zoos could potentially contribute more to conservation if they concentrated on smaller species, but acknowledged that the public prefers to see larger species. However, a study of data from 458 zoos in 58 countries found that zoos that keep a large number of animals from a large number of species – particularly large mammals – attract high numbers of visitors (Mooney et al., 2020).

The cost of keeping a breeding herd of Asian elephants (*Elephas maximus*) at Whipsnade Zoo (now known as ZSL Whipsnade Zoo), England, under optimum welfare conditions has been estimated by Sach et al. (2019). They estimated the annual cost to be £593,021–641,863 (US$749,697–811,443) per year, excluding indirect staffing costs, ground rent and the cost of contributions to field conservation projects. This represented up to 1.3 per cent of the funds spent on the ZSL's animal collections (London Zoo and Whipsnade Zoo) and conservation (excluding field conservation) in 2016–2017 (ZSL, 2018). In spite of the predicted future costs of implementing higher welfare standards, Sach et al. claimed that there was a clear argument for continuing to keep elephants at the zoo on commercial grounds and predicted that improvements would lead to 'increased secondary spend'.

A major zoo can play a significant part in the local economy. It creates employment and may attract tourists from beyond the local area. In assessing the economic impact of St Louis Zoo, Missouri, the St Louis Regional Chamber and Growth Association considered the direct impact, the indirect business impact and tourism and visitor spending impact (STL Zoo, 2014). This is summarised for 2014 in Table 2.3

AZA claims that its accredited institutions contributed US$22.5 billion to the economy of the United States in 2018, supporting 198,000 full-time jobs (AZA, 2022).

Little has been published concerning the economics of operating zoos. However, the economic value of the Central Zoo of Nepal – the country's only zoo – has been

Table 2.3 The economic impact of St Louis Zoo in 2014.

Full-time employees	330
Part-time employees	170
Seasonal employees	950
Capital improvement jobs created	274
Tourism spending jobs created	495
Total employment	2,139
Operating expenses/payroll	US$141,200,000
Capital improvements	US$42,700,000
Tourism spending	US$47,000,000
Total economic impact	US$230,900,000

examined by Mahut and Koirala (2006) in considerable detail, and Bandoli and Cavicchio (2021) have discussed the economic, ethical and operational challenges of recovering from the effects of the COVID-19 pandemic on Pistoia Zoo, a small private zoo in Italy.

2.12.1   Differential Pricing

Differential pricing has been used for some time to disperse visitors away from popular protected areas to those that are less well known. In Kenya, non-residents pay a daily fee to visit protected areas. In 2022, the adult fee for Amboseli National Park was US$60, but for the less popular Tsavo East it was US$52, and for Hell's Gate US$26. The fee for the Mombasa Marine Park was just US$17. 'Residents' and 'citizens' pay different (lower) rates for access to the same parks (KWS, 2022).

In parts of the world where there are distinct seasons, zoos commonly charge higher ticket prices when the weather is likely to be warm and dry (which generally coincides with school holidays) than when it is cold and wet (and children are at school on weekdays). Throughout the year, zoo attendance is generally higher at weekends than during weekdays. This results in overcrowding at weekends and underutilisation of facilities during the rest of the week. Some zoos employ a dynamic pricing model to spread visitors more evenly throughout the week, similar to that used by airlines that offer cheap seats to fill up their aircraft when demand is low.

Zoo visitor satisfaction falls when visitor number are very high. Dynamic pricing was implemented by Indianapolis Zoo, Indiana, to maintain visitor satisfaction prior to the expected spike in numbers resulting from the opening of the *Simon Skjodt International Orangutan Center*. The zoo adopted a ticket pricing structure based on years of attendance-impacting data, including weather data. Tickets were made less expensive on low-traffic days (weekdays and off-peak season) and when purchased further in advance (Herbert, 2014).

At Chester Zoo, in England, the ticket pricing structure was complex in 2017. Prices varied between months and between weekdays and weekends within a single week. On some days in August 2017 an online (advance) ticket for a single adult cost £23.63. This was during school holidays, when attendance should be high. In November of the same year, when school-age children would have been in school, a ticket for a single adult was £20.00 at the weekend but £16.36 between Monday and Friday. Tickets purchased online in advance were cheaper than those purchased on the day of the visit. In 2022 a single ticket costs £29.08 in August and £25.45 in November, with no difference between weekends and weekdays. In effect the zoo was operating a peak-season and off-season pricing system.

2.12.2 Effects of the COVID-19 Pandemic

The COVID-19 pandemic declared in March 2020 had major economic and societal effects on communities across the world. Visitor attractions have been particularly severely affected and many zoos and aquariums were closed for many months in an attempt to slow the spread of the disease (Fig 2.15).

The ZSL reported receiving a total of 901,115 visitors to London Zoo and Whipsnade Zoo in the 26.5 weeks they were open in the year May 2020 to April 2021. The Society estimated that it had seen a fall in visitor numbers of 713,000 visitors due to the COVID closures (ZSL, 2021a).

Fig. 2.15 Moving a resident at the zoo during the COVID-19 pandemic. The pandemic began in December 2019 and had a severe impact on zoos around the world. Institutions remained closed for long periods, depriving them of income from visitors, and then reopened with severe restrictions on the number of people who could visit each day.

The effects of the pandemic on one major zoo – St Louis Zoo, Missouri – and its ability to deliver its conservation activities were studied by surveying a selected subset ($n = 87$) of its staff members (Fine et al., 2022). This survey showed that the top three perceived challenges were the lack of funding (83.9 per cent), the reduction in zoo visitors – and concomitant reduction in communication with the public – (56.3 per cent) and reduced access to field sites and laboratories (55.2 per cent). At least half of the respondents indicated that the pandemic had decreased the ability of staff to perform the zoo's top three pre-pandemic activities: wildlife management and recovery programmes, caring for the animal collection and genetic breeding programmes. The authors concluded that zoo staff needed to find novel ways to connect with the public – considering the reduced number of visitors – and that continual funding would be essential to maintain its conservation activities.

An evaluation of the recovery in visitor numbers following the reopening of Czech zoos in spring 2020 that had been closed as part of measures to control the COVID-19 pandemic found that attendance was reduced the most by the requirement to buy tickets online, the closure of state borders and the restriction on the number of visitors allowed (Nekolný and Fialová, 2021). The authors suggested that controlling visitor numbers by restricting the number of people by surface area of visitor routes at one time would assist the recovery of zoos.

2.13   Conclusion

Zoos are important institutions of cultural, social, historical and scientific significance. They began primarily as places of entertainment and, to a very large extent, it is still this role that sustains them financially. However, the mission of zoos has changed over time so that the best zoos have evolved into international conservation organisations. They are places where families socialise and make memories, and they play an important role in keeping people connected to nature, especially those who live in urban areas.

3 Zoos and Education

3.1 Introduction

Education is considered to be a core activity of a modern zoo. Many studies of the value and efficacy of zoo education involve studies relating to single exhibits (e.g. Markwell et al., 2019), individual zoos (e.g. de White and Jacobson, 1994; Reade and Waran, 1996; Jensen, 2014) and self-reporting (Baechler et al., 2021). The majority are of short duration and many involve pre- and post-zoo visit tests of knowledge or attitudes, with few attempts to determine long-term effects or benefits. The educational value of providing information and testing knowledge shortly thereafter is limited. Most formal education tests knowledge and understanding months, and sometimes years, after students have been exposed to new facts and experiences.

Some journal editors encourage authors to summarise their findings in the title of their paper. This can be helpful to potential readers but can also be misleading. A paper entitled 'Staring at signs: biology undergraduates pay attention to signs more often than animals at the zoo' (Heim and Holt, 2022) would stimulate the curiosity of anyone interested in the educational value of zoo signage. However, on closer examination this work studied the behaviour of students from one institution while visiting one exhibit in a single zoo and, as the authors freely admit, it tells us nothing about the overall value of zoo interpretation to the general public.

Anyone who has ever visited a zoo on a busy day in summer cannot help but have noticed that most of the visitors are young people with small children, the large mammals attract most attention, visitors do not spend long looking at apparently empty enclosures or sleeping occupants and most visitors are intent on having an enjoyable day out rather than learning about environmental conservation. Scientists, of course, cannot rely on such subjective impressions but must draw their conclusions about zoo visitors and their behaviour from empirical evidence. Not surprisingly, numerous studies have discovered what most of us probably felt we already knew: people go on family-based trips to the zoo for fun; they want to see large, active animals; and they are not primarily interested in being educated (Turley, 2001; Ryan and Saward, 2004).

Unfortunately, this is not what zoos want to hear. Zoos claim to have an important conservation function – which often involves small, little-known species – and they claim to play an important role in educating the public about biodiversity conservation. The following discussion considers the extent to which they have been successful in educating the public and changing behaviour towards the environment.

3.2   Historical Perspective

Zoos have claimed to have an education function for many years. Indeed, in some countries zoos have a legal obligation to engage in education, for example in the Member States of the European Union (Rees, 2005a). Although some 700 million people visit zoos and aquariums worldwide each year (Gusset and Dick, 2011), even within zoos themselves – where information on exhibits is readily available – it is possible to hear visitors refer to chimpanzees as monkeys and call a jaguar a leopard. If some visitors ignore signage identifying an animal while standing next to its enclosure, are we really to believe that zoos are educating their public?

In 1967, Michael Boorer, who was at that time an education officer at the Zoological Society of London (ZSL), published an article in the first volume of the *Journal of Biological Education* entitled 'Zoos as teaching aids' (Boorer, 1967). In this article he stated that:

The [zoo] visitor gains more than mere amusement, and is both entertained and educated. The value of zoos as aids in the formal education of the young is therefore obvious . . .

London Zoo introduced a series of 'lecture demonstrations' for school parties in 1958 (Boorer, 1966). But the mere existence of educational activities does not guarantee that any significant learning actually occurred and the value of zoos as aids in formal education is most certainly not obvious.

Public lectures have always been an important part of the scientific and educational work of London Zoo. On 22 May 1875, *The Illustrated Sporting and Dramatic News* reported on a lecture given at London Zoo and emphasised the importance of making public lectures intelligible to the general public:

THE ZOOLOGICAL SOCIETY. – The fifth of the series of the Davis lectures was given at the gardens of the society on Thursday week. The audience did not number more than sixty, and we are afraid that the many unfamiliar scientific words used in the lecture frightened many people away. That a lecturer can explain the structure of animals by the use of English words was shown last Thursday by Professor Garrod.[1] Instead of using the words artiodactylate and perissodactylate, he spoke of the 'odd-toed group' and the 'even-toed group,' and in many places substituted English words for the Latin and Greek he had used in his previous lecture. (*The Illustrated Sporting and Dramatic News*, 1875)

In considering the role of the ZSL, Hatley (1986) noted that:

[It] may have had research as one of its aims but its early role in society was one of leisure provision and membership conferred some social status.

---

[1] Alfred Henry Garrod taught comparative anatomy at King's College London and was the Fullerian Professor of Physiology and Comparative Anatomy at the Royal Institution at the time.

Around the same time that Boorer published his article about zoos as teaching aids, others were published in the *International Zoo Yearbook* (IZYB) about education programmes at San Diego Zoo, Oklahoma City Zoo and Frankfurt Zoo (Brereton, 1968; Kinville, 1968; Kirchshofer, 1968). Other articles include that of Turkowski (1972), who described education at zoos and aquariums in the United States, Rammelmeyer and Porterfield (1978), who published an account of the use of the Smithsonian's National Zoological Park (also known as the National Zoo, Washington, DC) as an extension of the classroom for first-graders, and a progress report on keeper training at Santa Fe Community College (New Mexico) by Vandiver (1978). Now, as Santa Fe College, it continues to train keepers and runs one of a number of training zoos in the United States.

A quarter of a century ago, Broad and Weiler (1998) reviewed the literature on the history of zoo education and identified four key objectives or outcomes. These may be summarised as: an enjoyable, recreational and satisfying educational experience; the cognitive learning of facts about animals, zoos or exhibits; the development of a concern for wildlife conservation; and appropriate on-site behaviour and long-term environmentally responsible behaviour.

Andersen (2003) noted that zoo education had originally comprised the formal education of school groups. Towards the end of the twentieth century it had started to encompass all visitors. The author predicted that 'modern information technology' – allowing direct links to *in-situ* projects – would increasingly be used alongside signage, interpretive graphics, worksheets and staff presentations, and noted that zoo education departments were often consulted on the design of new exhibits.

3.3   Mission Statements and Education

Like many large organisations, zoos often use mission statements to define their purpose. These mission statements have changed over time, but the original purpose of at least some zoos was acknowledged when they were first founded, long before the term 'mission statement' had been created. In England, the Bristol, Clifton and West of England Zoological Society was established at a meeting in 1835 where Dr Henry Riley (a local physician) proposed that a Zoological Society in Bristol and Clifton

will tend to promote the diffusion of knowledge by facilitating observations of the habits, form and structure of the animal kingdom. (Warin and Warin, 1985).

This resulted in the establishment of the Bristol Zoological Gardens, which opened in 1836. The zoo currently lists the following as one of its aims:

To educate our audience, to encourage them to act in ways that benefit wildlife conservation. In this way, we can help create a sustainable future for both wildlife and people.

Patrick et al. (2007) analysed the mission statements of 136 AZA-accredited zoos in the United States and reported predominant themes of education and conservation in these statements. They concluded that zoos are uniquely placed to provide environmental and conservation education to large numbers of people

In a study consisting of an online questionnaire survey of 190 zoos located in 52 countries combined with an in-depth study at nine institutions (which included face-to-face interviews) Roe et al. (2014) found that the highest priority activity from the zoos' perspective and that of their visitors was educating visitors, including school children. In addition, visitors placed a high value on learning about actions they can take to help conservation efforts.

A survey of visitors to Wellington Zoo in New Zealand indicated that they believed that the single most important role of the zoo was education, but that recreation was also an important role (Mason, 2007). Conservation and the breeding of animals were recognised as important roles of the zoo by the majority of visitors, but 35 per cent were unaware of these roles. This lack of awareness of the zoo's conservation role raised issues relating to the future marketing and management of the zoo and concerns about the future of traditional zoos.

3.4   Do Zoo Visitors Want to be Educated?

Ryan and Saward (2004) surveyed 359 visitors to Hamilton Zoo in New Zealand and found that they were not generally interested in acquiring detailed knowledge about wildlife, so is there any evidence that zoo visitors engage with the education efforts of zoos?

How would we know if zoos have an educational function? Education may be defined as the act or process of acquiring knowledge. Three different types of knowledge acquisition may take place within a zoo environment: cognitive (intellectual) learning (e.g. learning facts about animals and conservation); affective (emotional) learning (e.g. changing attitudes towards animals or the environment); and behavioural learning (e.g. increasing the possibility that visitors will donate to conservation charities).

We can only know if education about a thing has been occurred if we measure knowledge, attitude or behaviour at least twice, such as before and after a visit to a zoo. Zoos may set objectives for an exhibit in relation to knowledge acquisition, attitude change and subsequent visitor behaviour (Seidensticker and Doherty, 1996). But is there any evidence that they impart knowledge or change the behaviour of visitors in relation to the environment?

A study of 3,347 zoo visitors in seven countries found that attitudes to species conservation in visitors were influenced by their age, the number of zoo visits they had made in the previous 12 months, their interest in animals and the country where they were surveyed, but not by their sex (Kleespies et al., 2021).

An identity-related motivational model was used by Schultz and Joordens (2014) to study the success of environmental education at Toronto Zoo in Canada. They proposed that the educational success of a zoo visit depends on visitors' dominant

Table 3.1 Categories of zoo visitors used by Schultz and Joordens (2014).

Identity	Characteristics
Explorers	Individuals who are curiosity-driven
Facilitators	Those who enable the learning of others
Professionals/hobbyists	Persons who have a close tie with the zoo's content
Experience seekers	Individuals who derive satisfaction from visiting an important site
Spiritual pilgrims	Those who seek a contemplative and/or restorative experience

motivation (Table 3.1). Their preliminary data suggested that 'spiritual pilgrims' and 'experience seekers' made greater knowledge gains than did 'facilitators' (the most numerous group). As 'facilitators' tended to be accompanied by more children than did those in the other groups, the authors suggested that this was why they appeared to show fewer knowledge gains themselves.

3.5   Does Participation Data Tell Us if Education Takes Place?

Schools, colleges and universities claim to have an educational function. They support this claim by providing structured educational programmes and assessing the knowledge and understanding of their students using examinations, coursework and other methods. The evidence of the effectiveness of their educational activities is available via published student pass rates and other measures of student performance.

Museums, art galleries, zoos, aquariums and other visitor attractions also claim to have an educational function, but these claims are much more difficult to substantiate. Zoos often use data on the numbers of school children that have visited the zoo and other participation data as evidence of their educational role.

In discussing its role in lifelong learning the ZSL listed the following achievements in 2015: 175,000 children visited the zoo on a school visit, 1,000,000 visitors attended a 'LIVE talk' (for example at the *Penguin Beach*); a conservation game was downloaded 40,000 times; 1,971 people attended 13 scientific meetings and conferences on conservation science; 175 papers were published in ZSL's academic journals; and 21 talks were given by ZSL scientists at London Zoo's summer 2015 Sunset Safari events (ZSL, 2015). In 2018–2019 (prior to the COVID pandemic), 153,851 school students visited the society's two zoos, and of these 60,201 took part in an additional educational workshop (ZSL, 2019).

Although all of this activity is to be commended, it tells us nothing about whether or not anyone learned anything or changed their behaviour as a result of contact with the ZSL. In many respects, this type of information is analogous to a school providing attendance data on its pupils and reporting on the number of classes provided by the teachers. No one would accept this as evidence that children at such a school had learned anything.

3.6   Educational Visits and Biological Education

Some modern zoos have well-equipped purpose-built education buildings. The facility at Detroit Zoo in Michigan has laboratory facilities containing educational aids for anatomical studies that replace the use of animal dissections, and a 4D cinema in which audiences can experience the sounds, smells and vibrations of going on safari (Fig. 3.1).

Most zoos provide education sessions for school children – and in some cases college students – but few measure the extent to which this education has been effective. Schools, colleges and universities have an education function because we can measure what they do. Most zoos make no attempt to assess their educational value so we must take their claim to be 'educational' at face value.

In addition to entrance fees, many zoos make additional charges for education sessions, and a cynic might argue that zoos encourage school visits to increase their visitor numbers and revenue rather than because they are serious about education. This is not to say that the two are mutually exclusive. However, it is clear that any zoo that did not make provision for school visits would be ignoring an important income stream while also failing to nurture the interest of potential adult visitors of the future.

From 1 July 2013 to 30 June 2014, Auckland Zoo in New Zealand was visited by 60,000 children on school trips (Auckland Zoo, 2014). The cost per child depended upon age, but assuming NZ$10 per child, this represented NZ$600,000 or around US$440,000.

The same cynic that might accuse a zoo of taking an interest in the education of children because of their importance as an income stream could also accuse a university of providing courses in zoology to take fees from students interested in learning

Fig. 3.1 Ford Education Center at Detroit Zoo, Michigan.

about animals. Even if most of the educational work of zoos has little real educational value and most zoology graduates do not find employment as zoologists, that does not invalidate the other good work done by both zoos and universities. If some of their activities cross-subsidise others, is that necessarily a bad thing?

Zoos have an important role to play in formal biological education, especially in relation to comparative vertebrate anatomy as it is taught at college and university level (Doolittle and Grand, 1995). Exotic species offer a greater diversity of forms from which to learn about anatomy, adaptive behaviour and evolution than do the species traditionally used to study anatomy in the laboratory (e.g. the frog, dogfish and rat). Students in the age range 18–22 years are able to integrate knowledge of gross anatomy with the functions of behaviour and relate behaviour to exhibit design and animal management. Furthermore, zoo visits provide students with opportunities to meet keepers and other zoo professionals following careers of which they may not have previously been aware.

My own zoological education included a field trip to Chester Zoo in 1974 as part of a first-year undergraduate module entitled 'Advanced Chordates', and throughout my teaching career I arranged numerous visits to zoos and aquariums for undergraduate students of zoo biology where they received lectures from keepers, curators, education officers, veterinarians, nutritionists and researchers, and fieldworkers engaged in *in-situ* conservation. For the keenest students these visits led to placements in zoos and, for some, employment opportunities.

In the 1980s, when some GCE 'A-level' Biology syllabuses in England included options such as 'Animal Behaviour' and 'Ornithology', some teachers who felt unable to offer any of these specialisms 'sub-contracted' this work to zoos. For example, staff at the Wildfowl and Wetland Trust at Martin Mere, Lancashire – which has a collection of captive wildfowl – offered a module in ornithology.

There is no doubt that for further and higher education students with an interest in biology and conservation a zoo visit provides learning opportunities that cannot easily be replicated elsewhere.

3.7 Studies of the Benefits of Zoo Visits and Teacher Education

Zoos have always made great efforts to attract and cater for children, including exhibiting models of animals alongside the real thing and building dinosaur exhibits (Fig. 3.2). However, there is evidence that interest declines with age.

An investigation of the biological topics of interest to German students ($n = 1,587$) found interest in animals and plants, and especially zoos, declined as students progressed through their high school education and that 'advanced' high school students showed little interest in either animals or zoos (Kleespies et al., 2020). However, for these older students, a one-hour guided tour of Opel Zoo in Kronberg increased their connection to nature.

Jensen (2014) studied children aged 7–15 years who participated in educator-guided and unguided visits to London Zoo, and found that conservation biology-related learning occurred in 41 per cent of educator-guided visits but only 34 per cent

Fig. 3.2 White stork (*Ciconia ciconia*) (a) and the species modelled in *Lego* building bricks (b) at Martin Mere Wetland Centre, Lancashire, England. (c) *Lego* model of a shark at Edinburgh Zoo, Scotland.

of unguided visits. He concluded that zoo visits have the *potential* to have an educational value for children, but that the zoo's unguided interpretive materials were insufficient to achieve good educational outcomes.

Fig. 3.3 A group of young children on a visit to Detroit Zoo, Michigan.

Wünschmann et al. (2017) compared the performance of a group of primary school children learning about amphibians and reptiles in a school setting with a second group learning the same material in an amphibian and reptile zoo. They concluded that students were more motivated and learned more in the zoo setting than in the school setting. The zoo environment also reduced sex differences in achievement; boys performed better than girls in the school group and girls reported feeling less pressure in the zoo group than in the school group (Fig. 3.3).

There is some evidence that providing pre-visit workshops for teachers leading zoo visits with young children may affect the knowledge and attitudes of the children. Children whose teachers participated in a zoo workshop in wildlife conservation targeted specifically at elementary school teachers prior to their visit to a zoo in Colombia achieved higher knowledge and attitude scores in a post-visit questionnaire than those who viewed a slide show prior to their visit and those who made an unstructured visit (de White and Jacobson, 1994).

After pre-service secondary school biology teachers were exposed to a three-week period of practical experience in an aquarium as part of their teacher education, all of the participants felt empowered as science educators and gained confidence in their teaching abilities (Anderson et al., 2006). They also developed a broader range of skills and views on education. This study illustrates the potential for improving teacher education by exposing student teachers to informal practical experiences.

3.8   Zoos Are 'Certain' That They Have an Educational Value

Zoos and zoo professionals make big claims for the educational work they do and published research on zoos routinely refers to their educational value:

Zoological institutions are powerful wildlife conservation entities that work to conserve wildlife and educate and inspire millions of people about animals and their habitat each year. (Gusset and Dick, 2010)

So begins the Introduction to a recent work, *Scientific Foundations of Zoos and Aquariums* (Kaufman et al., 2019). The study to which it refers is a three-page report in the journal *Zoo Biology*, conducted by the World Association of Zoos and Aquariums (WAZA) and showing that 700 million people visit zoos and aquariums worldwide and are thus 'potentially exposed to environmental education'.

In their recent review of zoo elephant research, Bechert et al. (2019) state that:

The purpose of zoos has changed over time from primarily providing entertainment for visitors to creating opportunities for public education and research (Barongi et al., 2015).

However, the citation in this quotation is not an academic source based on any evidence of an educational value, but a reference to WAZA's conservation strategy which makes the claim, unsupported by evidence. It is worth noting that this statement does not actually claim that any education takes place.

The website of the Association of Zoos and Aquariums (AZA) presents 'visitor demographics' and states that:

94% [of visitors] feel that zoos and aquariums teach children about how people can protect animals and the habitats they depend on. (AZA, 2020)

Of course, what zoo visitors feel is neither here nor there and is not evidence that any learning occurs.

In their discussion of the future costs of keeping Asian elephants at Whipsnade Zoo in the face of increased welfare costs, Sach et al. (2019) claimed that there was a clear argument for continuing to keep elephants on commercial grounds and because:

there is certainty that these animals provide substantial opportunities for education, conservation research and fundraising.

Again, the authors provided no evidence. There is absolutely no *certainty* that elephants kept in zoos have any clearly demonstrable educational value and, at the time this statement was made, there were very few studies that had attempted to address the question of such a value for these animals (Rees, 2021) (see Box 3.1).

3.9   Testing the Value of the Educational Impacts Claimed by Zoos

3.9.1   Inconsistent Outcomes and Poor Methodology

The educational value of zoo visits is equivocal. Studies are usually small-scale and often concerned with a single exhibit in one zoo. Some studies appear to demonstrate

Box 3.1 Do Elephants in Zoos Have an Educational Value?

A very small number of studies have looked specifically at elephant exhibits in zoos. When Smith and Hutchins (2000) claimed an educational function for elephants living in zoos the only study that appears to have been conducted was that of Swanagan (2000). He studied 350 visitors to Zoo Atlanta, Georgia (where he was deputy director at the time) and found that those who had an interactive experience with the zoo's elephant demonstration and 'bio-fact program' were more likely to support elephant conservation – by signing a petition or returning a solicitation card lobbying against the importation of elephant parts – than those who simply viewed the animals in their exhibit and read the associated graphics. Only 18.3 per cent of participants returned the solicitation cards: just 64 visitors. Smith and Hutchins did not mention this study or, indeed, any other study that was specifically concerned with the educational role of elephants in zoos.

Smith and Broad (2008) studied visitor behaviour at the *Trail of the Elephants* exhibit at Melbourne Zoo, Australia, but were concerned with determining dwell time – the time spent by visitors in the exhibit – rather than measuring the extent to which visitors improved their knowledge or changed their behaviour following their visit. Moss et al. (2008) studied the relationship between the size of the viewing area and visitor behaviour in the Asian elephant (*Elephas maximus*) exhibit at Chester Zoo, England. Neither of these studies addressed the question of whether the presence of living elephants affected attitudes towards their conservation or had any other specific educational function. Subsequently, Moss and Esson (2010) studied the popularity of different animal taxa – including Asian elephants – at Chester Zoo, and showed – not surprisingly – that zoo visitors are particularly interested in seeing large mammals.

Two recent studies have attempted to link close-up encounters with elephants in zoos and 'conservation intent' in visitors. Hacker and Miller (2016) reported that experiencing close-up encounters with elephants and observing active elephant behaviours resulted in the greatest changes in 'conservation intent' reported by visitors to the San Diego Zoo Safari Park in California. A similar study conducted by Miller et al. (2018) analysed 1,294 questionnaires completed by visitors following observations of elephants kept at nine zoos in the United States accredited by AZA. They concluded that 'up-close experiences' watching elephants engage in active species-typical behaviours correlated with a positive emotional experience and visitors' interest in getting involved in conservation. However, they also noted that most visitors arrived at elephant exhibits with very receptive predispositions towards wanting to learn more and get more involved in conservation. In response to the statement 'I now have a better understanding of what actions I can take that will help protect and preserve elephants and their habitats' the mean visitor response was 4.55 (s.d. = 1.61) on a scale from 1 (strongly disagree) to 7 (strong agree). Overall, the respondents, before they visited the elephant exhibits, reported that they already supported conservation organisations (mean = 4.29, s.d. = 1.74; where 1 = not at all, 4 = somewhat, 7 = very much) but their visit did not appear to have encouraged them to want to donate to an

Box 3.1 (*cont.*)

elephant conservation organisation, and the question asking about this received the lowest score of the 29 'questions' posed (mean = 3.93, s.d. = 1.69).

Asking the public about their future intentions is problematic. Recording a future *intention* to become involved in conservation on a questionnaire is not the same as recording an actual *change* in future behaviour. There is evidence that 'intention' in relation to the protection of the environment fails to predict future behaviour (Davies et al., 2002). Indeed, the very act of measuring intention may inflate the association between intention and behaviour: a phenomenon known as 'self-generated validity' (Chandon et al., 2005). Put simply, we cannot determine what someone will do in the future simply by asking their intentions, and research that implies we can is misleading.

ElephantVoices, the organisation run by the elephant biologist Joyce Poole, believes that elephants do not belong in zoos and that they are not necessary for zoos to educate the public about elephants:

zoos can offer high-end virtual educational exhibits that through animatronics and multimedia connect the public to the capabilities and lives of wild elephants, while stimulating the interest in their conservation. (*ElephantVoices*, 2019)

Elephants living in zoos may have an educational value. But the scant evidence available for this is unconvincing and, of the five studies concerned specifically with elephant exhibits discussed above, only Smith and Broad (2008) was written by authors who do not have a vested interest in zoos. Unless independent scientists are able to demonstrate conclusively that visitors exposed to elephants living in zoos are more likely to support elephant conservation as a result – and more likely to do so than if only exposed to other information about elephant conservation in the media and elsewhere, or to museum specimens – then it is difficult to see how zoos can continue to justify an educational function in keeping them.

In spite of the lack of evidence that elephants living in zoos have an educational role, zoo professionals continue to refer to its importance when discussing elephants:

Positive animal welfare is essential for a zoo to fulfil the underlying mission to raise awareness of conservation issues, public education and (re)-connecting people with animals. (Sach et al., 2019)

Reproduced with permission from Rees (2021). *Elephants under Human Care: The Behaviour, Ecology, and Welfare of Elephants in Captivity*. Elsevier, San Diego, CA.

an educational function, while others do not (Rees, 2011). Even some zoo educators have acknowledged that:

Zoos exude a certain self-confidence regarding their roles as education providers and that . . . [educational] outputs do not necessarily lead to outcomes. (Moss and Esson, 2013).

Many studies of the educational value of zoo exhibits simply compare the knowledge of visitors before and after visiting the exhibit. Lukas and Ross (2005) evaluated visitor knowledge and conservation attitudes towards chimpanzees and gorillas at Lincoln Park Zoo in Chicago. The 1,000 participants correctly answered 60 per cent of the knowledge questions asked. They performed better on exit than entrance surveys, and answered more chimpanzee- than gorilla-oriented questions correctly. But does this knowledge last and have any effect on behaviour? Broad (1996) found that knowledge about threatened species gained from a zoo visit had not influenced visitors in any way in 80 per cent of cases when they were contacted 7–15 months later. Some zoo visits have even apparently led to a decrease in wildlife knowledge and an increase in 'dominionistic' attitudes to conservation in visitors (Kellert and Dunlap, 1989).

An unusually large study was undertaken by Balmford et al. (2007). They examined the effects of a single informal visit by 1,340 respondents to one of seven zoos in the United Kingdom and found little evidence of any effect on adults' conservation knowledge, concern or ability to do something useful.

Clayton et al. (2009) studied the informal learning that occurs during a visit to a zoo, including the emotional dimensions and personal meaning-making that occur in the social context of a zoo visit. They surveyed 206 visitors and analysed 1,900 overheard conversations and found that animals in zoos facilitate topical interaction among social groups and allow them to make connections between people and animals. The authors concluded that a visit to a zoo is a positive emotional experience that leaves visitors wanting to learn more – irrespective of their reading exhibit labels – and these perceived positive connections may be related to conservation initiatives.

A recent study of the impact of a global biodiversity education campaign relating to the UN Convention on Biological Diversity's Aichi Targets examined almost 5,000 visitors to 20 zoos and aquariums located in 14 countries. It concluded that visitors to these institutions showed an increased understanding of biodiversity and actions to protect it (Moss et al., 2015). However, this study measured only the effects of exposure to campaign materials, not the effects of seeing captive animals.

A report by AZA considered the impact of a visit to a zoo or aquarium (Falk et al., 2007) and concluded that:

- Visits to accredited zoos and aquariums prompt individuals to reconsider their role in environmental problems and conservation action, and to see themselves as part of the solution.
- Visitors believe zoos and aquariums play an important role in conservation education and animal care.
- Visitors believe they experience a stronger connection to nature as a result of their visit.
- Visitors bring with them a higher-than-expected knowledge about basic ecological concepts. Zoos and aquariums support and reinforce the values and attitudes of the visitor.
- Visitors arrive at zoos and aquariums with specific identity-related motivations and these motivations directly impact how they conduct their visit and what meaning they derive from the experience.

However, an analysis of these conclusions by five academics at American colleges (Malamud et al., 2010) determined that AZA's claims were invalid:

> We conclude that Falk et al. (2007) contains at least six major threats to methodological validity that undermine the authors' conclusions. There remains no compelling evidence for the claim that zoos and aquariums promote attitude change, education, or interest in conservation in visitors . . .

A more recent study has expressed concern about the validity of studies that have evaluated zoo and aquarium conservation education. Mellish et al. (2019a) investigated the methods and reporting practices used in 47 peer-reviewed articles focusing on adult visitor samples used in zoo education evaluation. They assessed that the quantitative methods used in 83.3 per cent of these studies were weak, and in the remaining 16.7 per cent the methods were of moderate quality. In their examination of the quality and rigour of the qualitative methods and reporting practices used they found that 69.6 per cent of the articles discussed methods for identifying key themes from the data, whereas just 34.8 per cent reported how data verification was performed. Mellish et al. suggested a number of methodological improvements that need to be made in the evaluation of conservation education, including the use of intensive longitudinal methods to strengthen self-reported data, such as the keeping of a daily diary.

3.9.2   Do Zoos Provide General Environmental and Animal Welfare Education?

Poor zoos may have an unintended educational role in showing the public how not to keep and care for animals. Small, unsanitary cages and enclosures are unacceptable to modern zoo-goers, and even enclosures that are probably quite suitable for their occupants may be criticised by some animal lovers for being too small or appearing to provide little stimulation for the animals. Apart from being a legal requirement, in many – if not most – jurisdictions, the provision of good conditions for animals living in zoos can send an important subconscious message to visitors about how they should look after and respect animals, including their own pets. Other subconscious messages may be sent if zoos are obviously caring for the environment: recycling waste and water, generating electricity from solar panels, using electric vehicles and so on.

Some studies have measured the environmental knowledge, attitudes or concerns of zoo and aquarium visitors rather than their wildlife knowledge. Kelly et al. (2014) studied 3,594 visitors to zoos and aquariums in the United States and found that a greater proportion of these individuals were 'alarmed' and 'concerned' about climate change compared with the general public. The authors were seeking to gather information about the zoo and aquarium audience that could assist in the development of climate literacy resources. Although many educators in zoos and aquariums in the United States would like to say more about climate change to their audiences, they were inhibited by concerns about visitors' disinterest in the topic and lacked confidence in their ability to present climate change information (Swim and Fraser, 2014).

A conservation education programme at a zoo that focused on the harm done to wildlife by balloon litter found that visitors had a greater depth of understanding of the issue immediately after their visit but that post-visit recall was low (Mellish et al., 2019b). Completion of a pre-survey significantly influenced positive attitudes and reduced the likelihood that visitors intended to use balloons in the future.

3.9.3 Studies of the Educational Value of Individual Zoo Exhibits

Many of the studies of the educational value of zoos have been concerned with the assessment of a particular exhibit rather than an entire zoo. This is a perfectly valid approach but great care must be taken not to extrapolate conclusions drawn from a single exhibit to an entire institution or, indeed, zoos in general.

Markwell et al. (2019) examined the enablers and inhibitors of emotional engagement between visitors to Healesville Sanctuary in Australia and Tasmanian devils (*Sarcophilus harrisii*). They found that the factors that enabled visitors to form an emotional engagement with this species were clear views of the animals, first-person interpretation and an understanding of the conservation threats affecting the animals. The formation of an emotional connection was inhibited by the inability to view the animals and general misconceptions about the species.

Engaging visitors at chimpanzee (*Pan troglodytes*) exhibits was found to promote positive attitudes towards the species and their conservation by Craig and Vick (2021). A comparison of the interpretive techniques used at two tiger (*Panthera tigris*) exhibits was conducted by Broad and Weiler (1998) and the educational value of an elephant exhibit was assessed by Swanagan (2000).

The potential impact of a proposed immersive exhibit – *Grasslands* – with an African savannah theme on visitor behaviour and learning at Chester Zoo was examined by studying visitors to existing exhibits for giraffes (*Giraffa camelopardalis*), black rhinoceros (*Diceros bicornis*) and painted dogs (*Lycoan pictus*)(Smart et al., 2021). These three existing exhibits were considered to be 'traditional', 'intermediate' and 'immersive', respectively. The authors found that, although visitors spent more than five times as long at the traditional giraffe exhibit that at the immersive painted dog exhibit, they were more than twice as likely to engage (and spent more time engaging) with the interpretation at the painted dog exhibit than with that at the giraffe exhibit. They concluded that providing attractive and engaging interpretive elements should be central to the design of the new exhibit.

Heim and Holt (2022) used eye-tracking equipment to monitor those aspects of a Komodo dragon (*Varanus komodoensis*) exhibit attended to by undergraduates during a zoo trip. They found that the interpretive signs were the primary focus of these students – rather than the Komodo dragon – and that this was true for both novice and 'expert-like' students. They did not, however, determine whether or not the students learned anything from the signs.

3.9.4   Does a Zoo Visit Affect Visitors' Future Behaviour

Does a zoo visit have any subsequent positive effect on visitors' conservation knowledge or their behaviour in relation to the environment?

Smith et al. (2008) tested whether or not zoos can foster conservation behaviours by examining two behaviours communicated during a bird presentation at an Australian zoo (recycling and removing road kill from the road). Of the respondents, 81 per cent recalled hearing the actions during the presentation and 54 per cent stated an intention to start a new action or increase their commitment to an existing action. Six months later, 26 of 38 respondents (68.4 per cent) contacted by phone said that they had started or increased their commitment to a conservation action, but only 3 (7.9 per cent) had started a new action and these were actions previously known to them.

A study of 306 visitors to the National Aquarium in Baltimore, Maryland, found that they clearly absorbed the overall conservation message of the aquarium and their conservation knowledge, understanding and interests largely persisted over 6–8 weeks after their visit (Adelman et al., 2010). However, the visitors' conservation experiences at the aquarium rarely resulted in new conservation actions and the enthusiasm and emotional commitment to conservation inspired by their visit generally returned to their original levels.

Analysis of data collected from an ocean plastics exhibit at the Oregon Coast Aquarium in Newport, Oregon found that children (up to the age of 17 years) engaged more than adults (18 years of age and older) with the interactive elements of the exhibit (Baechler et al., 2021). Visitors between 18 and 29 years old were the group that expressed the greatest change in desire to address the ocean plastic problem, but this group had the lowest likelihood of engaging in the reduction of single-use plastic or plastic stewardship actions.

Clayton et al. (2017) assessed the impact of a visit to the Menagerie of the Jardins des Plantes in Paris on the conservation knowledge and engagement of visitors by surveying them on entering the zoo ($n = 88$) and on the way out ($n = 84$). They found that individuals who had completed their visit scored higher on conservation knowledge, general concern about threats to biodiversity and perceived self-efficacy – the perception that one can complete one's intended actions – than those entering the zoo. They concluded that by affecting self-efficacy a zoo visit has *potential* to influence future behaviour. The average education level of respondents was high, with an average of 3–5 years of education beyond 'high school'. Unfortunately, this study did not compare the same individuals before and after their visit because the researchers wanted to avoid the possibility that visitors would search for the answers to questions they had been asked on entering the zoo.

Smith et al. (2012) have suggested that one reason why there is limited evidence of post-visit behavioural change in zoo visitors is that they do not view the requested behaviours favourably. They reported the results of workshops including 152 staff members from zoos in Australia, at which over 500 behaviours were identified and prioritised based on criteria zoo visitors prefer. The participants favoured behaviours that promoted wildlife-friendly consumerism and donations.

3.9.5 The Value of Interactive Experiences and Observing Interesting Behaviour

Interactive experiences in zoos, such as birds of prey displays and encounters with reptiles, have been shown to increase awareness and learning, and to promote positive attitudes towards animals (White and Barry, 1984; Morgan and Gramann, 1989; Yerke and Burns, 1991) (Fig. 3.4). Short- and long-term cognitive learning occurred when students handled either live or dead specimens of sea stars (*Asterias forbesi*) and horseshoe crabs (*Limulus polyphemus*) during a visit to an aquarium (Sherwood et al., 1989). Gains in affective learning (changes in attitude) only occurred when the students handled live animals.

Active experiences with animals at zoos are more likely to encourage visitors to support conservation than are passive experiences. Swanagan (2000) has shown that zoo visitors who had an active experience with an elephant show were more likely to support elephant conservation than those who simply viewed the animals and read the graphic displays, although sample size was small.

A study of the effects of performing animal-training sessions with Asian small-clawed otters (*Aonyx cinereus*) at Zoo Atlanta's otter exhibit while zoo visitors watched were examined by Anderson et al. (2003). They found that public animal training and public animal training with interpretation resulted in longer visitor exhibit stay times and produced more positive zoo experiences, training perceptions, exhibit size and staff assessments, when compared with passive exhibit viewing and interpretation-only sessions.

The conversations of visitors at three bear exhibits were recorded and analysed by Altman (1998): polar (*Ursus maritimus*), sloth (*Melursus ursinus*) and spectacled (*Tremarctos ornatus*) bears. She coded conversation as 'human-directed' or 'animal-direct' and used information about what visitors attended to as a proxy for what information visitors were consciously processing or learning when bears were active, inactive, pacing and not visible. Altman found that what bears were doing affected what visitors attended to, with animated activity eliciting most attention to behaviour. She noted that conversation about bear behaviour was highest and human-directed conversation was lowest in the presence of highly animated polar bears. In contrast, behaviour content was limited in the presence of the less animated sloth and spectacled bears.

A study of the effects of giant panda (*Ailuropoda melanoleuca*) play on visitors to exhibits at the Chengdu Research Base of Giant Panda Breeding, and the Chengdu Zoo, in China and Zoo Atlanta in Georgia in the United States, found that, although those who witnessed panda play were no more likely to request a flyer on panda protection than those who did not, visitors who observed play on the day of their visit were more satisfied with their visit than those who did not (Bexell et al., 2007).

Fig. 3.4 Interactive experiences in zoos have been shown to increase awareness and learning, and promote positive attitudes towards animals. (a) and (c) Burmese python (*Python bivittatus*); (b) curlyhair tarantula (*Tliltocatl albopilosus*). (A black and white version of this figure will appear in some formats. For the colour version, please refer to the plate section.)

3.10   Educational Outreach Work

Some zoos undertake a considerable amount of outreach work in their local community, especially in schools. Moss et al. (2017) evaluated the educational impact of an 'in-school zoo education outreach programme' run by Chester Zoo in England for 199 students aged 7–11 years. As a result of the programme, they concluded that students exhibited an increase in conservation-related knowledge and a more positive attitude towards conservation and zoo-related issues. However, this research did not include a zoo visit and tells us nothing about the educational value of zoos themselves. Indeed, Moss et al. draw attention to a previous study which attempted to measure the educational value of zoo visits (Moss et al., 2015) in conveying an understanding of biodiversity which appeared to have a smaller beneficial effect than the 'in-school' education programme, stating

Whether this suggests that an in-school zoo education programme is in some way more beneficial when compared to a standard zoo visit is open to debate.

Staff at the Monterey Bay Aquarium in California began using theatre techniques, such as the use of puppets of aquatic animals and pantomiming, as part of their outreach programme to teach children about animals in 1985 (Rutowski, 1990).

Esson and Moss (2016) have acknowledged that the evaluation of educational outcomes in the field of conservation is 'newish' and suggest this may be due to lack of expertise and because differences in cultures require a flexible approach to evaluation. The authors suggest that one way of transcending the boundaries of culture, language and literacy when evaluating the effectiveness of conservation education is to employ drawings in the style of personal meaning mapping. This method has been used by others to assess the educational value of experiences, such as a visit to a science centre (Philipps et al., 2019) and a planetarium (Faria et al., 2020).

3.11   Interpretation and Conservation Messages

Zoos use interpretation displays to communicate information to visitors. The type and quantity of information undoubtedly affects the efficacy of such displays, and even their orientation may be important in determining how well they are received by members of the public. In an environment where visitors spend just a few minutes looking at the living exhibits and static displays, they are unlikely to read the verbose interpretation boards produced by some zoos (Fig. 3.5).

Dove and Byrne (2014) explored the public's understanding of animal biology and the conservation of biodiversity in a zoo setting to determine whether or not visitors need zoological knowledge to understand conservation messages. They found that although their understanding of basic animal biology was 'fairly secure', their comprehension of complex concepts such as ecological interdependence and the physiological needs of animals was unsophisticated and sometimes confused. However, visitors' understanding of some concepts was greater than expected from the existing literature, particularly in relation to the effect that humans can have on the natural world.

Conservation issues are often extremely complex and zoos may not be the most appropriate organisations to educate the public about them. Many species are

Fig. 3.5 Too much information. People visit zoos to see animals, not read signs. Elaborate signs may simply be ignored. (A black and white version of this figure will appear in some formats. For the colour version, please refer to the plate section.)

threatened by the bushmeat trade, but in some parts of the world it provides an important source of food for local people. When photographs relating to this trade were shown to zoo visitors, a study of their responses suggested that a static display of photographs and text may not be the most effective method of educating the public about bushmeat and other complex environmental issues (Stoinski et al., 2002). Surprisingly, 83 per cent of the visitors surveyed had never heard of the bushmeat trade, but 98 per cent felt that zoos should be educating people about the issue.

An assessment of four Australian aquarium displays of the largetooth sawfish (*Pristis pristis*) – a Critically Endangered species – involved the calculation of conservation impact scores for research, education and conservation-related activities (Buckley et al., 2020). Sawfish-related education was evaluated before and after exposure to the displays. Although visitors to all four aquariums demonstrated significant positive attitudinal changes, no significant changes in behavioural intentions were detected. Knowledge was gained by visitors to just one of the four sites. The authors found that the educational messages that addressed public attitudes and behaviours were mostly generalised and untargeted and concluded that research projects and conservation activities were unlikely to contribute substantially to sawfish conservation because they received limited support from the aquariums.

Ichino et al. (2013) examined the effect of display angle on the communication of information to museum visitors. They compared responses to large, flat interactive displays that were mounted horizontally, vertically and tilted. Users rated tilted

Fig. 3.6 It has been shown that young people prefer vertical displays and older people prefer displays that are tilted. (A black and white version of this figure will appear in some formats. For the colour version, please refer to the plate section.)

displays as best with respect to attracting attention (measured by approach rate), ease of perusal and understanding of the content (measured by dwell time) and interaction (measured by touch rate), irrespective of age. The display angle at which the displayed content was easiest to understand and remember differed depending on age, with young people preferring vertical displays and older people preferring displays that were tilted (Fig. 3.6).

Some exhibits are able to provide useful scientific information to visitors by enhancing their senses so that they are able to experience the world as experienced by the animals. Some fish species (Gymnotidae, Mormyridae and Gymnarchidae) possess weak electric organs and are able to use their electric sense to detect species-specific signals in the form of electric organ discharges and in the rhythm of the discharge. Lücker (2003) has discussed how the world of weakly electric fishes can be made accessible to aquarium visitors by using silver electrodes, an amplifier, an oscillograph and loudspeakers. When the effectiveness of electronic and graphic displays created for the electric eel exhibit at Belle Isle Aquarium in Detroit, Michigan, were examined by Safadi and Ram (2020), they found that the electronic enhancements were associated with an increase in knowledge and enjoyment among visitors and an increase in time spent at the exhibit.

Studies have shown that visitors are more likely to read short passages than long passages (Borun and Miller, 1980; Bitgood and Patterson, 1993) and the quantity of text that is read is determined more by the perceived costs of reading than by the visitor's interest in the subject matter (Bitgood, 2006). The likelihood that text will be read is influenced by where it is placed (Bitgood et al., 1989) and it is more likely to be read if it asks provocative questions (Hirshi and Screven, 1990; Litwak, 1996).

A set of principles for interpreting and displaying animals in captive environments has been developed by Woods (1998) and the potential role of interpretation in wildlife tourism settings has been discussed by Moscardo et al. (2004). The development of augmented reality signage for zoos that allows greater personalised connections to be made between animals and visitors has been described by Kelling and Kelling (2014).

Working in cooperation with Cincinnati Zoo, Ohio, Debevec and Kernan (1984) examined the effect of the physical attractiveness of male and female presenters (played by models) on the outcome of slide show presentations aimed at recruiting volunteers to assist with a (fictitious) campaign for a zoo levy to appear on the ballot in an upcoming election. Males responded significantly more favourably towards the attractive female model and she was more persuasive than the message alone in persuading males to favour passage of the levy. This cross-gender effect was not observed in female participants.

3.12   Accessibility

3.12.1   Making Zoos More Accessible for the Visually Impaired

To increase the accessibility of an aquarium to visually impaired visitors, fishes and other animals were tracked with computer vision by Walker et al. (2007) and the data collected on their movements was used to create music that varied with the number and types of fishes, their locations and behaviours. The soundscapes created communicated the dynamic aspects of the exhibit and were also designed to convey an emotional content such as 'amazement and wonder at the massive whale shark gliding by'. This work was part of the Accessible Aquarium Project at the Georgia Institute of Technology in the United States.

Jennings (1996) has discussed the use of focus groups with visitors who are visually impaired in order to examine the usefulness of various aids in communicating with visitors to Brookfield Zoo, Illinois. She noted that only a small minority of visually impaired people read Braille and that wayfinding within the zoo was a major problem. Participants identified that 3D maps could be useful and, although some adults were reluctant to touch sculptures depicting animals, when they did they learned a great deal about the animals depicted.

3.12.2   Ethnic Minorities, Culture and Education

Many zoos make a great effort to engage with local communities by developing outreach projects with local schools. This may involve taking small animals into classrooms and more generally talking to children about conservation and local biodiversity.

Aboriginal education officers at Taronga Zoo in Sydney provide specific Aboriginal educational programmes that have been created in partnership with local Aboriginal communities (Taronga Zoo, 2017). An evaluation of indigenous wildlife interpretation at captive wildlife attractions in Australia has been conducted by Zeppel and Muloin (2007) using telephone interviews of staff. Staff believed that tourists benefited from the inclusion of indigenous content in many ways, including dispelling myths, learning about indigenous cultures, increasing cultural awareness and developing more positive attitudes towards indigenous people. The employment of indigenous guides to provide interpretive talks and tours at wildlife attractions was supported by managers and staff.

Arab and Jewish visitors to the Tisch Family Zoological Gardens in Jerusalem perceived it as an educational institution, but neither regarded conservation as a prominent message (Tishler et al., 2020). The two communities differed in their animal preferences and in their interpretations of the message the zoo was conveying.

3.12.3   Zoo Libraries, Archives and Access to Information

The history of zoo and aquarium libraries began in the early nineteenth century with the development of modern zoos. However, zoos were not established in America until much later in the nineteenth century and, according to Kisling (1993), few of them had libraries before 1960 or archives prior to 1970.

Rohr (1989) noted that archives are a low priority for most zoos and aquariums due to limited space and a lack of both funds and knowledgeable personnel. She has suggested that librarians and archivists should become knowledgeable about the importance of zoo records and more assertive about their maintenance. More recently, Barr (2005) found that zoo and aquarium libraries had increased their holdings, resources and scope in facilities of a wide range of sizes, in spite of the general economic hard times. They also acknowledged that these libraries helped their parent institutions fulfil their modern missions relating to education, conservation and research.

An international biography of works written about animal health libraries, including zoo and aquarium libraries, has been published by Croft (2008).

3.13   Animal Shows, Tricks and Fundraising

Animal shows in zoos have evolved from degrading circus-like displays of parrots riding miniature tricycles and roller-skating chimpanzees into performances of naturalistic behaviours such as climbing, flying, feeding or swimming (Fig. 3.7). Apart from the opportunity these shows provide for keepers and trainers to show off the natural behaviour of their animals, they also provide a chance for the zoo to collect donations towards the *in-situ* conservation of these animals or closely related species.

Visitors that have watched the California sea lion (*Zalophus californianus*) show at Blackpool Zoo in England are asked to donate money to assist in the conservation of the Mediterranean monk seal (*Monachus monachus*); South Lakes Wild Animal Park

Fig. 3.7 A sea lion show can be used to demonstrate the natural behaviours of animals, but unnatural behaviours send confusing messages to visitors.

(now Safari Zoo) in England collected donations for the conservation of wild Sumatran tigers (*Panther tigris sumatrae*) following a talk and the feeding of the animals with meat attached to the top of a telegraph pole.

There is some evidence that this approach to fundraising is both acceptable to visitors and generates a greater response than simply giving individuals an opportunity to donate at a donation point. At Melbourne Zoo in Australia, a donation request to help wild fur seals was introduced into an interactive Australian fur seal (*Arctocephalus pusillus doriferus*) presentation and the response of visitors was surveyed as they exited the zoo. Researchers found that the introduction of the request did not negatively impact on visitors' satisfaction with the presentation, and a higher proportion of visitors reported that they either had donated or intended to donate than was the case for those who were only exposed to the seal exhibit and the donation point (Mellish et al., 2017).

The impact of a sea lion and mixed-species bird show on audience knowledge of facts about animals found that visitors demonstrated more knowledge after the show than before the show (Spooner et al., 2021). However, conservation action awareness showed only a weak increase post-show. The study also found that audiences were confused about natural adaptations after seeing 'trick-type behaviours'. The authors of the study concluded that live animal shows are an important potential source of knowledge transfer – over one-quarter of visitors attended some form of animal show during their visit –and that priority should be given to natural behaviours with a focus on conservation action.

3.14 Zoos and Higher Education

3.14.1 Zoo Biology and Management as a Discipline in Higher Education

Although zoos may have difficulty demonstrating that they have an educational function in relation to influencing the general public or significantly contributing to the education of school children, there is no doubt that many zoos assist in the training and education of future generations of zoo scientists. In the UK, programmes in zoo biology have become popular in the last decade or so from college level up to Master's degree level (Table 3.2). In some cases higher education institutions cooperate directly with zoos to provide courses, while in others, although they are not directly involved with teaching, zoos provide research opportunities for student projects, including doctoral projects.

3.14.2 Zoos and Veterinary Training

Most veterinarians deal with a narrow range of species. Clearly, this is not the case for wildlife and zoo vets. Zoos provide an important opportunity for veterinary students to encounter species and clinical conditions that they might not otherwise experience. Internships in zoos in the United States are considered particularly valuable (Cushing, 2011).

The history of the field of zoo and wildlife pathology has been discussed by Lowenstine and Montali (2006) along with a consideration of training opportunities available to veterinary students and qualified veterinarians. They noted that much of the initial emphasis in the field was on the management of game animals but that this gradually changed when veterinarians became involved and attention turned to the improvement of captive care and conservation strategies.

Liptovszky et al. (2021) have assessed the educational value of a five-day zoo placement for final-year veterinary students ($n = 200$) using an end-of-placement questionnaire. Before the placement only 35 per cent of students stated that their understanding of modern zoos was good or excellent compared with over 90 per cent of students after the placement. In addition, 43 per cent reported that the placement had a positive impact on their attitudes towards zoos.

Although zoos provide useful experiences for vets, employment opportunities are limited by both the small number of zoos (compared with the opportunities to work in general practice) and the size and budgets of these zoos. Of the 96 zoos (including six aquariums and one primate centre) that responded to a survey of American Association of Zoological Parks and Aquariums (AAZPA) institutions (now AZA) in 1987, 40 (45 per cent) employed at least one full-time veterinarian and of these 12 zoos (13 per cent) had more than one full-time veterinarian (Gentz, 1990). Most of these vets were male (76 per cent) and only 14 per cent of the total held a Master's degree. The zoos that employed more than one full-time veterinarian had large collections – with an average of over 2,000 animals – and large annual operating budgets, averaging over US$14 million.

3.15 Zoos, Technology and Education

There is some evidence that engaging with technology during a visit can interfere with children's informal learning in a zoo or aquarium environment. A study of the

Table 3.2 Universities in England and Wales offering undergraduate courses in which zoo and/or aquarium studies are a significant component.

Institution	Location	Programme
Bangor University	Bangor	BSc Zoology with Animal Management
Cornwall College (University of Plymouth)	Newquay	FdSc Zoological Conservation
Derby College (University of Derby)	Derby	FdSc/BSc Animal Management (Zoo and Wildlife)
Kingston Maurward College	Dorchester	FdSc Animal Behaviour, Welfare and Conservation
		BSc Animal Behaviour, Welfare and Conservation (Animal Collections)
Merrist Wood College (Kingston University)	Guildford	FdSc/BSc Zoo Management
Nottingham Trent University	Nottingham	BSc Zoo Biology
South Gloucestershire and Stroud College (University of Gloucestershire)	Bristol	FdSc Zoological Management
		BSc Zoological Management and Conservation
South Staffordshire College/Rodbaston College (Staffordshire University)	Stafford	FdSc/BSc Zoo Animal Management with Conservation
University Centre Askham Bryan	York	FdSc Management of Aquatics and Conservation of Oceans
		FdSc Management of Animal Collections with Conservation
		BSc Zoo Management
University Centre Myerscough	Lancashire	BSc Animal Science and Welfare (Zoo Conservation Biology)
University Centre Reaseheath (University of Chester)	Nantwich	FdSc/BSc Zoo Management
University Centre Sparsholt	Winchester	BSc Zoo Biology
University of Greenwich	London	BSc Zoo Husbandry and Management
University of Salford	Greater Manchester	BSc Wildlife Conservation with Zoo Biology
Writtle University College	Chelmsford	BSc Animal Management (Zoo and Wildlife Conservation)
		BSc Animal Science (Zoo and Wildlife Conservation)

Accurate at 5 June 2022. BSc, Bachelor of Science; FdSc, Foundation Degree in Science.
Source: extracted from the database of the Universities and Colleges Admissions Service.

relationship between smartphone use and parent–child conversations during visits to an aquarium found that 55 per cent of families used a phone during their visit, primarily to take photographs and record videos (Kelly and Ocular, 2021). More phone-using adults reported off-topic talk than did those who did not use their phones, and the more parents used their phones the less contingent were their conversations.

The collecting and sharing of location-based content on mobile phones by children during a visit to London Zoo were studied by O'Hara et al. (2007). The 80 subjects consisted of 33 children on an organised zoo visit and 47 children who were members of the public on a family visit (mean age 10.7 years). The children used mobile camera phones to read 2D barcodes on signs at the animal enclosures to access related information. Some aspects of the phone application used encouraged cooperation and sharing among the participants, while others encouraged competition. After their visit the participants were able to see what they had 'collected' on a dedicated website. O'Hara et al. suggested that this was important to how the zoo visit contributed to family and group relationships and extended the value of the outing, becoming the basis for ongoing conversations about the visit back home between family members who participated in the visit and those who did not.

Hirskyj-Douglas et al. (2021) claim that young zoo visitors largely ignore interactive signs and videos. They explored the use of an educational card and role-playing game called *ZooDesign* by children to investigate how animals can use technologies to cater for their welfare needs and human involvement. Some of the cards contained information about individual species, including their size, weight, life span, diet and other aspects of their biology and habitat requirements. Other cards provided information about different types of environmental enrichment: dietary, play, sensory and social. The authors concluded that the approach constituted a 'good process' but not all of the designs produced by the children were appropriate for use with animals.

Many zoos have established their own YouTube channels, giving them an opportunity to send important conservation messages to their visitors and those who are unable to visit. But are they making good use of this platform? Llewellyn and Rose (2021) examined the YouTube videos made by 20 zoological institutions – most of which were located in Europe or North America – and found that they focused more on entertainment than education and did not reflect the taxonomic diversity of the zoos' animals. The most popular videos were of mammals, especially charismatic taxa such as elephants and bears. Videos with a conservation content made up only 3 per cent of the most viewed videos, but there was some evidence that more conservation-focused videos are being produced over time.

3.16 Are Zoos Educationally Redundant?

It is often claimed that zoos have become 'educationally redundant' because of the many nature documentaries on television (Margodt, 2000). However, the footage used in such programmes is highly edited and over-emphasises the amount of time many species are active during the day, reinforcing the view that many animals living in zoos appear 'bored'. In reality, animals such as big cats do not spend large amounts of time engaging in unnecessary activity in the wild. In his study of lions in the Serengeti, Schaller (1972) found that they were largely inactive for 20–21 hours per day, although he could not assess how much of this time was spent asleep. Some authors have suggested that behaviour observed in some species in zoos is more natural – and consequently of greater educational value – than that seen in television documentaries, especially in modern exhibits (Burgess and Unwin, 1984; Andersen, 2003).

Compared with other sources of information about wildlife, zoos do not appear to be as important as television. Broad and Smith (2004) interviewed 125 zoo visitors and found that 42 per cent had obtained their previous knowledge of animals and habitats from the television, 29 per cent had obtained it from documentaries and 21 per cent from books, newspapers and magazines. Only 14 per cent had gained this prior knowledge from previous visits to zoos.

Contrast the conclusions drawn from many of the studies discussed above regarding the educational value of a zoo visit with those of Fernández-Bellon and Kane (2020). They analysed Twitter and Wikipedia big data activity following the airing of Sir David Attenborough's television series *Planet Earth 2* by the BBC. They found that, despite lacking specific conservation themes, the series generated species awareness and stimulated audience engagement and desire for information at magnitudes comparable to other conservation-focused campaigns. The authors concluded that natural history films can generate durable shifts in audience awareness beyond the period of broadcast of the programmes and provide the public with vicarious connections with nature.

3.17   Conclusion

Asking whether or not zoos have an educational function is rather like asking if drugs cure disease. Drug trials consider the efficacy of one drug – or a combination of a small group of drugs – at a time; studies of education in zoos usually examine the efficacy of a single exhibit, single activity or a single zoo in educating the public. For this reason, there is no single comprehensive answer to this question. Children and many adults enjoy a day at the zoo. Whether or not they learn anything while they are there or change their behaviours after their visit is unclear. Some zoo professionals appear certain that zoos have an educational value, but some professional zoo educators are more circumspect. Now, over 60 years after London Zoo introduced its lectures for school groups and Boorer (1967) wrote about zoos as teaching aids, empirical studies on the educational impact of environmental education programmes in zoos are still rare (Kleespies et al., 2020).

4 Anthrozoology and Visitor Behaviour

4.1 Introduction

Ethnobiology is the science that examines the interactions between human cultures and plants and animals. Anthrozoology is the branch of ethnobiology that considers the interactions between humans and other animals in a range of cultural, historical and geographical contexts. This chapter considers the relationships between people – especially zookeepers and zoo visitors – and the animals kept in zoos and aquariums. It also includes a brief consideration of visitors' behaviour when exploring zoos and viewing exhibits.

4.2 The Effect of Visitors on the Behaviour of Zoo Animals

Zoos and aquariums cannot survive without visitors, but the presence of visitors has the potential to alter the behaviour of animals and compromise their welfare. The relationship between visitors and the animals they visit has become a fruitful area of research in recent years.

Studies of the effect of visitors on the behaviour of animals generally combine an analysis of activity budgets or frequencies of particular behaviours (e.g. aggression) and the location of individual animals within the enclosure with a measure of visitor numbers or some other visitor metric (e.g. group sizes, visitor noise). Studies have shown that characteristics such as visitor presence, density, activity, size and position are associated with behavioural and, to a lesser extent, physiological changes in some zoo animals (Davey, 2007b). However, in some cases the effects on the animals are positive (enriching) and in other cases they are negative (undesirable), so the extent to which visitors affect animals' welfare is inconsistent between and within both species and zoos.

In a study of 25 little penguins (*Eudyptula minor*) Sherwen et al. (2015) showed that when the birds were exposed to visitors they exhibited increased levels of aggression, huddling and visitor-avoidance behaviours (e.g. spending time behind enclosure features and increased distance from the visitor viewing area). They suggested that the presence of visitors or some aspect of their behaviour may have provoked fear in these birds.

A study of 15 primate species found that in the presence of visitors, animals were less affiliative, more active and more aggressive (Chamove et al., 1988). These effects were particularly marked in arboreal species, especially smaller species. Lowering the height of the spectators (by asking them to crouch) reduced these adverse effects by 50 per cent. In the same study, as visitor numbers increased a group of mandrills (*Mandrillus sphinx*) exhibited a linear increase in attention to visitors, activity and stereotypical behaviour. The researchers concluded that, rather than a source of enrichment, visitors are a source of stressful excitement for the primate species studied.

A study of orangutans (*Pongo* sp.) living in two naturalistic enclosures in Singapore Zoo found that visitor numbers had little effect on the animals except at one of the exhibits where the likelihood of looking at visitors and food soliciting increased when visitor numbers exceeded 40 (Choo et al., 2011). Visitor activity did not appear to cause stress in the animals, and visitors with food provided enrichment. However, orangutans played less and looked at visitors more when they were in close proximity. Birke (2002) found that adult orangutans (*P. pygmaeus*) used paper sacks to cover their heads more, and infants held onto adults more, during periods of high visitor density (Fig. 4.1).

Although there are many instances where visitors have been shown to influence the behaviour of animals living in zoos, this is not always the case. Sherwen et al. (2014) found that variation in the intensity of visitor behaviour – visitor noise level and the extent to which visitors attempted to interact with the animals – did not affect the

Fig. 4.1 In the wild, orangutans (*Pongo*) cover their heads with large leaves. Zoos often provide these animals with paper bags, sacks or old clothes. They have been recorded covering their heads more when visitor density is high.

behaviour of captive-born meerkats (*Suricata suricatta*) at three zoo exhibits. Signage requesting visitors to be quiet and not to interact with the meerkats – the regulated treatment – reduced noise level by around 32 per cent and reduced the intensity of visitor behaviour compared with that in the unregulated treatment where there was no such signage. However, the regulated treatment had no effect on the distance that meerkats positioned themselves from visitors or the proportion of time they engaged in vigilant behaviours.

There was no association between visitor numbers and the performance of stress-related behaviours (pacing and aggression) or critical behaviour (feeding, nesting and resting) in birds living in a walk-through mixed-species, free-flight aviary, suggesting that the birds were not experiencing substantial negative welfare (Blanchett et al., 2020).

Persson et al. (2018) examined spontaneous imitation in interactions between chimpanzees (*Pan troglodytes*) and visitors at Furuvik Zoo/Lund University Primate Research Station, Furuvik, Sweden. They found that humans and chimpanzees imitated each other at a similar rate (about 10 per cent of all actions) and imitation was promoted by physical proximity (i.e. indoors). The authors suggested that imitation accomplished a social-communicative function because interactions between the two species that contained imitative actions lasted longer than those without imitation. The four chimpanzees – out of a total of five – who engaged in imitation engaged in 'imitative games'.

The COVID-19 pandemic resulted in the closure of zoos around the world and gave scientists a unique opportunity to compare the behaviour of animals in the presence and absence of visitors during normal opening hours. Williams et al. (2021a) studied the impact of zoo closures (visitors absent) and reopening (visitors present) on the behaviour of individuals of eight mammal species at two zoos in the UK. Behaviour change and enclosure use varied across species, but most changes were not significant. They noted that some species have the potential to take longer than others to re-habituate to the presence of visitors.

In a second study, Williams et al. (2021b) examined groups of slender-tailed meerkats (*Suricata suricatta*) at three institutions in the UK (Knowsley Safari Park, *Plantasia* and Twycross Zoo) and a single group of African penguins (*Spheniscus demersus*) in a South African facility (uShaka Sea World). They found no differences in the behaviour of the penguins between the COVID closure and the period after the facility was reopened. However, during the COVID closures the meerkats used more of their enclosures and performed more environmental interactions and less alert behaviour than during the reopening periods. After the zoos were reopened, the meerkats performed more positive social interactions than during the closure. Unfortunately, the sizes of the three meerkat populations were small ($n = 2$, 7 and 10) and the authors concluded that it was not possible to determine whether the presence of visitors was 'stressful' or 'enriching' for meerkats after a long period of visitor absence or whether the meerkats were exhibiting naturally inquisitive behaviour when visitors returned.

Close contact between non-domesticated animals and visitors is permitted in some zoos and aquariums and this has the potential to exacerbate the negative

Fig. 4.2 Handling koalas (*Phascolarctos cinereus*) causes physiological stress in some individuals.

effects of visitors on animals. Handling koalas (*Phascolarctos cinereus*) causes physiological stress in some individuals (Narayan et al., 2013) (Fig. 4.2). Handled males had faecal cortisol metabolite (FCM) levels 200 per cent higher than non-handled males, but females were only affected by handling if they were lactating. A study of the effect of photographing koalas on their stress levels at Sydney Zoo, Australia was conducted by Webster et al. (2017). Koalas were housed in a multiple-bay enclosure for photography sessions with visitors, but no touching of the animals was permitted. Faecal cortisol metabolite levels were higher in intensive photography conditions compared with controls, but returned to basal levels during standard photography conditions. In the experimental bays males had higher mean FCMs than females.

Anderson et al. (2002) have suggested that improvements in exhibit design and management techniques may improve human–animal interactions and increase animal wellbeing in a petting zoo environment (Fig. 4.3). They examined the relationship between the behaviour of sheep and goats towards humans in a petting zoo and the effect of their environment. Five African pygmy goats (*Capra hircus*) and two Romanov sheep (*Ovis aries*) were observed in an environment that contained a 'retreat space' which the animals could use to avoid contact with humans. The study measured the animals' behaviour in three conditions: no retreat space, semi-retreat space and a full-retreat space. In both species, undesirable behaviours were lowest in the full-retreat condition.

Fig. 4.3 Interactions between animals and visitors are unavoidable in petting zoos as animals harass visitors for food and visitors approach animals to pet them. These inquisitive and perpetually hungry goats were residents of Berlin Zoo, Germany.

Many aquariums allow contact between fishes, starfishes and other aquatic taxa in open-topped tanks known as touch pools. The ethical issues arising from these interactions have been discussed by Biasetti et al. (2020). Animals kept in touch pools are held in shallow water – so often unable to exhibit species-typical behaviours – and cannot escape from visitors and aquarium staff (Fig. 4.4). The authors draw attention to the fact that some fishes kept in touch pools exhibit evasive behaviours and may exhibit stereotypic behaviours. Visitors handling organisms may cause them physical damage or may be bitten, and there is the possibility of exposure to pathogens.

Although it is clear to anyone that visits a zoo than some visitors behave in a manner which is likely to disturb animals, little attention has been paid by researchers to what is acceptable behaviour for members of the public. Visitors vary in their empathy for animals kept in zoos. A study conducted at the Henry Doorly Zoo and

Fig. 4.4 Aquatic animals kept in shallow pools are at risk of harm from visitors. (a) Beadlet anemone (*Actinia equina*). (b) Thornback ray (*Raja clavata*). Fishes like this ray frequently appear at the surface of touch pools.

Aquarium in Omaha, Nebraska, found that visitors' views regarding the purpose of zoos and the similarities they perceived between humans and animals influenced the behaviour they considered appropriate around zoo animals (Mullar et al., 2021).

Fig. 4.5 Noisy visitors can cause stress for some species in zoos and alter their behaviour. Signs asking visitors to be quiet when visiting (a) okapis, (b) gibbons, (c) monkeys and (d) gorillas.

Those who reported seeing other visitors behaving inappropriately – for example, hitting glass barriers and trying to provoke a response from animals – more often reported similarities between animals and humans (Fig. 4.5). Overall, visitors ranked education and conservation as the most important functions of zoos, and entertainment and research as the least important. However, those who identified entertainment as the most important zoo function reported fewer similarities between humans and animals and also reported observing fewer inappropriate behaviours in other visitors.

Concerns regarding the negative effect of the presence of visitors on the behaviour of some species has resulted in a number of studies of the effects of reducing animals' exposure to visitors by obscuring windows and screening off the perimeters of enclosures. These studies are discussed in Chapter 8.

4.3 Keeper–Animal Relationships

Zookeepers form strong bonds with the animals for whom they care, and may spend more time with them than they spend with their own family. This is especially true for long-lived and intelligent species that have strong personalities, such as elephants, cetaceans and primates (Fig. 4.6).

Fig. 4.6 Keeper Hicks at London Zoo, 1885. Many keepers form close and long-lasting relationships with the animals in their care (source: An Amateur Photographer at the Zoo. *The Graphic*, 5 September 1885. p. 271).

The relationship between chimpanzees (*Pan troglodytes*) and their keepers can affect their welfare. In a study conducted at Zoo Northwest Florida in Gulf Breeze, United States, Jensvold (2008) showed that chimpanzees engaged in more friendly behaviours (e.g. play) when keepers communicated with them using typical chimpanzee behaviours and vocalisations, than when the keepers communicated with human speech and human behaviours. The apes also interacted more with keepers when they used chimpanzee behaviours.

A strong bond has welfare benefits for both elephants and their keepers. A survey of 427 keeper–elephant pairings in North American zoos found that dissatisfaction with the job was associated with weaker keeper–elephant bonds and that stronger bonds were more common with Asian elephants (*Elephas maximus*) than African elephants (*Loxodonta africana*). When keepers are treated by Asian elephants as part of the herd, the animals experience lower serum cortisol levels (Carlstead et al., 2018).

In a study of keeper–animal dyads (pairs) involving black rhinoceros (*Diceros bicornis*), Chapman's zebra (*Equus burchellii*) and Sulawesi crested macaques (*Macaca nigra*) in six zoos across the UK and the United States, Ward and Melfi (2015) found that there was a significant difference in the animals' latency to respond appropriately to commands and cues from different keepers. This indicated that unique keeper–animal relationships had been formed. Stockmanship styles varied between keepers, determined by their attitude towards the animals and their knowledge and experience of them.

The presence of a keeper may have adverse effects on the behaviour of some animals. The effect of the close presence of a keeper on the undesirable behaviours exhibited by African pygmy goats (*Capra hircus*) and Romanov sheep (*Ovis aries*) in a contact yard at Zoo Atlanta, Georgia, was studied by Anderson et al. (2004). Undesirable behaviours were defined as aggressive, avoidance and escape behaviours directed at humans, that is behaviours which were incompatible with the purpose of visitors gaining a positive experience through contact with the animals. They found that the rate of undesirable behaviours increased as the rate of visitor touches increased and that – contrary to expectations – the close presence of a keeper increased the rate of undesirable behaviours in the animals.

Melfi et al. (2021) used a modified Lexington Attachment to Pets Scale (LAPS) to measure the strength of keeper–animal relationships among 187 keepers working in 19 collections in three countries, extending an earlier study which used just 22 keepers (Hosey et al., 2018). They found that female keepers scored more highly than males for their relationships with both zoo animals and pet animals. Pet animal scores were comparable with those obtained for the general public, while zoo animal scores, although reflecting attachment, were influenced by differences in culture between the institutions.

Zoo staff can feel loss when an animal dies that is identical to that of losing a family member. On 28 May 2016, *Harambe*, a 17-year-old Western lowland gorilla kept at Cincinnati Zoo, Ohio, was shot by zoo staff after displaying aggression towards a three-year-old boy who had climbed a fence and fallen into the moat around his enclosure. The trauma and distress suffered by onlookers, his keepers and the director of the zoo where he was raised have been described by Cohen and Clark (2019).

Bonds between keepers and 'their' animals may have adverse animal welfare consequences. Richardson (2000) has suggested that zoos sometimes delay making decisions about euthanising animals that are suffering due to staff sentiment. There is some evidence from necropsies performed on euthanised animals to suggest that pathological changes of organs and/or the musculoskeletal system were already advanced in many cases (Föllmi et al., 2007). The adverse effects of euthanising animals on animal care staff have been discussed by Reeve et al. (2004).

4.4   Keeper Education and Motivation

Semi-structured interviews were used by Birke et al. (2019) to investigate the relationships between zoo professionals and animals across five institutions. Keepers entered the job because of their 'love of animals', but accepted that they had to maintain a physical and emotional distance from them; they often perceived themselves as the best person to deal with a particular animal; they considered the animals in the zoo 'wild' while recognising they were captive; and they formed bonds with particular individuals.

The amount of training received by zoo staff varieties widely. Some have no formal qualifications while others may be educated to degree level or beyond. Zookeepers, in particular, tend to train on-the-job. A study of felid keepers employed in North

American zoos demonstrated that job satisfaction was directly linked to access to training on felid welfare and the perceived fulfilment of the Five Freedoms (see Section 9.3) (DeSmet and Ogle, 2022). However, job satisfaction did not appear to influence the bond keepers felt with felids (Fig. 4.7).

Fig. 4.7 Some keeper tasks are more pleasant than others. (a) Keepers with white tiger cubs (*Panthera tigris*) and (b) cleaning an antelope paddock.

In a survey of animal welfare educational opportunities and barriers conducted among zoo staff in European and Chinese zoos, respondents reported that motivation and enthusiasm for the job were important alongside formal training (Bacon et al., 2021). Staff considered contextual information – including wildlife ecology – to be important in zoo animal welfare education. They valued learning opportunities but did not always feel supported by management. Other barriers to zoo staff education were identified as the lack of time and funding, and the English language bias in the published literature. The authors of the study noted that stockmanship was lacking in some keepers and they were unaware of the possible effects of their behaviour on the welfare and behaviour of the animals with which they worked. Although interesting, it should be noted that the sample sizes in this study were very small: only eight zoo staff were interviewed from European zoos and eight from Chinese zoos.

The relationship between zookeepers and their work was examined by Bunderson and Thompson (2009). They surveyed keepers in the United States and Canada and found that those with a strong sense of calling found broader meaning and significance in their work but were also more likely to see this work as a moral duty, to sacrifice pay, personal time and comfort for their work and to hold their zoo to a higher standard. The results of this study revealed that the meaningful work done by zookeepers can become a double-edged sword.

4.5 Using Caregivers and Visitors to Collect Data

4.5.1 Using Untrained Observers to Collect Behavioural Data?

Keepers and other zoo staff do not usually have much time to engage in research. The possibility that untrained visitors could be used to collect animal behaviour data in a zoo environment was examined by Williams et al. (2012). They were asked to collect behavioural data for a group of captive North American river otters (*Lontra canadensis*) at Slimbridge Wetlands Centre, in England, using a simple questionnaire, and these data were then compared with baseline activity budget data collected by a trained biologist. Williams et al. concluded that the visitors were unable to collect accurate data mainly because they failed to comply with the time limit set for recording observations (30 seconds), resulting in sampling errors. Many visitors recorded for much longer than the specified time, resulting in more behaviours being recorded than would have been seen in the 30-second sampling period they were asked to use.

4.5.2 Keeper Assessments of Animals

Developing methods for monitoring the welfare of animals in zoos has received considerable recent attention and keeper and animal trainer ratings are increasingly being used in zoos as a welfare tool (Whitham and Wielebnowski, 2009). This is particularly important in cross-institutional studies where it is not possible for one or a small group of scientists to undertake time-consuming behavioural studies of each animal.

Experienced keepers know their animals well and may be important repositories of information about the behaviour of particular animals, such as knowledge of past relationships between individuals or the husbandry techniques in their last home, which may have led to particular behavioural abnormalities.

Carlstead et al. (1999) studied the relationship between behaviour and breeding success in black rhinoceros (*Diceros bicornis*) by asking keepers at 19 zoos in the United States to rate 60 animals on 52 behaviour elements using a questionnaire. Subsequently they surveyed 70 black rhinoceros at 24 zoos and concluded that keeper ratings of the behaviour and temperament attributes of this species can be used as reliable and valid cross-institutional descriptions of individual differences between animals. Keeper questionnaires were also used by Yasui et al. (2012) to study the personalities of 75 elephants, and keeper ratings of behaviours were used in a study by Wielebnowski et al. (2002) of the effects of stress on captive clouded leopards (*Neofelis nebulosa*). Keepers have also assisted in assessments of personality in snow leopards (*Uncia uncia*) (Gartner and Powell, 2012) and in cheetahs (*Acinonyx jubatus*) (Chadwick, 2014).

4.6   Health and Safety: Accidents and Incidents

The close proximity between humans – especially keepers and veterinarians – and non-human animals in zoos inevitably results in accidents, including human fatalities, from time to time.

Accidents involving animals in zoos are relatively rare. Most of the accidents in British zoos reported to the Health and Safety Executive in Britain in 2016 did not involve animals but were the result of children falling off playground equipment (*The Times*, 2017) (Fig. 4.8).

A number of studies have examined the risks associated with zoo visits and disaster planning in zoos. A study of the public's awareness of safety instructions in relation to crowd management after the birth of a cub at the Giant Panda Conservation Centre at Zoo Negara, Malaysia found that the level of awareness was affected by visitor background (Siew et al., 2018).

Ginsburg and Miller (1982) found that significantly more 3- to 11-year-old boys than girls visiting San Antonio Zoo, Texas, engaged in risk-taking behaviour: rode the elephants, petted the burro, fed the animals and climbed the river embankment. Older children of both sexes were more likely to take these risks than were younger children.

A number of serious incidents in zoos that have led to human deaths and the subsequent destruction of animals have been described by Barreiros and Haddad (2016). They noted that awareness of the dangers in zoos appeared to be poor among the general public and suggested that those who live in urban areas have a romanticised view of wildlife that bears little resemblance to reality.

Fig. 4.8 (a) Safr-Kali giving rides to children at London in 1885 (source: An Amateur Photographer at the Zoo. *The Graphic*, 5 September 1885. p. 270). (b) 'The latest tip': an artist's impression of an accident at London Zoo (source: *The Illustrated Sporting and Dramatic News,* 4 June 1881).

In 2014 a two-and-a-half-year-old girl was attacked by an adult female Brazilian tapir (*Tapirus terrestris*) when she entered the animal's enclosure during a behind-the-scenes visit at Dublin Zoo, Ireland. The tapir had recently calved. The child sustained multiple injuries, including a deep laceration of the left forearm, haematoma on the right forehead and multiple abdominal lacerations, the deepest of which resulted in evisceration of the large bowel. Fortunately, she had recovered completely after six weeks (O'Grady et al., 2014).

The death of a man found inside the cage of two Himalayan black bears (*Ursus thibetanus*) at Belgrade Zoo in Serbia has been described by Mihailovic et al. (2011). The man was intoxicated as a result of visiting a nearby beer festival and the autopsy revealed that he had died of exsanguination as a result of multiple internal and external injuries.

In April 2000 an outbreak of Legionnaires' disease – caused by *Legionella pneumophilia* – at the Melbourne Aquarium in Australia affected 125 visitors. Of these, 76 per cent were hospitalised and four died. The infection was associated with poorly disinfected cooling towers at the aquarium (Greig et al., 2004).

Modern zoos should have plans in place to deal with a wide range of eventualities such as storms, power failures and outbreaks of disease (Fig. 4.9). Current and anticipated plans for the protection of zoo animals in the event of a thermonuclear catastrophe were examined in a survey of 115 zoos in 41 countries by Maroldo (1982). Of these, 78 zoos located in 23 countries responded and the analysis indicated that 91 per cent of these zoos had no protection plans for such a catastrophe. What is perhaps more surprising is that 9 per cent of the zoos had a plan.

Zoos have legal responsibilities for the safety and welfare of their visitors, animals and staff. The legal aspects of the safety of animals and humans in Malaysian zoos have been considered by Hassan (2014).

Fig. 4.9 Zoos need to be prepared for natural disasters. A toilet block at Detroit Zoo, Michigan, doubles as a tornado shelter (inset).

During natural disasters and wars, animals may escape from zoos and become a threat to the local human population. During the Second World War the Japanese government destroyed many zoo animals, ostensibly to protect civilians if zoos were bombed by the Allied forces. However, Itoh (2010) has argued that this was done for political reasons to impress upon the population the gravity of the war and to focus the anger of civilians on the Allied forces, since the killings began long before the first air raids occurred.

The deaths of animals and the destruction of buildings at Berlin Zoo in Germany as a result of bombing by Allied forces in the Second World War have been discussed by Kinder (2013). In total, some 90 per cent of the animals were killed. The author describes the manner in which a memorial to the seven elephants killed when the elephant house was bombed transformed the animals into emblems of the German nation-state: innocent victims of outside invasion. Kinder also notes that Berlin Zoo played a significant historical role in validating Nazi ideology. Another view of the effect of war on zoos and their animals has been provided by Malamud (1998).

In 2009, Newquay Zoo in England recreated a 'dig for victory' allotment garden based on those created in many European zoos during and after the Second World War (Norris, 2014). The purpose of the garden was to educate the public in the roles of zoos and botanical gardens in wartime and to provoke discussion about future sustainability challenges by considering past crises.

4.7 Accidents and Illnesses among Zookeepers, Veterinarians and Other Zoo Staff

Incidents involving animals attacking people, especially keepers, regularly attract the attention of the media. In December 1874 the front page of *The Penny Illustrated Paper and Illustrated Times* carried a drawing depicting 'The gallant rescue of keepers from the infuriated rhinoceros at the Zoological Gardens [London Zoo]' (*The Penny Illustrated Paper and Illustrated Times*, 1874) (Fig. 4.10).

In zoos, aggression towards keepers may be the result of the failure to establish long-lasting keeper–animal relationships. This may be due to the high rate of keeper turnover in some facilities and is a particular problem with long-lived dangerous animals such as elephants.

Lehnhardt (1991) documented 15 elephant-related deaths between 1976 and 1991 – approximately one death per year. Data on 36 Asian elephants responsible for serious accidents involving keepers have been analysed by Benirschke and Roocroft (1992). They found that bulls were significantly more dangerous than cows. The mean age of bulls involved in accidents was 18 years (corresponding with the approximate age when they first experience musth) and the mean age of cows was 25.3 years (corresponding approximately with the age when they become matriarchs) (Kurt, 1995). Gore et al. (2006) have reviewed the factors responsible for injuries caused by captive elephants. Litchfield (2005) has suggested that lower-ranking elephant keepers may be at greater risk than the head elephant keeper and their deputy because their ranks are not recognised by herd members. However, he presented no empirical evidence to support this theory.

In a study of occupational fatalities due to mammal-related incidents in Japan during the period 2000–2019, Hioki and Inaba (2021) found 12 fatal accidents affecting persons working with mammals in zoos. The taxa involved were bears (five) elephants (three), tigers (two), lion (one) and rhinoceros (one). In many cases of fatal attacks in zoos it is difficult to determine the precise circumstances leading up to the attack, especially when there are no witnesses.

A fatal attack by a cow elephant on a male elephant keeper has been reported by Hejna et al. (2011). The keeper tripped over a foot chain while the elephant was receiving medical treatment. The elephant repeatedly attacked him with her tusks and he later died in hospital. The keeper sustained multiple penetrating injuries to the groin and abdomen, laceration of the abdominal aorta, contusion of both lungs, laceration of the liver, prolapse of the small intestine, multiple rib fractures and comminuted fractures of the pelvic arch and left femoral body. A detailed reconstruction of an incident involving the killing of a keeper by a musk ox (*Ovibos moschatus*) in a zoo has been provided by Schalinski et al. (2008).

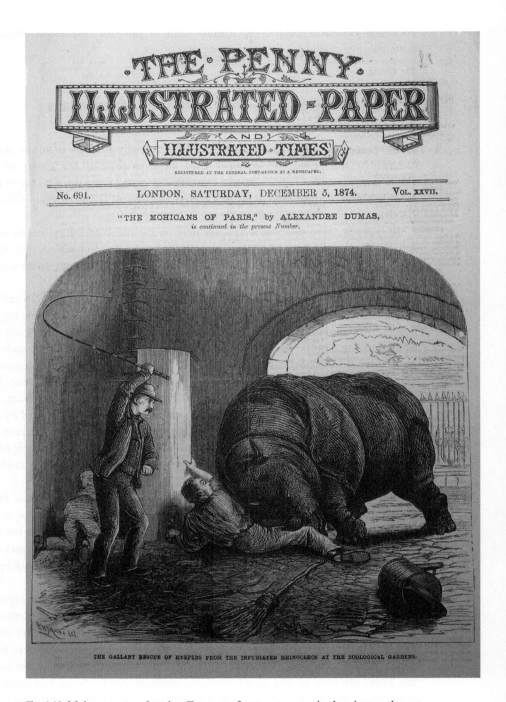

Fig. 4.10 Major events at London Zoo were front-page news in the nineteenth century. The front page of *The Penny Illustrated Paper and Illustrated Times* of 5 December 1874 shows the rescue of keepers who were attacked by a rhinoceros at London Zoo.

A neck injury caused by a male Sumatran tiger (*Panthera tigris sumatrae*) that resulted in the death of a keeper in a European zoo has been described by Szleszkowski et al. (2017). They described the nature of the attack, based on eyewitness testimony, and concluded from autopsy results that death was caused by a combination of penetration of tissues by the canines, crushing and distension caused by the tiger jerking its head after biting the victim's neck.

Some descriptions of non-fatal attacks appear in the scientific literature. An adult Komodo dragon (*Varanus komodoensis*) – a large varanid lizard that produces an anticoagulant from glands in its lower jaw – attacked a female keeper at a zoo in the United States, causing serious tendon and neurovascular injuries to the left forearm and the right lower leg (Boyd et al., 2021). The authors warned of the possibility of clinically important envenomation – exposure to venom or toxins – in humans following varanid lizard bites.

A survey of occupational injuries and illnesses reported by zoo veterinarians in the United States analysed questionnaires received from 315 veterinarians (Hill et al., 1998). A summary of their responses is provided in Table 4.1.

The number and type of incidents encountered in practice were affected by the length of experience and sex of the veterinarian and the practice type. Female veterinarians reported a higher rate of zoonotic infection, insect allergy and adverse exposure to anaesthetic gas, formalin and disinfectants/sterilants than did males. More experienced zoo veterinarians were more likely to receive major animal-related injuries and associated hospitalisation, back injury and lost work time associated with back injury than those with fewer years of experience. Full-time zoo veterinarians were more likely to report back injury and inadequate knowledge of occupational hazards.

A study of occupational accidents involving animals at the National Zoo Park of Cuba found that 81.4 per cent of biologists, medical and veterinary experts experienced accidents (Echarte and Vasallo, 2016). A study of employees at the National Zoological Park, New Delhi, India found that among individuals handling animals 59.2 per cent had been injured at least once, of which 79.3 per cent required the attention of a doctor (Bagaria and Sharma, 2014). Only 24.5 per cent of the staff had attended a training programme on zoonotic diseases.

Table 4.1 Occupational injuries and illnesses reported by veterinarians in the United States.

Injury/illness reported	Percentage ($n = 315$)
Major animal-related injury	61.5
Back injury	55.0
Necropsy injury	44.1
Adverse formalin exposure	40.2
Animal allergy	32.2
Zoonotic infection	30.2
Insect allergy	14.2

Source: based on data in Hill et al., 1998.

A number of studies have examined occupational hazards of working with particular taxa in zoos. The risk of infection of zookeepers with intestinal parasites from mammals and birds kept at Beni-Seuf Zoo, Egypt, has been investigated by Kamel and Abdel-Latef (2021), and work-related respiratory symptoms in bird zookeepers have been examined by Świderska-Kiełbik et al. (2009).

A useful source of records of dangerous animal incidents involving the death or injury of zookeepers and visitors and the deaths of animals for bears, big cats, elephants, giraffes and primates is the data compiled by People for the Ethical Treatment of Animals (PETA, 2021).

Zoonoses affecting animals in zoos are discussed in more detail in Chapter 13.

4.8   Visitor Behaviour: Dwell Time, Orientation and Circulation

The way visitors behave in areas where items are exhibited, such as art galleries and museums, is a legitimate area of academic study and is supported by a significant body of published work. Research conducted in zoos and aquariums on visitor behaviour and the evaluation of exhibit design has been accumulating since the early 1980s (Bitgood, 1987). An understanding of the implications of this research can assist in the design of animal enclosures and zoo layouts, and the design and positioning of signage and other interpretation materials.

4.8.1   Dwell Time

Zoo designers want visitors to be attracted to their exhibits and to spend time watching the animals and reading the associated interpretive materials (Fig. 4.11). However, zoo visitors spend very little time looking at individual exhibits. This phenomenon is not unique to zoos, but also occurs in museums and other facilities where many 'items' must be viewed in a relatively short time. The length of time a visitor spends at an exhibit is known as the 'dwell time', although this term has also been used to refer to the total time spent in a zoo or aquarium during a single visit.

In a zoo with many exhibits a short dwell time (at each exhibit) is inevitable. If a zoo is open for eight hours a day (10.00–18.00) and has 100 exhibits, a visitor could only spend an average of 4.8 minutes at each exhibit assuming she arrived when the zoo opened and left when it closed, and no time was spent walking from one exhibit to the next or doing anything else.

On 1 January 2022, the stocklist at London Zoo recorded 14,926 individual animals (227 of which were mammals) from 396 species/subspecies (including 57 mammals) (ZSL, 2022). In an eight-hour visit – assuming a visitor did nothing but view animals and it took no time to move between each species – an average of just 73 seconds could be spent looking at each species. If a visitor spent just three minutes looking at each species they would have to spend about 2.5 days in the zoo from opening time until it closed.

Fig. 4.11 Visitors interacting with giraffes (*Giraffa camelopardalis*) at South Lakes Safari Zoo in England. Dwell time increases when animals are active and when visitors are able to interact with them.

Bitgood et al. (1988) recoded the percentage of visitors who stopped at selected exhibits at 13 zoos in the United States and the duration of viewing time. They found that visitor behaviour was correlated with both the animals' characteristics (animal activity, size of species, presence of an infant) and the architectural characteristics of the exhibits (presence of visually competing exhibits, proximity of visitors to the animals, the visibility of animals and the physical features of the exhibit).

In his study of the factors affecting viewing time, Johnston (1998a) studied visitors at 10 exhibits in six zoos and identified more than 50 variables with potential effects. In a second study he used viewing time as a means of estimating demand for wildlife viewing in zoos and demonstrated that variables representing exhibit design, zoo design and exogenous conditions all have significant impacts on this (Johnston 1998b).

Some studies have recorded extremely short dwell times. A study of visitors to the reptile house in the Smithsonian's National Zoological Park (also known as the National Zoo, Washington, DC) showed that they spent less than 30 seconds in front of each enclosure (Mullan and Marvin, 1999). In an earlier study of visitor behaviour in the reptile house of the National Zoo, visitors were found to spend a mean of 14.7 minutes in the building and only about 8 minutes looking at the exhibits, regardless of age, sex or group type (Marcelli and Jenssen, 1988). Visitors spent more time looking at larger animals than smaller ones, regardless of taxon, and crocodiles were looked at for the longest time; snakes and turtles were looked at for longer than lizards and amphibians. The time spent looking at the exhibits was negatively correlated with visitor density.

Ross and Gillespie (2008) studied visitor behaviour in the *Regenstein African Journey*, an exhibit at Lincoln Park Zoo, Chicago, that was designed to take visitors on a simulated African safari. The median visit was just over 11 minutes. Of this, 41 per cent of the time was spent looking at animals and 9 per cent spent engaged with interpretive elements. Visitors in groups without children spent more of their time engaged with signage than those who had children. Visitors who spent more of their visit interacting socially spent less time engaged with signage.

Long dwell times are problematic for zoos as they create congestion at popular exhibits and reduce visitor satisfaction. There are few giant pandas (*Ailuropoda melanoleuca*) in zoos outside China so, in those zoos that keep them, demand will always be high. When a new panda exhibit opened at Edinburgh Zoo in Scotland in 2011, visitors were required to pre-book a 15-minute time slot in advance of their visit sometime between 10.00 and 11.45 or 12.30 and 17.15.

Staff at Chester Zoo in England have made a detailed study of the behaviour of visitors to their jaguar exhibit. *Spirit of the Jaguar* has an indoor visit area of 315.6 m^2 and is designed so that visitors travel through it in one direction (Fig. 4.12). The exhibit consisted of 39 elements at the time of the study. The living elements consisted of two 'flagship species' enclosures housing jaguars (*Panthera onca*) and three 'integral species' enclosures housing poison arrow frogs (*Dendrobates* sp.), butterfly goodeids (*Ameca splendens*) and leafcutter ants (*Atta cephalotes*). In addition, there were 18 interactive interpretation elements (including video monitors, soundboards, flip panels, scent signs and signs with tactile models) and 16 non-interactive interpretation elements (signs consisting only of text and images). The effectiveness of the various elements was measured by tracking individual visitors through the exhibit (Francis et al., 2007). The median dwell time in the exhibit was 5 minutes 41 seconds, and ranged from 29 seconds to almost 48 minutes. Dwell time was heavily dependent upon the visibility of the flagship species and was almost twice as long when at least one jaguar was visible (median dwell time = 419 seconds) than when no jaguars were visible (median dwell time = 249 seconds). The average visitor spent just 109 seconds at the jaguar enclosures. On average, visitors stopped at just seven of the 39 exhibit elements (18 per cent), and even the most interested visitors only stopped at a maximum of 21 (54 per cent). Most of the visitor groups (92.5 per cent) stopped at the flagship species (jaguar enclosures) but only 36 per cent stopped at the integral species. The interactive exhibits attracted 17 per cent of visitors but just 5.9 per cent stopped at the non-interactive exhibits. Although visitors spent 61 per cent of the time engaged with the exhibit elements, they were passive for 39 per cent of the time.

Moss and Esson (2010) examined visitor interest in zoo animals by measuring the proportion of visitors that stopped and the dwell time at exhibits of 40 species at Chester Zoo. They found taxonomic grouping to be the most significant predictor of visitor interest, with visitors showing more interest in mammals than in other taxa. Visitors also showed more interest in larger species, those exhibiting increased activity and species that were the primary or 'flagship' species within an exhibit. Moss and

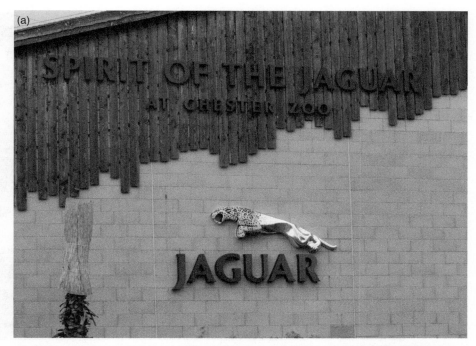

Fig. 4.12 The *Spirit of the Jaguar* exhibit at Chester Zoo, England. (a) The exhibit was funded by the makers of *Jaguar* cars. (b) A pair of jaguars (*Panthera onca*) in their outdoor enclosure.

Fig. 4.13 In safari parks the concept of dwell time at an exhibit is complicated by the fact that visitors often have to queue in their cars to get close to the animals. In this case, these visitors are eager to move on before the baboons damage their car.

Esson suggest that information on visitor interest should be used by zoos to inform collection planning.

In safari parks, visitors usually move around the exhibits by driving through them in their own vehicles (Fig. 4.13). The concept of dwell time, as used in traditional zoos to measure the popularity of particular exhibits, is inappropriate in a safari park context because of differences in, for example, road length between exhibits. Instead Lloyd et al. (2021) used average speed and found that primates and felids elicited slower speeds while bovids and cervids elicited faster speeds. This is similar to differences found in dwell times in traditional zoos.

Davey (2006a) reported a previously overlooked variable that may influence the effect of visitors on animal welfare in zoos: hourly variation in visitor interest. He noted that visitor interest wanes during the course of the day and this should be taken into account in research design. Visitor effects on animal welfare are likely to be more intense in the morning or in earlier visits as this is when visitor interest is greatest.

4.8.2   Visitor Orientation, Wayfinding and Circulation

Visitor orientation and circulation are important areas of the study of visitor behaviour because they influence whether or not people actually visit a zoo, whether or not they see a particular exhibit and what they learn (Bitgood, 1988). They also affect what visitors tell their friends and relatives, and whether or not they return.

Visitor orientation and circulation has three major elements. The first is conceptual (or thematic) orientation, which is concerned with an awareness of the themes and the organisation of the subject matter of the zoo. The second is wayfinding (or topographical or locational) orientation: the manner in which visitors find places in the zoo. The third is circulation: the pathways that visitors take while exploring the zoo.

Zoo visitors need to know where they are (orientation) and how to get to where they want to go (wayfinding). A study of students that visited the National Zoo in the United States found that those that were given pre-visit orientation information – such as what they would see and when they would eat lunch – learned more than other groups of students, even when they were given more learning-oriented pre-visit preparation (cited in Bitgood, 1988).

Zoos offer a number of ways for visitors to find their way around, including signposts, maps and zoo guides. They also provide information to visitors prior to their visit. More than half (55.6 per cent) of the visitors to the Arizona-Sonora Desert Museum (a zoo) in the United States wanted information on conceptual orientation before their visit – information about what they could see – and only 18.5 per cent wanted wayfinding information. During their visit almost 65 per cent of visitors reported using a hand-held map to find their way (Shettel-Neuber and O'Reilly, 1981) (Fig. 4.14).

Fig. 4.14 Zoo maps are an important aid to wayfinding in zoos and add to the excitement of exploring an unfamiliar place.

Providing visitors with hand-held maps can increase the number of exhibits visited. In a study at the Birmingham Zoo, Alabama, 77 per cent of visitors who received hand-held maps were observed using them and those with a map viewed a greater percentage of the exhibits (86 per cent) than those without (78 per cent) (Bitgood and Richardson, 1986).

Most zoos have signs illustrating the direction in which visitors must walk to find a particular exhibit. The system of trails marked by signs using pictograms of animal heads introduced in the National Zoo in Washington in 1974 is considered a classic in zoo graphics (Yew, 1991). Pictograms are particularly useful for children who cannot read well and for overseas visitors (Fig. 4.15). In the bird collection at Martin Mere, West Lancashire (England) there are signs indicating the time it will take to walk to particular places, which is useful for visitors with mobility problems or those accompanying small children.

There is evidence that visitors prefer a suggested path (Shettel-Neuber and O'Reilly, 1981). This is particularly important in congested areas of zoos where one-way systems may be used, and in dark areas such as indoor aquariums and nocturnal houses, to prevent visitors from walking into each other in the dark. Visitors generally appear to comply with traffic flow instructions. When the flow in the reptile house in the Birmingham Zoo was changed from two-way to one-way less than 1 per cent of visitors violated the signs (Bitgood et al., 1985).

Knowledge of visitors' preferred paths can assist a zoo in selecting a suggested route, because people are more likely to follow a route that has been freely chosen

Fig. 4.15 A wayfinding sign in Barcelona Zoo, Spain. Multilingual signs containing silhouettes of animals help children and foreign visitors to find their way.

Fig. 4.16 Arrows on the paths in London Zoo help visitors to find their way.

(Fig. 4.16). Knowledge of visitor circulation can help a zoo to plan exhibits and time interpretive talks to coincide with visitors' circulation patterns.

The visitor experience in a zoo is determined as much by the overall design of the facility as it is by the individual exhibits. Bitgood (2006) argues that the value of an experience is unconsciously calculated by visitors using the general value principle: the ratio between the benefits and costs. Essentially, a visitor will only do something if the benefits are perceived as greater than the costs. Visitors tend to walk in a straight line unless something catches their attention and pulls them away, and they are reluctant to backtrack.

Often visitors to an exhibit do not follow the intended traffic pattern around a zoo and they tend to stay on the main paths. In a study of visitors to North Carolina Zoo, Bitgood et al. (1990) observed that visitors were reluctant to walk down a path to overlook an exhibit that took them away from the main path; they call this phenomenon 'dominant path security'.

The most common visitor circulation pattern observed at the Reid Park Zoo in Arizona involved turning right and circulating anti-clockwise on the periphery of the zoo, but not using paths that connected one part of the outer circle to another (Deans et al., 1987). This outer path was probably perceived by visitors as the main path.

Visitors tend to move only along one side of a path and are reluctant to move from one side to the other. They will only approach objects that are attractive or interesting and the cost of this approach must be perceived as low. Visitors tend to walk on the right side of paths, and when they reach a 'choice point' they tend to turn right (although this finding is not confirmed in all studies).

Bitgood et al. (1988a) compared exhibits of the same species in different zoos. In some institutions exhibits were placed on both sides of a walkway and in others they were placed on just one side. They found that visitors stopped fewer times at exhibits when displays were on both sides of a path. In a study of visitors to the Steinhart Aquarium in San Francisco, California, Taylor (1986) found that visitors were unwilling to backtrack during their visit. The only visitors who saw the whole aquarium were repeat visitors who were familiar with the layout.

Many zoos – and especially aquariums – have exhibits that contain dark areas such as nocturnal houses and free-flying bat exhibits. Loomis (1987) has shown that visitors are reluctant to enter dark areas and are more likely to follow lighted pathways in buildings.

The general value principle has several important implications for zoo design. A zoo should be designed so that visitors do not have to walk more than necessary and are not required to backtrack through areas they have already visited. Multiple-choice points – where visitors must choose which path to follow – should be avoided in the design as should two-sided designs where exhibits on one side of a path compete with those on the other.

4.9   Conclusion

Studying the relationships between animals and people is a relatively new area of academic endeavour. For many species it is important to their husbandry that keepers form a relationship with the animals in their care, and this relationship can benefit the wellbeing of both. Increasingly, keepers are being used to collect data on animals and it is evident that they can make important contributions to scientific research by providing information that is otherwise difficult for scientists to obtain. The study of the behaviour of visitors in zoos is important because it can inform zoo and exhibit design and improve the visitor experience.

5 Zoo Organisation and Regulation

5.1 Introduction

Zoos and aquariums need to be regulated by legislation to ensure that animals are properly cared for and safely contained. In recent years legislation has begun to focus on the purpose of zoos, and in some jurisdictions – notably within the European Union (EU) – this has led to an increased emphasis on their potential roles in education, research and conservation. This chapter examines the role of national, EU and international laws in the regulation of zoos and private collections of dangerous animals.

5.2 Regulation, Legislation and Inspection

5.2.1 Introduction

In 2003, Cooper surveyed zoo legislation in various countries and pointed out that, although its form and content are diverse, there are many common elements such as licensing and inspection systems (Cooper, 2003). This remains true today. Zoo legislation generally makes provision for the licensing of collections of animals that are open to the public, their inspection by veterinarians and others with knowledge of animal welfare and the proper operation of zoos, and their closure if licensing provisions are breached. In addition, zoos are required to keep records of their animals, prevent their escape, educate the public and engage in other conservation activities.

5.2.2 Zoo Governance, Standards and Policies

The peculiar legal status of animals in American zoos has been examined by Braverman (2011). She argues that animal living in zoos are almost extralegal by virtue of the fact that, although there is a significant body of law that relates to animals in various settings, many exceptions and exemptions are made for those in zoos. She also notes variances and exceptions in the law that apply to the physical facilities of zoos, and draws attention to zoos' self-regulatory industry standards.

Table 5.1 List of BIAZA policies (March 2021).

BIAZA Health and Safety Guidelines 2020	BIAZA Governing Document: Memorandum and Articles of Association
BIAZA Conservation Translocations Policy	Requirements for Membership of BIAZA
BIAZA Close Contact Policy	BIAZA Animal Transfer Policy
BIAZA Ethical Acquisition Policy	BIAZA Euthanasia Policy
BIAZA Animal Welfare Policy	BIAZA Disposal of Dead Stock Policy
BIAZA Privacy Policy	BIAZA Sanctions Policy
BIAZA Surplus Animals Policy	BIAZA Animal Inventories: Guidelines for Full Member Applicants

Zoos operate within a quasi-legal regulatory framework at national or regional level, European level and international level. For example, a zoo in England may be accredited by its regional zoo organisation (the British and Irish Association of Zoos and Aquariums (BIAZA)), at the European level by the European Association of Zoos and Aquaria (EAZA) and at the international level by the World Association of Zoos and Aquariums (WAZA). Each of these organisations produces its own regulatory guidelines and policies, and zoos must abide by these in order to retain their accreditation.

If a member of BIAZA violates any of the association's policies, articles or requirements of membership (Table 5.1) the BIAZA Council can impose four levels of sanction. In order of severity, these are: (1) Warning; (2) Under Mentorship; (3) Suspension; and (4) Termination.

At the European level, EAZA publishes governing documents, standards, guidelines, strategies, position statements and best practice guidelines (Table 5.2).

In addition to the documents listed in Table 5.2, EAZA also makes available best practice guidelines for a range of taxa, produced by EAZA Taxon Advisory Groups (TAGS) (Table 5.3). These include guidelines for keeping De Brazza's monkey (*Cercopithecus neglectus*) (Fig. 5.1).

5.3   Zoo Registrars and Animal Assets

The zoo registrar is the person responsible for keeping records of each individual animal in a zoo, with the exception of small species kept in large numbers, such as ants and other small invertebrates, and perhaps small fishes. These records include details such as the species, sex, age, sire, dam, date and place of birth, or date of transfer to the zoo and origin. They are essential for the genetic management of zoo populations and for the effective operation of captive breeding programmes. Zoo registrars, along with other zoo staff, may be responsible for the operation of studbooks that maintain a detailed record of all of the individuals held in a particular breeding programme.

Table 5.2 List of EAZA documents (March 2021).

Governing documents	Constitution of the European Association of Zoos and Aquaria (2018)
	EAZA Code of Ethics (2015)
	EAZA Population Management Manual (2020)
	EAZA Membership and Accreditation Manual (2020)
	Sanctions in the case of a violation of the EAZA Code of Ethics or EEP procedures (2016 updated 2019)
	European Union Zoos Directive (1999)
	The Code of Ethics also requires members to comply with the following:
	IUCN Species Survival Commission Guidelines on the Use of Ex Situ Management for Species Conservation (2014)
	IUCN Guidelines for Reintroductions and Other Conservation Translocations (2013)
	IUCN Guidelines for the Management of Confiscated, Live Organisms (2019)
	WAZA Strategy Committing to Conservation
	WAZA Strategy Caring for Wildlife
Standards	EAZA Standards for the Accommodation and Care of Animals in Zoos and Aquaria (2020)
	EAZA Research Standards (2003) – should be read in conjunction with the Research Strategy
	EAZA Conservation Standards (2016)
	EAZA Conservation Education Standards (2016)
EAZA guidelines	EAZA Palm Oil Guidelines (2019)
	EAZA Guidelines for Ethical and Environmental Policies for Suppliers and Contractors (April 2017)
	EAZA Guidelines on the Definition of a Direct Contribution to Conservation (2015)
Strategies	EAZA Strategic Plan 2017–2020: Progressive Zoos and Aquariums Collaborating to Lead on Conservation
	Strategic Plan 2013–2016: EAZA Moving Forward in the UN Decade of Biodiversity
	EAZA Research Strategy (2008)
	EAZA Plant Conservation Strategy (2013–2016)

Table 5.2 (*cont.*)

Position statements	EAZA Response to the EU Biodiversity Strategy for 2030 (June 2020)
	EAZA Elephant Taxon Advisory Group Position Statement on Management Systems (July 2019)
	EAZA Position Statement on the European Commercial Trade in Tigers and Tiger Parts (November 2018)
	EAZA Position Statement on Songbird Trafficking (April 2018)
	EAZA Position Statement on the EU Regulation on the Prevention and Management of the Introduction and Spread of Invasive Alien Species (1143/2014) (September 2017)
	EAZA Position Statement on Circus Membership of the Association (April 2017)
	EAZA Position Statement on the EU Zoos Directive 1999/22/EC (March 2017)
	The Application of a Considered Culling Policy (EAZA Culling Statement, 2015)
	Intentional Breeding for the Expression of Rare Recessive Alleles (May 2013)
	EAZA Position Statement on Bears in Commercial Entertainment (2012)
	Council Regulation 1/2005: Protection of Animals during Transport (December 2010)
	Proposed Regulation on Food Information to Consumers (revised November 2010)
	Developing EU Strategy for Invasive Alien Species (September 2010)
	Animal Health Strategy for the European Union (August 2009)
Other	EAZA Information Sheet on Animal Management during the Coronavirus Crisis (2020)
	EAZA Contribution to the Birds and Habitats Directives Fitness Check (2014–2016)
	EAZA Manifesto (European Parliamentary Elections (2019))
	EAZA Contribution to the Consultation on the EU Approach against Wildlife Trafficking (April 2014)
	European Code of Conduct on Zoological Gardens and Aquaria and Invasive Alien Species (October 2012)
	EU Zoos Directive Good Practices Document

Table 5.3 EAZA best practice guidelines (1 March 2021).

Invertebrates:
Desertas Wolf Spider (*Hogna ingens*)
Montserrat Tarantula (*Cyrtopholis femoralis*)
Polynesian tree snails (*Partula* spp.)

Amphibians:
Midwife toad (*Alytes* sp.)
Lake Oku frog (*Xenopus longipes*)
Mountain chicken frog (*Leptodactylus fallax*)
Sardinian brook salamander (*Euproctus platycephalus*)
Typhlonectid caecilians (including *T. compressicauda, T. natans* and *P. kaupii*)

Reptiles:
Ploughshare tortoise (*Astrochelys yniphora*)
Egyptian tortoise (*Testudo kleinmanni*)

Birds:
Vietnam Pheasant (*Lophura edwardsi*)
Turacos (Musophagidae)
Red-Crested Turaco (*Tauraco erythrolophus*)
Ecuadorian Amazon Parrot (*Amazona lilacina*)
EAZA Reference Document: Virus Management for Parrots
Javan Green Magpie (*Cissa thalassina*)
Rufous-fronted Laughing thrush (*Garrulax rufifrons*)
Straw-headed bulbul (*Pycnonotus zeylanicus*)
Sumatran Laughing thrush (*Garrulax bicolor*)
Husbandry and Management Guidelines For Demonstration Birds
Parrot supplement to the Bird Demonstration Guidelines

Mammals:
Primates
Capuchin monkeys (*Sapajus* and *Cebus* spp.)
Callitrichidae

De Brazza's monkey (*Cercopithecus neglectus*)
Mangabey (*Cercocebus* spp., *Lophocebus* spp. and *Rungwecebus* spp.)
Gorilla (*Gorilla gorilla gorilla*)
Orangutan (*Pongo* sp.)
Bonobo (*Pan paniscus*)
Carnivores
Asiatic golden cat (*Catopuma temminckii*)
Cheetah (*Acinonyx jubatus*)
Demonstration Guidelines for Felid Species used in Public Demonstrations
Dhole (*Cuon alpinus*)
European otter (*Lutra lutra*)
Red Panda (*Ailurus fulgens*)
Marine species
Antillean manatee (*Trichechus manatus manatus*)
Pinnipeds (Otariidae and Phocidae)
Marine Mammal Demonstration and Public Interaction Guidelines
Ungulates
Greater one-horned rhinoceros (*Rhinoceros unicornis*)
White rhinoceros (*Ceratotherium simum*)
Black rhinoceros (*Diceros bicornis*) 2nd Edition
Babirusa (*Babyrousa*)
Pygmy hippopotamus (*Choeropsis liberiensis*)
Burmese brow antlered deer (*Rucervus eldii thamin*)
Tufted deer (*Elaphodus cephalophus*)
Lesser kudu (*Tragelaphus imberbis*)
Elephants (Elephantidae)
Other mammals
Anteater demonstration guidelines

Note: EAZA lists the taxa in inconsistent styles.

Fig. 5.1 De Brazza's monkey (*Cercopithecus neglectus*).

Braverman (2010) discussed the role of zoo registrars and concluded that their rising importance highlighted two significant changes that have occurred in zoos in recent times. First, their transition from places of entertainment and education into organisations focusing on conservation through the management of information. Second, she highlights the increased legalisation of zoos. Braverman has characterised animal governance in zoos in North America as surveillance and coined the term 'zooveillance' to describe this. She claims that the notions of care, stewardship and conservation underlie three modes of surveillance: elementary surveillance (naming, identifying and recording animals at an institutional level); 'dataveillance' (the computerised management of populations at a global level); and collective reproductive control (Braverman, 2012).

Abbott and Tan-Kantor (2022) have argued that a zoo's greatest asset is its animals and its balance sheet should reflect their value. They noted that there is no internationally recognised accounting framework and guidance on asset management techniques for zoo animals. The authors argue that the asset measurement of zoo animals underpins their ability to account for their animals in their annual financial reporting.

Modern zoos rarely pay for their animals, but in the past zoos maintained stock books that indicated the origin of individual animals and the prices paid to dealers (Fig. 5.2). In its 1970 annual report, the North of England Zoological Society valued the animals kept at Chester Zoo at £117,082 on 31 December 1970 (NEZS, 1970). However, in its annual report of 1980 the animals were given a nominal value of just

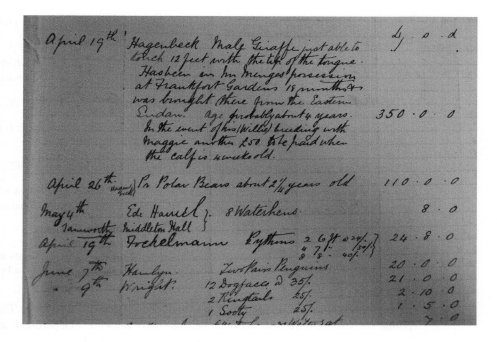

Fig. 5.2 Entries in the stock book of the Belle Vue Zoological Gardens, Manchester, England, for 1905 (courtesy of Chetham's Library, Manchester). Entries include a giraffe costing £350, a pair of polar bears obtained for £110 and two pairs of penguins purchased for £20.

£1,000 (NEZS, 1980). It is common practice for modern zoos to attach no monetary value to their animals.

5.4 Zoo Legislation

5.4.1 Introduction

Zoo legislation exists at national, EU and international levels. Most countries have legislation concerning the regulation of zoos open to the public and many have laws that restrict and control the keeping of dangerous wild animals by members of the public in what are effectively private zoos.

5.4.2 Defining Zoos in the Law

The term 'zoo' is an abbreviation for zoological gardens. It is a place where wild animals are kept for exhibition to the public, entertainment, breeding, study and conservation purposes. However, the legal definition of a zoo varies between jurisdictions.

In the United States, the Animal and Plant Health Inspection Service (APHIS) 9 CFR, Ch.1, §1.1 defines a zoo very broadly as:

Zoo means any park, building, cage, enclosure, or other structure or premise in which a live animal or animals are kept for public exhibition or viewing, regardless of compensation.

In Arizona State, in the United States, Arizona Revised Statutes, 17-101, A23:

'Zoo' means a commercial facility open to the public where the principal business is holding wildlife in captivity for exhibition purposes.

In New South Wales, Australia, under the Zoological Parks Board Act 1973, s.4(1):

'zoological park' means a zoological garden, aquarium or similar institution in which animals are kept or displayed for conservation, scientific, educational, cultural or recreational purposes.

In India, under the Wildlife (Protection) Act 1972, s.2(39):

'Zoo' means an establishment, whether stationary or mobile, where captive animals are kept for exhibition to the public but does not include a circus and an establishment of a licensed dealer in captive animals.

Zoos in India must be licensed by the Central Zoo Authority (CZA). In Thailand a public zoo may only be operated with permission of the Forest Department under the Wild Animal Reservation and Preservation Act B.E. 2535 (1992):

s29 Whoever is desirous of establishing and conducting a public zoo operation shall obtain permission from the Director-General [of the Royal Forest Department].

5.4.3   Regulation of Zoo Building Design and the Housing of Animals

Some states have sought to closely control the design of zoo buildings and the manner in which certain species are housed. India published its National Zoo Policy in 1998. The Preamble to the policy noted that:

1.6 In India, many well-designed zoos were set up in some of the states but for the most part, zoos have not been able to meet the challenges imposed by the changing scenario and still continue with the legacy of past i.e. displaying animals under conditions which are neither congenial to the animals nor educative and rewarding to the visitors.

1.7 The amendment of the Wildlife (Protection) Act, in 1991, provided for the enforcement of mandatory standards and norms for management of zoos through the Central Zoo Authority. However, it is realised that the objectives of the Act can be achieved only through co-operation and participation of various government agencies, nongovernmental organisations and people at large.

1.8 The National Zoo Policy aims at giving proper direction and thrust to the management of zoos by mustering co-operation and participation of all concerned.

The policy requires zoos to, among other things, prepare a masterplan, provide suitable housing, avoid keeping single animals and surplus animals, prioritise the keeping and breeding of endangered species and only take animals from the wild to support approved breeding programmes.

MINISTRY OF ENVIRONMENT & FORESTS RESOLUTION New Delhi, the 28th October, 1998 NATIONAL ZOO POLICY, 1998

OBJECTIVES

3.1 GENERAL POLICY ABOUT ZOOS

[. . .]

3.1.4 Zoos shall give priority to endangered species in their collection and breeding plans. The order of preference for selection of species shall be (in descending order) locality, region, country and other areas.

3.1.5 Zoos shall regulate the number of animals of various species in their collection in such a way that each animal serves the objectives of the zoo. For achieving this objective, a detailed management plan of every species in the zoo shall be prepared.

3.1.6 Every zoo shall endeavour to avoid keeping single animals or non-viable sex ratios of any species. They shall cooperate in pooling such animals into genetically, demographically and socially viable groups at zoos identified for the purpose.

3.1.7 Zoos shall avoid keeping surplus animals of prolifically breeding species and if required, appropriate population control measures shall be adopted . . .

3.3 ANIMAL HOUSING

3.3.1 Every animal in a zoo shall be provided housing, upkeep and health care that can ensure a quality of life and longevity to enable the zoo population [to] sustain itself through procreation.

3.3.2 The enclosure for all the species displayed or kept in a zoo shall be of such size that all animals get adequate space for free movement and exercise and no animal is unduly dominated or harassed by any other animal.

3.3.3 Each animal enclosure in a zoo shall have appropriate shelters, perches, withdrawal areas, wallow, pools, drinking water points and such other facilities which can provide the animals a chance to display the wide range of their natural behaviour as well as protect them from extremes of climate.

The CZA licenses and regulates the activity of zoos in India and has published *Guidelines on Minimum Dimensions of Enclosures for Housing Exotic Animals of Different Species* (Bonal et al., 2012).

In New South Wales, Australia, in 2009 the state government approved a *Policy on the Management of Solitary Elephants in New South Wales* under the Exhibited Animals Protection Act, 1986 (NSW Department of Primary Industries 1986). After exploring the relevant issues, the author of the policy concluded that:

All reasonable efforts should be made to integrate solitary elephants into other groups unless compelling reasons can be provided that warrant the retention of a solitary elephant. Only in the event that all avenues for integration have been exhausted should the maintenance of a solitary elephant be contemplated.

5.4.4   Zoos and International Law

Parties to the United Nations Convention on Biological Diversity 1992 are required, under Article 9, to adopt measures for the *ex-situ* conservation of biodiversity. This may be defined as 'keeping components of biodiversity alive away from their original habitat or natural environment' (Heywood, 1995). However, these measures are not to be taken as an end in themselves but predominately for the purpose of complementing *in-situ* measures:

Article 9

(a) Adopt measures for the ex-situ conservation of components of biological diversity, preferably in the country of origin of such components.

However, the major captive breeding programmes for many species are undertaken outside their countries of origin because most of the more advanced zoos in the world – and consequently the largest cooperative breeding programmes – are located in North America and Europe. For example, in 2018 Zoological Management Information System (ZIMS) – a global system for recording the holdings of the world's major zoos – listed 750 Asian elephants (*Elephas maximus*), of which 305 (40.7 per cent) were distributed among 76 European zoos and 100 (13.3 per cent) were spread among 30 North American zoos (Rees, 2021). The Convention also requires Parties to conserve variation *within* species:

Article 2

"Biological diversity" means the variability among living organisms . . . this includes diversity within species, between species and of ecosystems.

It is important that the genetic integrity of species and subspecies is protected wherever possible. Orangutans were originally classified as a single species (*Pongo pygmaeus*), but now (2022) three species are recognised: the Bornean (*P. pygmaeus*), the Sumatran (*P. abelii*) and the Tapanuli (*P. tapanuliensis*). Consequently, orangutan populations kept in zoos contain hybrid individuals who cannot be used for breeding if the species are to be kept distinct. The problem of hybridisation between subspecies of the Asian elephant (*Elephas maximus*) has been discussed by Schmidt and Kappelhof (2019). They found 20 hybrids within EAZA's Asian elephant Ex-Situ Programme (EEP; formerly known as the European Endangered Species Programme) in 2018.

In domesticated animals, organisations such as the Rare Breeds Survival Trust in the UK and the American Livestock Breeds Conservancy play an important role in conserving rare breeds of horses, cattle, pigs, goats, chickens and other farm animals (Kendall, 2003). Some zoos also keep rare breeds of farm animals (Fig. 5.3). In Germany, Tierpark Berlin keeps spotted Alpine sheep, Skudde sheep, Rouge du Roussillon sheep, Girgentana goat, Thuringian goat, Fjäll cattle, Mangalica pig and the Cröllwitzer turkey, all of which are listed on Germany's Red List of Endangered Livestock Breeds (Bundesanstalt für Landwirtschaft und Ernährung, 2021). The educational value of keeping domestic animals in zoos has been discussed by Jewell (1976).

The Convention on International Trade in Endangered Species of Wild Fauna and Flora (CITES) 1973 established a system of import and export licences that strictly

Fig. 5.3 British Saddleback. Rare breeds of farm animals contribute to intraspecies genetic diversity.

monitors, among other things, the international movements of live animals, including those taken from the wild, transfers between zoos and from private sources to zoos, such as the acquisition of elephants from Asian logging camps.

Although CITES appears to restrict the activities of zoos, Schmitt (1988) has argued that the opposite is true and that it reinforces the interdependence of the remaining wild populations of natural areas and zoos in their role as stewards acting on behalf of wildlife. He argues that the legislation safeguards the future of rare species and enables zoos to exhibit them for future generations.

It is rare to find published data on the numbers of animals imported into zoos for an entire country in the academic literature. However, Huang et al. (2022) have analysed data from China that identify the destiny of living animals imported into Chinese zoos, aquariums and breeding centres between 1996 and 2015. China began importing animals duty-free for conservation purposes in 1996. The study used a database of almost 300 vertebrate species (mammals, bird, fishes and reptiles) compiled from information relating to 123 institutions. It showed that the number of imported living animals decreased significantly year by year over the period 1996–2015 (the number of birds and reptile individuals decreasing the most). Importations were suspended in 2003 in response to an outbreak of severe acute respiratory syndrome (SARS). Between 1996 and 2015, 64,843 individuals were imported from 278 species. Of these, globally threatened species were represented by 47,096 animals from 85 species. Included in these totals were 43,465 juvenile Siamese crocodiles (*Crocodylus siamensis*) sent to five institutions, representing 67 per cent of all

Fig. 5.4 Siamese crocodile (*Crocodylus siamensis*).

imported animals (Fig. 5.4). No explanation was provided by the authors of the study for this extraordinarily high number of crocodiles.

Huang et al. noted that by 2017 only eight of the threatened species imported into China between 1996 and 2015 had failed to breed and survive captivity, and that the numbers in zoos increased for 50 per cent of the imported threatened species. They emphasised the need for more cooperative breeding between zoos in China to reduce the need for imports and the more effective use of trade licences, quotas and sustainability assessments for threatened species.

Some of the animal trade involves supplying animals that are kept by private individuals. The opportunities for public aquariums to increase the sustainability of the aquatic animal trade – aquarium fishes and invertebrates – has been discussed by Tlustry et al. (2013). They suggested that working with the home aquarium industry could transform it from a threat to a positive force for conservation.

5.4.5   Zoos and European Law

In the EU, zoos located within the territories of the Member States must comply with Council Directive 1999/22/EC of 29 March 1999 relating to the keeping of wild animals in zoos (the Zoos Directive). The purpose of the Directive was to improve the welfare of animals living in zoos and to strengthen the role of zoos in biodiversity conservation:

Article 1

Aim

The objectives of this Directive are to protect wild fauna and to conserve biodiversity by providing for the adoption of measures by Member States for the licensing and inspection of zoos in the Community, thereby strengthening the role of zoos in the conservation of biodiversity.

Under the Directive, 'zoos' are widely defined – and include aquariums – and exceptions are allowed so that small collections of animals do not require a licence and are therefore exempt from the obligations imposed by the Directive:

Article 2

Definition

For the purpose of this Directive, 'zoos' means all permanent establishments where animals of wild species are kept for exhibition to the public for 7 or more days a year, with the exception of circuses, pet shops and establishments which Member States exempt from the requirements of this Directive on the grounds that they do not exhibit a significant number of animals or species to the public and that the exemption will not jeopardise the objectives of this Directive.

All zoos covered by the Directive must engage in education, provide appropriate accommodation and husbandry (including enrichment), prevent animal escapes (and pest intrusions) and keep records. In addition, zoos must engage in at least one of the following: conservation research, training, information exchange or captive breeding/ reintroduction to the wild:

Article 3

Requirements applicable to zoos

Member States shall take measures under Articles 4, 5, 6 and 7 to ensure all zoos implement the following conservation measures:

– participating in research from which conservation benefits accrue to the species, and/or training in relevant conservation skills, and/or the exchange of information relating to species conservation and/or, where appropriate, captive breeding, repopulation or reintroduction of species into the wild,

– promoting public education and awareness in relation to the conservation of biodiversity, particularly by providing information about the species exhibited and their natural habitats,

– accommodating their animals under conditions which aim to satisfy the biological and conservation requirements of the individual species, inter alia, by providing species specific enrichment of the enclosures; and maintaining a high standard of animal husbandry with a developed programme of preventive and curative veterinary care and nutrition,

– preventing the escape of animals in order to avoid possible ecological threats to indigenous species and preventing intrusion of outside pests and vermin,

– keeping of up-to-date records of the zoo's collection appropriate to the species recorded.

I have argued elsewhere that the Zoos Directive was a lost opportunity to implement the Convention on Biological Diversity because it did not *require* zoos to engage in captive breeding and reintroductions (Rees, 2005a). These activities are merely some of the ways in which zoos can fulfil their conservation obligations under the Directive, but other options are available. I have also argued that, at least in the past, most of the research undertaken and published by zoos was not related to conservation so did not comply with the requirement of Article 3 of the Directive that refers specifically to 'participating in research from which conservation benefits accrue to the species' (Rees, 2005b). This situation has been steadily improving over the last two decades.

The European Commission conducted an evaluation of the Directive between 2015 and 2018 (European Commission, 2018). This concluded that the Directive was fit for purpose and played a small but useful and necessary role within a wider legislative framework and did not cause disproportionate costs. However, improved implementation of the Directive was required to realise its full potential.

5.5   National Zoo Licensing Systems

5.5.1   Introduction

A number of studies have examined the nature and efficacy of national zoo inspection systems. Those operating in the UK, Sweden and the United States are considered below, along with examples of countries listed in a ranking system created using an Animal Protection Index produced by World Animal Protection.

5.5.2   United Kingdom

The licensing and legal regulation of zoos and their activities is a relatively recent development. Zoos in the UK have been required to hold a licence to operate from the local (government) authority in which they are located since 1984 under the Zoo Licensing Act 1981. Zoo inspections are undertaken by government-appointed zoo inspectors under the authority of the Act.

The importation of specimens of rare species into the UK has been strictly controlled since the coming into force of the Endangered Species (Import and Export) Act 1976 (which implemented CITES), thereby effectively preventing zoos and animal traders from taking animals from their natural habitats in the uncontrolled manner that was previously the case, and controlling the international movements of rare species between zoos. This was later superseded by European legislation when the UK became part of the EU.

Although the UK left the EU in January 2020, at the time of writing (December 2022) UK legislation in relation to zoos and the wildlife trade continued to reflect EU law.

Draper and Harris (2012) criticised the quality of zoo inspections in their analysis of zoo reports for 192 British zoos made between 2005 and 2008. They concluded that the requirement for a zoo to be inspected by two inspectors was not met in at least 11 per cent of full licence inspections. Animal welfare assessments were made predominantly on inputs rather than outcomes and it was not possible to determine how welfare was assessed and whether assessments considered all animals or only a sub-sample. A total of 9,024 animal assessments were made. Of these 7,511 (83 per cent) were described as meeting the standards while 782 (9 per cent) were deemed to be substandard. The remainder were not graded. Only 47 zoos (24 per cent) out of the 192 examined were assessed as meeting all the animal welfare standards. Draper and Harris found that zoos classified as 'farm parks' and 'other bird collections' performed least well and that membership of a zoo association was not linked with a higher overall assessment of welfare standards.

When Draper et al. (2013) examined two consecutive inspection reports for each of 136 zoos in Britain produced between 2005 and 2011 they found no conclusive evidence of overall improvement in compliance. They found inconsistencies between the application of inspection criteria between inspectors; when the same inspector performed both inspections for the same zoo, animal welfare was more likely to be assessed as unchanged than if the second inspection was made by a different inspector. No significant difference was found in the overall number of criteria assessed as substandard on the second inspection between institutions that were members of BIAZA and those that were not. However, BIAZA members were more likely than non-members to meet the standards on both inspections and less likely to have criteria remaining substandard at the second inspection.

The effectiveness of licensing laws in guaranteeing animal welfare have been discussed at length by Tyson (2021).

5.5.3 Sweden

An epidemiological approach has been taken to an analysis of the results of the regulatory inspections of zoos in Sweden by Hitchens et al. (2017). They examined all inspections for 2010–2014, comprising 318 inspections of 179 zoos. For animal-based measures of welfare (ABMs) they found that 14.3 per cent did not comply with general care requirements for hooves/claws and coat, 8.6 per cent for body condition, and 1.7 per cent for animal cleanliness. In addition, 17 per cent of inspections found that animals were not kept in appropriate social groups. When resource- and management-based measures were examined, 29.4 per cent of inspections did not comply with space requirements and 28.8 per cent did not comply with enrichment requirements. Hitchens et al. also found that zoos with inadequate or unsafe housing and space design, inadequate bedding or that did not meet nutritional requirements were more likely to be non-compliant with at least one ABM. The authors recommended that more ABMs should be included in inspections and that risks and trends should be benchmarked over time.

5.5.4 United States

The efficacy of the Animal Welfare Act of 1966 (AWA) in protecting the welfare of animals living in zoos in the United States has been scrutinised by Jodidio (2020) and found wanting. She cites a lack of enforcement of the Act by APHIS, noting that there were only approximately 130 inspectors conducting yearly inspections of over 8,000 licensees and registrants under the Act and investigating complaints. Jodidio also notes that the AWA requires only bare minimum standards which are frequently vague and often open to interpretation. Zoos are rarely closed for violations of the Act and, as there is no citizen's suit provision, a concerned citizen cannot sue on behalf of a zoo animal. Jodidio suggests that species-specific guidelines should be added to the AWA; contact between animals and the public should be prohibited; licensing procedures should be amended; and the US Department of Agriculture (USDA) should provide facilities to house confiscated animals from non-compliant zoos.

5.6   The Animal Protection Index

The Animal Protection Index is produced by World Animal Protection and ranks 50 countries according to their legislation and policy commitments to protecting animals. The scoring bands range from A (best) to G (worst). Examples are provided in Table 5.4. The World Animal Protection website provides additional information for each of the 50 countries with respect to the existence and efficacy of national laws formally recognising animal sentience, laws against animal cruelty, the legal protection of wild, companion and farm animals and laws protecting those used for recreation and research. With respect to zoos and private collections of animals, there is information regarding whether or not national zoo legislation exists, the efficacy of law enforcement and recommendations for improvement. The overall grade assigned to each country is assigned based on a consideration of the individual grades assigned for each element of the assessment. So, for example, although Kenya has an overall grade of 'D' it received a grade of 'A' for laws protecting the welfare of wild animals but only an 'F' for protecting animals in captivity because the assessors could not

Table 5.4 Animal Protection Index scoring bands of selected countries (A–G).

Country	API scoring band	Country	API scoring band
Austria	B	Canada	D
Netherlands	B	Kenya	D
Sweden	B	Russia	D
UK	B	United States	D
France	C	China	E
Germany	C	Japan	E
India	C	Nigeria	E
Malaysia	C	South Africa	E
Mexico	C	Egypt	F
New Zealand	C	Morocco	F
Australia	D	Vietnam	F
Brazil	D	Iran	G

Note: data from www.api.worldanimalprotection.org (accessed 4 April 2022).

identify any legislation dealing specifically with the welfare of animals kept in captivity in zoos or privately (WAP, 2022).

5.7 Private Animal Collections: The Consequences of Poor Regulation

The keeping of exotic animals in private collections has been popular for millennia. Some countries have very strict laws regulating the keeping of exotic species, especially those that are dangerous and pose a threat to the local ecology if they escape. In many jurisdictions this practice is controlled by licensing.

In Great Britain anyone who keeps dangerous wild animals on their property (other than in a zoo, pet shop, circus or scientific laboratory) requires a licence under the Dangerous Wild Animals Act 1976:

s1(1). Subject to section 5 of this Act, no person shall keep any dangerous wild animal except under the authority of a licence granted in accordance with the provisions of this Act by a local authority.

This Act applies to anyone who keeps animals in a private zoo that is not open to the public but does contain at least one dangerous wild animal. The Act does not define a 'dangerous wild animal', but the Schedule to the Act provides a list of the taxa of animals to which it applies and includes hybrids of the listed species. Part of the Schedule as defined for England and Wales is shown in Table 5.5 and includes a wide range of taxa from large mammalian predators and monkeys (Fig. 5.5) to small invertebrates.

In Britain a number of felids are thought to have been released by private owners as a result of the passing of the Dangerous Wild Animals Act in 1976, possibly because they were unwilling or unable to comply with the restrictions imposed by the new law. The Department for the Environment, Food and Rural Affairs (DEFRA) received 27 reports of non-native cats that had escaped into the wild between 1975 and 2001, including a Eurasian lynx (*Lynx lynx*). Eight of these animals were recaptured and 12 were shot. The remaining animals were either found dead or their fate was never established (DEFRA, 2007).

A review of the effectiveness of the Dangerous Wild Animals Act was undertaken by DEFRA in 2001. The authors noted that the Schedule to the Act had been modified on a number of occasions and that in 1984 the listing policy changed to include animals that were potentially dangerous without requiring any evidence for this (Greenwood et al., 2001). The welfare and public health risks associated with the keeping of dangerous pets have been considered by Loeb (2020). It has been noted by Loeb and Leeming (2020) that once poisonous snakes are in private possession access to veterinary care is limited as few vets are willing to treat them and that many private owners evade inspection by not possessing a licence to keep them.

Private owners of dangerous animals often underestimate the risk of contact with these animals. Nyhus et al. (2003) analysed data on attacks on humans by captive tigers and found that six out of seven fatal attacks between 1998 and 2001 in the United States occurred where tigers were privately owned or held in private facilities.

Table 5.5 Examples of taxa listed in the Schedule to the Dangerous Wild Animals Act 1976 – Kinds of dangerous wild animals. This list was amended by the Dangerous Wild Animals Act 1976 (Modification) (No.2) Order 2007 and applies specifically to England and Wales. This list reflects the version of the schedule that extends to England and Wales only.

Scientific name of kind	Common name or names
Family Macropodidae: The species *Macropus fuliginosus*, *Macropus giganteus*, *Macropus robustus* and *Macropus rufus*.	The western and eastern grey kangaroos, the wallaroo and the red kangaroo.
Family Cercopithecidae: All species.	Old-world monkeys (including baboons, the drill, colobus monkeys, the gelada, guenons, langurs, leaf monkeys, macaques, the mandrill, mangabeys, the patas and proboscis monkeys and the talapoin).
Family Felidae: All except— (a) the species *Felis silvestris*, *Otocolobus manul*, *Leopardus tigrinus*, *Oncifelis geoffroyi*, *Oncifelis guigna*, *Catopuma badia*, *Felis margarita*, *Felis nigripes*, *Prionailurus rubiginosus* and *Felis silvestris catus*; (b) a hybrid which is descended exclusively from any one or more species within paragraph (a); (c) a hybrid of which— (i) one parent is *Felis silvestris catus*, and (ii) the other parent is a first generation hybrid of *Felis silvestris catus* and any cat not within paragraph (a); (d) any cat which is descended exclusively from any one or more hybrids within paragraph (c) (ignoring, for the purpose of determining exclusivity of descent, the parents and remoter ancestors of any hybrid within paragraph (c)); (e) any cat which is descended exclusively from *Felis silvestris catus* and any one or more hybrids within paragraph (c) (ignoring, for the purpose of determining exclusivity of descent, the parents and remoter ancestors of any hybrid within paragraph (c)).	All cats including the bobcat, caracal, cheetah, jaguar, leopard, lion, lynx, ocelot, puma, serval and tiger. The following are excepted: (i) the wild cat, the pallas cat, the little spotted cat, the Geoffroy's cat, the kodkod, the bay cat, the sand cat, the black-footed cat, the rusty-spotted cat and the domestic cat; (ii) a hybrid cat which is descended exclusively from any one or more species within paragraph (a); (iii) a hybrid cat having as one parent a domestic cat and as the other parent a first generation hybrid of a domestic cat and any cat not within paragraph (a); (iv) any cat which is descended exclusively from any one or more hybrids within paragraph (c); (v) any cat which is descended exclusively from a domestic cat and any one or more hybrids within paragraph (c).
Family Hyaenidae: All except the species *Proteles cristatus*.	Hyænas. The aardwolf is excepted.
Family Struthionidae: All species.	The ostrich.
Family Helodermatidae: All species.	The gila monster and the (Mexican) beaded lizard.
Family Theridiidae: The genus *Latrodectus*.	The widow spiders and close relatives.

Until recently Ohio's animal ownership laws were among the weakest in the United States. That changed after Terry Thompson, the owner of the Muskingum County Animal Farm in Zanesville, Ohio, released his private collection of dangerous animals before committing suicide in October 2011. Police were forced to shoot almost all the animals to protect the public. Over 50 large animals were released, including 18 tigers,

Fig. 5.5 Black and white colobus monkeys (*Colobus guereza*) are considered to be dangerous by the Dangerous Wild Animals Act 1976 in Great Britain and as such a licence is required by any member of the public who keeps them. (A black and white version of this figure will appear in some formats. For the colour version, please refer to the plate section.)

17 lions, six black bears and two grizzly bears (Rees, 2021). Thereafter the state revised its animal laws to control the ownership of exotic animals.

Ohio Dangerous Wild Animals and Restricted Snakes Act

Ohio Revised Code §§ 935.01–935.99

§935.02 Possession of dangerous wild animal prohibited.

(A) No person shall possess a dangerous wild animal on or after January 1, 2014 . . .

[. . .]

§935.04 Registration of dangerous wild animals.

(A) A person that possesses a dangerous wild animal on the effective date of this section shall register the animal with the director of agriculture in accordance with this section not later than sixty days after the effective date of this section . . .

[. . .]

§935.08 Restricted snake possession permit.

(A) (1) A person that possesses a restricted snake in this state prior to January 1, 2014, that wishes to continue to possess the restricted snake on and after that date, and that does not intend to propagate, sell, trade, or otherwise transfer the snake shall obtain a restricted snake possession permit under this section not later than January 1, 2014 . . .

'Dangerous wild animals' are listed in §935.01(C) and 'restricted snakes' are listed in §935.01(L). Exceptions are allowed, for example, for bona fide accredited zoos and aquariums, accredited research establishments, animal shelters, circuses and veterinarians providing temporary care. 'Restricted snakes' are listed in the Act as follows:

(L) "Restricted snake" means any of the following:

(1) All of the following constricting snakes that are twelve feet or longer:
(a) Green anacondas;
(b) Yellow anacondas;
(c) Reticulated pythons;
(d) Indian pythons;
(e) Burmese pythons;
(f) North African rock pythons;
(g) South African rock pythons;
(h) Amethystine pythons.

(2) Species of the following families:
(a) Atractaspididae;
(b) Elapidae;
(c) Viperidae.
(3) Boomslang snakes;
(4) Twig snakes.

In the state of Missouri the keeping of certain dangerous species is restricted to certain facilities and, outside such facilities, by a registration system:

Missouri Revised Statutes

Chapter 578

Miscellaneous Offenses

Beginning January 1, 2017 – Keeping a dangerous wild animal – penalty.

578.023. 1. A person commits the offense of keeping a dangerous wild animal if he or she keeps any lion, tiger, leopard, ocelot, jaguar, cheetah, margay, mountain lion, Canada lynx, bobcat, jaguarundi, hyena, wolf, bear, nonhuman primate, coyote, any deadly, dangerous, or poisonous reptile, or any deadly or dangerous reptile over eight feet long, in any place other than a properly maintained zoological park, circus, scientific, or educational institution, research laboratory, veterinary hospital, or animal refuge, unless he or she has registered such animals with the local law enforcement agency in the county in which the animal is kept.
2. The offense of keeping a dangerous wild animal is a class C misdemeanour.

5.8   Legal Cases

Although some progress has been made in protecting animals using the law – especially in relation to cruel treatment – courts have been reluctant to recognise animals as having a legal personality and allow lawyers or NGOs to act as 'guardians' in legal proceedings.

A West German court, in 1988, refused to recognise a group of environmental lawyers as having standing as guardians of approximately 15,000 seals poisoned by marine pollution in the North and Baltic Seas. However, other cases have been brought to the courts in attempts to release captive animals or prevent them being imported by zoos.

The San Diego Zoo, California and the Lowry Park Zoo, Florida successfully imported 11 elephants from Swaziland to the United States after the US Fish and Wildlife Service issued the necessary permits, in spite of attempts by Born Free USA to obtain a preliminary injunction to prevent the importation in the District Court of Columbia (*Born Free USA v. North* 2003).

An attempt by the Free Morgan Support Group to use a Dutch court to prevent a female killer whale (named *Morgan*) from being transferred from the dolphinarium in Harderwijk (Netherlands) to Loro Parque in Tenerife failed in 2011. The orca had been rescued from shallow waters in Waddenzee. The group argued that the transfer would breach EU wildlife trade laws and argued that she should be returned to the wild.

People for the Ethical Treatment of Animals (PETA) brought a lawsuit in the US District Court for the Southern District of California against *SeaWorld* in 2012 on behalf of five wild-caught orcas. They claimed that the animals were held in violation of Section One of the Thirteenth Amendment to the Constitution of the United States, which prohibits slavery and involuntary servitude. They further argued that the law did not require the defendant to be a person. This case also failed.

In 2015, The Nonhuman Rights Project, Inc. petitioned the Supreme Court of New York State, in the United States, for a writ of *habeas corpus* on behalf of *Hercules* and *Leo*, two chimpanzees. The purpose of the writ was to attempt to secure the animals' release from the State University of New York to a sanctuary in Florida (*The Nonhuman Rights Project, Inc. v. Stanley* 2015). Such a writ would normally be addressed to the custodian of a prisoner to appear before a court to determine if he has lawful authority to detain him. The petition was denied because the court did not recognise the chimpanzees as legal persons and only a legal person is entitled to bring a writ of *habeas corpus*.

Advances in our scientific knowledge of animals, changes in public attitudes and developments in legal thinking may one day improve the legal position of animals, especially sentient, long-lived species that do not thrive in captivity. In 2010 a *Declaration of Rights for Cetaceans: Whales and Dolphins* was formulated by experts during a conference entitled *Cetacean Rights: Conference on Fostering Moral and Legal Change* at the Helsinki Collegium for Advanced Studies, University of Helsinki, Finland. The Declaration covers 10 rights, including the following:

We believe that:

1. Every individual cetacean has the right to life.
2. No cetacean should be held in captivity or servitude; be subject to cruel treatment; or be removed from their natural habitat.
[...]
4. No cetacean is the property of any State, corporation, human group or individual.
[...]
5. Cetaceans have the right not to be subject to the disruption of their cultures.

The Declaration asserts that all cetaceans have the right to life, liberty and wellbeing. It opposes the keeping of cetaceans in captivity and their use as a resource, but has no legal force.

In a court decision of 2015, an orangutan, *Sandra*, appeared to have been awarded nonhuman personhood rights by a court in Argentina and was sent from Buenos Aires

Fig. 5.6 A female chimpanzee (*Pan troglodytes*) with her infant. Will chimpanzees and other great apes one day soon have the right not to be kept in captivity? In law the principle of *habeas corpus* requires that a person who imprisons another be brought before a court to prove that the detention is lawful. (A black and white version of this figure will appear in some formats. For the colour version, please refer to the plate section.)

Zoo to live at the Center for Great Apes in Wauchula, Florida. On further investigation it became clear that, although *Sandra* was sent to Florida, these rights had not been awarded by the court (Wise, 2015). In the same year The Nonhuman Rights Project failed in its bid to have chimpanzees recognised as legal persons (Fig. 5.6). However, the judge noted that this may not be the case in the future, concluding that:

The similarities between chimpanzees and humans inspire the empathy felt for a beloved pet. Efforts to extend legal rights to chimpanzees are thus understandable; some day they may even succeed. (*The Nonhuman Rights Project, Inc. v. Stanley* 2015).

5.9   Conclusion

Modern zoos must not only work within the constraints of the law, they also play an important role in helping countries fulfil their international obligations to conserve biodiversity. Licensing laws are important to ensure that – through an inspection system – animals living in zoos are appropriately housed, contained and cared for, and that members of the general public cannot keep dangerous animals on their property without adequate facilities. The international obligations that most states have undertaken to protect biodiversity are to some extent fulfilled by zoos' conservation efforts through captive breeding and their *in-situ* work. However, although recent attempts to use the law to have animals such as elephants, great apes and cetaceans released from zoos to sanctuaries or the wild have had limited success, this law is still developing and changes in public and legal attitudes may ultimately restrict which species zoos are able to keep.

6 Ethics, Zoos and Public Attitudes

6.1 Introduction

The keeping of animals in zoos raises complex ethical questions. Many of these are concerned with balancing the interests of individual animals with the desire of humans to protect species from extinction.

Animals have no interest in conservation. An individual animal has interests in its own welfare and, at least in the more socially complex species, in that of its close relatives and companions, but not in its species *per se*. The idea of species conservation is a human construct. Chimpanzees have no interest in the survival of their *species*, but they have an interest in their own survival and that of their close relatives. People, or at least some people, care about the long-term future of chimpanzees, and therein lies the dilemma. If we keep chimpanzees in captivity we may be infringing the 'rights' of these animals – by keeping them in unnatural conditions – to satisfy the needs of humans to conserve wildlife. Whether or not an animal born and reared in a zoo can be successfully released to the wild is another question and is addressed in Chapter 12.

6.2 Zoos and Ethics

6.2.1 Is It Ethical to Keep Animals in Zoos?

People who object to zoos in principle on animal welfare grounds often have a naïve appreciation of what well-managed modern zoos do – or how they do it – and can see no further than the individual animals in cages and enclosures. They may, for example, see a single animal in an enclosure and be unaware that she is part of a captive breeding programme and will have companions at some time in the future when transfers from other zoos have been completed. Some animals are solitary in the wild but many visitors do not know this and assume all animals need the company of others of their species. Unfortunately, many zoos do not display appropriate signage explaining why lone animals are alone or explain that the individuals are part of a breeding programme involving other zoos, leaving their visitors to draw their own conclusions.

Peter Singer's *Animal Liberation* was published in 1975 and became the 'bible' of the animal liberation movement (Singer, 1975). However, its focus was the plight of animals living in laboratories and factory farms, not animals living in zoos. In recent years the ethics of keeping animals in zoos have been discussed at length by philosophers, zoo professionals and others (e.g. Bostock, 1993; Wemmer and Christen, 2008; Donahue, 2017).

In my view the best discussions have been provided by philosophers specialising in ethics rather than the proponents of the cause of animal welfare, who predominantly argue 'against' zoos, or those employed by zoos, who inevitably argue 'for' zoos. I do not want to repeat the many 'for' and 'against' arguments here. Instead, I offer a slightly edited version of some of the discussion about the ethics of keeping elephants in zoos that I have published elsewhere (Rees, 2021) in Box 6.1. To a very large extent, whether or not animals should be kept in zoos should probably be considered on a species-by-species basis, and elephants are a particularly contentious group.

Box 6.1 Is It Ethical to Keep Elephants in Captivity?

Peter Singer published *Animal Liberation* in 1975 (Singer, 1975). The work was revised in 1995, yet even this updated version largely focuses on the treatment of animals living in factory farms and laboratories, making little reference to animals in zoos and circuses, and no reference at all to elephants in captivity (Singer, 1995).

In 1975 scientists knew little about the complexity of the social and mental lives of elephants, but by 1995 this was most certainly not true. Iain Douglas-Hamilton's work on the Lake Manyara elephants in Tanzania was published in popular form in 1975 (Douglas-Hamilton and Douglas-Hamilton, 1975), and Cynthia Moss's book on the elephants of Amboseli, in Kenya, appeared 13 years later (Moss, 1988).

In his book *Animal Rights*, DeGrazia suggests that in considering whether or not an animal should be kept in a zoo two tests should be applied (DeGrazia, 2002). First, can a zoo provide for the basic physiological and psychological needs of the animal (the basic needs test)? Second, can a zoo provide a life at least as good as the life the animal could expect in the wild (the comparable life test)? DeGrazia suggests that there should be a strong presumption against keeping great apes and dolphins in zoos, but makes no specific reference to elephants.

In 2003 a symposium entitled *Never Forgetting: Elephants and Ethics* was held at the Conservation and Research Center in Front Royal, Virginia (now the Smithsonian Conservation Biology Institute). This resulted in the publication of *Elephants and Ethics: Towards a Morality of Coexistence* (Wemmer and Christen, 2008). In spite of its title, this book contains little serious philosophical discussion of the ethics of keeping elephants in captivity, and of the 37 contributing authors only one appears to have held an academic post in a university philosophy department at the time the work was published.

Box 6.1 (*cont.*)

Unfortunately it is not uncommon for biologists and others to write about ethics without using the terminology and principles of the discipline of ethics developed by philosophers. In a discussion of the animal rights–conservation debate, Hutchins (2007) suggested that:

A conservationist's work is analogous to that of an emergency room doctor's, and emergency room ethics apply. Sometimes desperate acts are going to be necessary to preserve life, even if it means causing the patient short-term harm.

He acknowledged that the 'patients' are endangered species and the harm is done to a subset of captive individuals, thereby demonstrating that this is not, in fact, an analogy. In an emergency room doctors may sometimes need to cause short-term suffering in their patients in their future long-term interests, but they do not harm one patient for the benefit of others. Elephants do not care if they become extinct, only (some) people do. Any welfare compromises suffered by captive elephants in the name of conservation are ultimately for the benefit of people, not other elephants.

Doyle (2014), an anthrozoologist, has recently argued against the keeping of elephants in captivity as a contributor to *The Ethics of Captivity* (Gruen, 2014), but, like many of the contributors to *Elephants and Ethics*, she largely reiterated well-rehearsed arguments, out-of-date statistics and her account contained no ethical – philosophical – arguments whatsoever (Fig. 6.1).

Some authors, however, have attempted a serious philosophical analysis of the ethics of keeping elephants. Varner (2008) discussed the extent to which elephants may be considered to be persons and how this should inform our attitude to the keeping of elephants in captivity. He considered the extent to which elephants can be shown to have a 'robust, conscious sense of their own past and future' and to have the 'biographical life' described by Rachels (1986), which some ethicists believe is necessary if an animal species is to be afforded a special moral status similar to that given to humans. After reviewing the evidence from studies of elephant memory, mirror self-recognition and theory of mind, Varner concluded that holding elephants in captivity could be justified if they are both born in captivity and their keepers treat them like 'domesticated partners'. He defines such a partner as 'a companion animal who works with humans in ways that emphasize and exercise [their] mental and/or physical faculties in a healthy way'. Varner recognises that adult elephants 'sometimes pose special difficulties' but concludes that keeping captive-born elephants in a circus or zoo environment may be consistent with respecting them as near-persons provided that their management is 'enlightened'. Furthermore, he suggests that a well-treated working elephant or an elephant performing in an idealised circus could be better off than one in a zoo.

A different approach to analysing the ethics of keeping elephants in captivity has been taken by Alward (2008). The 'capabilities approach' – when applied to

Box 6.1 (*cont.*)

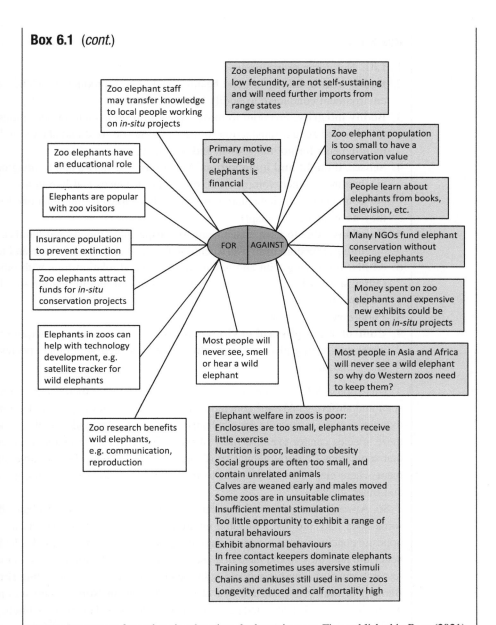

Fig. 6.1 Arguments for and against keeping elephants in zoos. First published in Rees (2021). Reproduced with permission from Elsevier.

animals – analyses ethical dilemmas by asking what an individual animal in a given situation is able to do and to be; whether they are able to live a full life. This is defined by first developing a list of the central functional capabilities of the species. Alward proposes the following as a list of central elephant functional capabilities:

Box 6.1 *(cont.)*

1. **Life.** Not dying prematurely.
2. **Bodily health**. Living in good health – including reproductive health – in an environment conducive to bodily health.
3. **Bodily integrity**. Freedom to move, freedom from assault, opportunities for self-directed sexual satisfaction, freedom to choose food and shelter and other requirements for normal growth and wellbeing.
4. **Senses, thought and imagination.** Able to use the senses and to imagine in an elephantine manner, communicate over long distances and live in an environment to which elephants are naturally adapted (rather than one to which they have learned to adapt).
5. **Emotions.** Able to form attachments to conspecifics (friends and family), places and things. Freedom from overwhelming fear and anxiety, and abuse and neglect.
6. **Affiliation.** Living with and having normal social relationships with others.
7. **Other species** Being able to live with concerns for and in relation to other elements of nature.
8. **Play.** Freedom to play.
9. **Control of the physical environment**. Being able to seek and acquire food, water and resting places. This control must be shared with other species and humans.

This approach focuses on autonomy and freedom, rather than health, normal life span and suffering. Alward concludes that the capabilities approach demonstrates that circuses are unsuitable for elephants – and cannot be made suitable – because these species cannot exercise most of their abilities in a circus environment, but that some zoos ('natural habitat zoos') may be able to provide a suitable environment where elephants are allowed to roam freely and associate with conspecifics and other species.

So, Alward's analysis concludes that circuses are unsuitable for elephants but that some zoos may provide a suitable environment; but Varner's analysis led to the opposite conclusion that working in an 'enlightened' circus may provide elephants a better life than they would have in a zoo. Indeed, Dennis Schmitt, a veterinarian and research scientist who has worked with the elephants owned by Ringling Bros. and Barnum & Bailey, has argued that elephants belong in North American circuses, where they are well-treated and perform a useful educational purpose (Schmitt, 2008).

Conservationists work to prevent the extinction of species. So, from an ethical perspective, would the extinction of elephants be a harm? Should elephants have a right not to become extinct? If so, this could help to justify the keeping of elephants in captivity for captive breeding purposes, but only if we can show that these programmes have a realistic prospect of assisting the future survival of elephants.

An individual elephant clearly has interests and may suffer harm; elephant *species* have no interests and thus may not suffer harm. Russow (1994) proposes that we protect animals because of their aesthetic value, their rarity, their

Box 6.1 (*cont.*)

adaptations and for many other reasons, not because they belong to a particular species. Russow would argue that it is not the species *Elephas maximus* or *Loxodonta africana* that we admire, but individual elephants. We value encounters with rare animals because they are less frequent than encounters with common animals. Russow says we should preserve these animals because we value possible future encounters with other individuals of the same species.

The material in this box was first published in Rees (2021) and is reproduced here with the permission of Elsevier.

6.2.2 Aesthetics, Pornography and Ethics

Tafalla (2017) has considered the extent to which we can appreciate the aesthetics of animals living in zoos by employing the theories concerning the aesthetic appreciation of nature propounded by Saito (1998) and Carlson (2000). After considering reasons why we can and why we cannot appreciate the aesthetic qualities of wild species in zoos, she concluded that, on balance, a serious and deep appreciation of their aesthetic qualities cannot be attained because of the effects of captivity on the health, appearance and behaviour of the animals (Table 6.1). She claims that because zoos impose their story on animals, we are prevented from appreciating the animals on their own terms.

Contrary to Tafalla's claim, some zoos rescue animals from appalling conditions and tell their stories to the public. Detroit Zoo, Michigan, rescued a polar bear (*Ursus maritimus*) from a Puerto Rican circus; Blackpool Zoo in England took in two Asian elephants (*Elephas maximus*) from the bankrupt Berlin Circus; and an Asian elephant called *Anne* was retired from Bobby Roberts' Super Circus and rehomed by Longleat Safari Park in England after her owners were convicted of causing unnecessary suffering (Rees, 2021). In 2010 the Yorkshire Wildlife Park, England, rescued 13 African lions (*Panthera leo*) from Oradea Zoo in Romania, where they were poorly fed and held in small concrete pens. Their new home covers an area of 7 acres (2.83 ha) divided into three grass reserves. A sign at the zoo tells their story (Fig. 6.3).

Acampora (1998) has likened the exhibition of animals in zoos to pornography and has called for their abolition. He claims that zoos make the natures of their subjects disappear by overexposing them, and they degrade and marginalise them through the marketing of their visibility. The exhibition of animals in zoos extinguishes their 'existential reality' while claiming to preserve their biological existence; it takes away the wildness we admire. In his analogy with pornography Acampora asks us to imagine an apologist for pornography taking the position that we should permit the institution because it excites or inspires us to esteem the subjects on display.

Table 6.1 A summary of Tafalla's analysis of the aesthetic appreciation of animals living in zoos.

Reasons why we *can* appreciate them	Reasons why we *cannot* appreciate them
Zoos present animals as ambassadors of their species, not as fantasies or metaphors.	Animals cannot be properly appreciated outside their natural environment.
Modern zoos exhibit animals in naturalistic enclosures with better viewing conditions than in the wild.	The ubiquity of 'star species' (e.g. orangutans) in zoos around the world instils in people the belief that they know these species and creates the illusion that they do not belong in any specific environment. It also exacerbates the tendency for many people to neglect their native fauna.
Appreciating animals in a zoo is a more multisensory and interactive experience than contemplating them in photographs and films.	Limited space in zoos makes it impossible for us to appreciate natural behaviour and causes the development of 'artificial behaviour'. Zoos modify behaviour by training.
At a zoo we can appreciate animals better than in any other type of exhibition of live animals (e.g. a circus).	Some of the aesthetic qualities of wild species cannot be appreciated in a zoo: strength, ferocity, speed, aggression. Zoo gift shops sell toys and other merchandise that are sentimental and neutralise the force and power of wild species (Fig. 6.2). Zoo play facilities depict animals as playthings.
Zoos offer a good opportunity to understand animals because they present real animals rather than artistic representations.	Animals are subjects, not objects; they look back at us. Animals are not ambassadors of their species; each is an individual with a subjective life and a personal story. Being exhibited causes them stress.
Modern zoos offer a scientific framework within which to appreciate animals.	Information boards offer information about species but rarely about the individual animals, such as their origin, history, personality and health. In contrast, some sanctuaries and rescue centres focus on the stories of particular abused individuals.

Source: based on Tafalla, 2017.

6.3 Public Attitudes

6.3.1 Introduction

Zoos cannot survive without public approval. Many famous zoos have closed in the past because they were not financially viable. Even London Zoo, perhaps the most famous zoo in the world, came close to closure when visitor numbers fell from over two million in 1970 to around 866,000 in 1995 (Rees, 2011). Without the financial support provided by visitors, zoos as we know them would cease to exist and this support depends heavily on what members of the public think about the standards of animal welfare in zoos.

Fig. 6.2 Toys representing *Snowflake* in the gift shop at Barcelona Zoo, Spain, in 2018. He lived at the zoo for 37 years and was believed to be the only albino gorilla in captivity. Snowflake died as a result of skin cancer in 2003. (A black and white version of this figure will appear in some formats. For the colour version, please refer to the plate section.)

6.3.2   Public Attitudes to Zoos

Studies of public attitudes to, and visitor perceptions of, zoos are usually based on a relatively small sample of individuals, often selected from visitors to a single zoo or a small number of zoos. It is, therefore, very difficult to make generalisations from their conclusions because those who choose not to visit zoos are usually excluded. Nevertheless, such studies provide a useful starting point for understanding the views and attitudes of the public.

The setting in which an animal is exhibited determines the public's attitude towards it. When slides of animals in natural, semi-natural and zoo settings were shown to American college students, they perceived zoo animals as significantly less dignified than those in the other settings and also confined, unhappy, unnatural, tame and dependent (Rhoads and Goldsworthy, 1979). In a similar study, Finlay et al. (1988) showed subjects eight species of animals in wild, naturalistic and caged zoo environments. They found that zoo animals were seen as restricted, tame and passive, while wild animals were described as free, wild and active.

Fig. 6.3 (a) In February 2010 the Yorkshire Wildlife Park, England, rescued 13 African lions (*Panthera leo*) from Oradea Zoo in Romania. (b) Two of the lionesses resting in the sunshine.

Zoos that display animals in authentic environments and have an educational focus have a positive influence on visitor attitudes towards wildlife. Conversely, visitors' fear of, or indifference to, wildlife increases after visits to more traditional zoos (Kellert and Dunlap, 1989).

Zoo and animal sanctuary visitors value interacting with animals and enjoy seeing a wide range of species. However, they are also concerned about welfare standards. Woods (2002) asked zoo visitors to describe their best and their worst wildlife tourism experiences. Captive environments were responsible for 363 of the worst experiences reported and 323 of the best experiences. The best experiences commonly involved interactions with wildlife, learning and viewing large numbers and varieties of wildlife. The worst experiences were dominated by poor animal management, poor service or facility management and experiencing threatening behaviour from animals.

Learmonth (2020) has noted that human–animal interactions in a zoo context can be 'fraught with ethical and welfare perils' but may also be rewarding and beneficial for both parties. It may foster ideas about inappropriate pet ownership, but also lead to pro-conservation attitudes. He suggested that zoos should adopt three ethical frameworks when deciding whether or not to incorporate human–animal interactions in a zoo: compassionate conservation, conservation welfare and duty of care. The World Association of Zoos and Aquariums (WAZA) has published guidelines on human–animal interactions for its member institutions (WAZA, 2020).

A comparison of the perceptions of zoo animals reported by visitors to Edinburgh Zoo, Scotland, with those of the general public – from a street survey – found that zoo visitors considered the main role of zoos to be conservation, while the general public had a number of negative perceptions of zoo animals, such as them being 'bored and sad' (Reade and Waran, 1996). Zoo visitors appeared to have a more positive perception of zoo animals and a greater awareness of the value of environmental enrichment, and appeared to be influenced by visual messages as they moved through the zoo.

6.3.3   Public Attitudes to the Dissection of Zoo Animals

Zoos are an important source of animals that can be dissected for scientific and educational purposes. From its earliest days London Zoo provided animals for dissections, the results of which were published in the Society's *Proceedings* (see Section 7.3.2). Attitudes to animals have changed considerably since the nineteenth century and zoos now receive much more scrutiny of their decisions and actions as a result of increased public access through the media.

A young giraffe called *Marius* was culled at Copenhagen Zoo, Denmark, in February 2014, in spite of a public campaign to save him, to prevent inbreeding because his genes were already well-represented in the captive population (*The Independent*, 2014). Yorkshire Wildlife Park, in England, had offered to take the animal, but this offer was declined. The giraffe was dissected in front of members of the public and some of the meat was later fed to the zoo's lions.

This incident caused an international outcry and was reported in the press around the world, from the *New York Times* to the *Sydney Morning Herald*. The incident attracted considerable multidisciplinary academic interest, including from philosophers (Levin, 2015; Gunasekera, 2018; Palmer et al., 2019), those who engage in research on tourism and journalism (Cohen and Fennell, 2016; Einwiller et al., 2017; Mkono and Holder, 2019), zoo veterinarians who must balance the medical interests of animals kept in zoos with those of conservation (Braverman, 2018) and others (Zimmerman et al., 2014; Gingrich-Philbrook, 2016; Mäekivi and Maran, 2016).

The Danish attitude to the use of animals is rather less squeamish than that of many nationalities, especially those protected from the harsh reality of meat production. The Danish Crown slaughterhouse at Horsens allows visitors to watch the entire pork production process from a viewing platform: the arrival of the pigs, their slaughter and butchery, and the packaging of the meat. In contrast to this open attitude towards the killing of animals for food, some zoos will not even allow visitors to see inside the indoor accommodation of their animals. Although animals need to be able to escape the attention of visitors, this lack of access may give rise to doubts about the quality of the housing and standards of animal welfare.

6.3.4   Public Attitudes to Euthanasia and Culling

Zoos may cull healthy animals because they cannot be used in captive breeding programmes or because zoos are unable to continue to house them. Some zoos allow animals to continue breeding because animal births bring in visitors. In some cases these animals are culled once they have served their purpose.

In December 2000, Edinburgh Zoo euthanised four Arabian oryx (*Oryx leucoryx*) because they could not find another zoo that would take them (BBC, 2000). In February 2017 Takagoyama Nature Zoo in Japan culled 57 of its 164 Japanese macaques or snow monkeys (*Macaca fuscata*) because they were the result of hybridisation with rhesus macaques (*M. mulatta*) (Demetriou, 2017).

Killing healthy animals causes a dilemma for zoo staff who are employed to care for them, especially when the welfare of the animals conflicts with the aims of captive breeding programmes and the ability of zoos to house them.

The options open to zoos to prevent the birth of surplus animals that will put pressure on scarce accommodation are to sterilise them or use contraception. If the surplus animals already exist they may be disposed of by euthanasia or transfer to other zoos. The ethical, practical and welfare implications of these options have been discussed – with particular reference to primates – by Glatston (1998), who concluded that a combination of contraception and euthanasia seemed to be the most acceptable solution from an animal welfare viewpoint.

A nationwide survey of staff in zoos and aquariums in the United States found that managers and animal programme leaders were more supportive of euthanasia as a population management tool than keepers (Powell and Ardaiolo, 2016). Regardless of their role, men were more supportive of euthanasia than women, but education, tenure in the position and taxonomic expertise were not strong predictors of attitudes. However, support for culling varied with taxon. Overall, the surveyed population split approximately evenly between those in favour, those against and those who were unsure.

A similar study of the attitudes of zoo veterinarians working in zoos in the United States found that they were aware of population management euthanasia and knowledgeable about the practice (Powell et al., 2018). Attitudes varied depending upon the type of animal being considered for euthanasia, and female veterinarians and those not aware of the practice being used at their facilities were less supportive of it than males and those familiar with the practice. Veterinarians tended to be either more population-focused or more individual-focused when considering culling, whereas other zoo personnel often rated themselves as being more equally focused on the needs of populations and individuals. The authors of the study suggested that veterinarians were more likely to be supportive of management euthanasia if they were provided with information on the rationale for using it.

Animal *rights* advocates have traditionally argued that the use of euthanasia for population management is a violation of the animals' rights. The argument from an animal *welfare* perspective is that if the animal does not suffer there is no harm in the practice and it may be justified by the potential benefits to the population. However, Browning (2018) has suggested that if the welfare view is expanded to include longevity and opportunities for positive welfare, this weakens the argument for management euthanasia unless greater benefits can be identified to justify its use.

6.3.5 Public Attitudes to Live Feeding

Modern zoos are expected to encourage normal behaviour in the animals they keep, but this is difficult for predatory species that catch and kill their prey in the wild. In many jurisdictions the feeding of live vertebrate prey is illegal and in any event watching animals being killed in a zoo is unpalatable to many visitors, especially those attending with children.

A survey of visitors to Edinburgh Zoo found that all agreed with feeding live insects to lizards off-exhibit and only 4 per cent objected if this took place on-exhibit (Ings et al., 1997). When asked if penguins should be fed live fish, 72 per cent agreed with this on-exhibit and 84.5 per cent off-exhibit. Only 32 per cent found it acceptable to feed a live rabbit to a cheetah on-exhibit, but 62.5 per cent found this acceptable off-exhibit. Female interviewees were more likely than males to object to live feeding vertebrate prey. Those that agreed with live feeding perceived it as 'natural', while those that objected said they did so because 'it would upset them or their children'.

The majority of visitors to Zurich Zoo, Switzerland, agreed with feeding live prey (vertebrates and invertebrates) both off- and on-exhibit, except that (especially) women and frequent visitors did not agree with feeding live rabbits to tigers on-exhibit (Cottle et al., 2010). Better-educated visitors were more likely to agree with feeding live prey off-exhibit. The authors of this study noted that zoo visitors in Switzerland were more often in favour of live feeding of vertebrates than those in Scotland studied by Ings et al. (1997).

In the context of feeding live vertebrates to other animals in zoos, Keller (2017) has argued that zoos have an ethical responsibility to take the welfare of invertebrates into consideration because there is evidence of higher cognitive functions – such as emotion and learning – in some invertebrate taxa and the possibility of their experiencing suffering exists.

6.3.6 Public Attitudes to Keeping Cetaceans and Elephants in Zoos

The keeping of elephants in zoos and cetaceans in marine theme parks has been controversial, but the attitudes of the public to these practices have attracted little attention from researchers.

Fig. 6.4 The dolphinarium at Barcelona Zoo was closed in 2020 and the three bottlenose dolphins (*Tursiops*) moved to Attica Zoological Park, Greece. Inset: One of the former occupants.

Public attitudes to the keeping of whales and dolphins in marine theme parks and aquariums vary between cultures, but an international online survey conducted in 2015 ($n = 858$) found that there was little support overall (Naylor and Parsons, 2019). Overall, 54.4 per cent of respondents were opposed to the captive display of dolphins and whales, but only 5 per cent of respondents from the United States stated they 'strongly support' this compared with 21 per cent of respondents from India. These figures increased to 38.3 per cent and 64.4 per cent, respectively, when respondents who 'supported' captive display were added. Some 86 per cent of respondents said they would prefer to watch cetaceans in the wild; only 9 per cent of US respondents preferred to watch them in a marine theme park, compared with 26 per cent of respondents from India. Participants in the survey that supported keeping cetaceans in captivity were significantly more likely to consider cetacean conservation unimportant. Some institutions have stopped keeping cetaceans in recent years as a result of welfare concerns and public option (Fig. 6.4).

The attitudes of Australian ($n = 101$) and Indian ($n = 101$) respondents towards keeping elephants in zoos have been examined by Gurusamy et al. (2015). Elephants are indigenous in India, whereas the importation of eight Asian elephants (*Elephas maximus*) into Australia from Thailand (via the Cocos Islands) in 2006 had been controversial and received international attention. Australian respondents were more concerned about the husbandry conditions for animals in zoos and sanctuaries than were Indian respondents, and this concerned increased with education level.

Australian men were less concerned about husbandry than were Australian women. Over 42 per cent of Australians were prepared to pay more to visit a zoo with elephants, but only 7.9 per cent of Indians were prepared to do this. More Indian than Australian respondents believed it was important for zoos to display elephants, and wanted to feed, touch and ride them. Indian respondents considered elephants to be of religious, cultural and historical importance, while Australians acknowledged that captive elephants have a scientific value.

I have listed elsewhere (Rees, 2021) over 30 zoos around the world that no longer keep elephants, having kept them in the past. In addition, in 2009 the Central Zoo Authority in India announced that 140 elephants living in 26 Indian zoos would be transferred to wildlife parks and sanctuaries.

6.4   Sentience and Animal Dignity

Whether or not we recognise animals as sentient is central to the discussion of how we should treat them. There is no single universally accepted definition of sentience (Proctor et al., 2013). It may be defined as:

having the awareness and cognitive ability necessary to have feelings. (Broom, 2014)

In an earlier definition, Broom (2007) considered a sentient being as

one that has some ability to evaluate the actions of others in relation to itself and third parties, to remember some of its own actions and their consequences, to assess risk, to have some feelings and to have some degree of awareness.

By the time of the Renaissance the concept of sentience was accepted by lay people in mammals and birds, but it was not until the Enlightenment of the eighteenth century that it was acknowledged by philosophers (Duncan, 2006). By the end of the nineteenth century a fairly sophisticated concept of sentience had been developed by scientists and philosophers, but the influence of Behaviourism inhibited any further progress throughout most of the twentieth century. Animal sentience received a surge of interest from animal welfare scientists in the last quarter of the twentieth century as they came to appreciate that tackling welfare problems is facilitated by an understanding of how animals feel.

Science-based discussions of sentience in animals depend upon our scientific knowledge of their behaviours and capacities available at the time. It was not long ago that it was generally accepted that fish did not feel pain (Fig. 6.5). There is now ample evidence to suggest that fish experience pain in a similar manner to that experienced by other vertebrates. In his review of the research conducted on fish perception and cognition, Brown (2015) concluded that their abilities often match or exceed those of other vertebrates.

However, Key (2016) takes a different view. He has made an anatomical comparison of the brains of humans (who can report feeling pain) and fish (who cannot report feeling pain) and concluded that:

Fig. 6.5 Lionfish (*Pterois volitans*). Until recently many people believed that fishes did not feel pain.

fish lack the necessary neurocytoarchitecture, microcircuitry, and structural connectivity for the neural processing required for feeling pain.

As we extend our knowledge of the nervous systems and sensory capabilities of different animal taxa – taxa other than vertebrates – it is inevitable that the law will eventually provide protections for a wider range of animals. In a consideration of which aquatic animals should be protected, Broom (2007) noted that:

There is evidence from some species of fish, cephalopods and decapod crustaceans of substantial perceptual ability, pain and adrenal systems, emotional responses, long- and short-term memory, complex cognition, individual differences, deception, tool use, and social learning. The case for protecting these animals would appear to be substantial.

Birch (2017) has argued that there is some justification for giving an animal the 'benefit of the doubt' and applying the precautionary principle when the evidence for sentience is inconclusive. However, Jones (2013) noted that:

even the most progressive current welfare policies lag behind, are ignorant of, or arbitrarily disregard the science on sentience and cognition.

Blattner (2019) has discussed the role of sentience in determining which animals receive protection from the law and which do not. She argues that most states recognise that animals must be protected by law because, and to the extent that, they are sentient. Blattner discusses the nature and scope of the legal recognition of animal sentience, highlights the remaining prejudices in society and science, and notes that some states refuse to commit themselves to the recognition of animal sentience.

In the European Union (EU), the Lisbon Treaty (2007) amended the Treaty on the Functioning of the European Union and introduced the recognition that animals are sentient beings in Article 13 of Title II:

In formulating and implementing the Union's agriculture, fisheries, transport, internal market, research and technological development and space policies, the Union and the Member States shall, since animals are sentient beings, pay full regard to the welfare requirements of animals, while respecting the legislative or administrative provisions and customs of the EU countries relating in particular to religious rites, cultural traditions and regional heritage.

On 13 May 2021 the Animal Welfare (Sentience) Bill was introduced in the UK Parliament as part of its *Action Plan for Animal Welfare* (DEFRA, 2021). The new law was intended to recognise that vertebrate animals can experience feelings such as joy or pain and will ensure that animal sentience is taken into account when developing policy across the government through the establishment of an Animal Sentience Committee:

2 Reports of the Committee

(4) The purpose is that of ensuring that, in any further formulation or implementation of the policy, the government has all due regard to the ways in which the policy might have an adverse effect on the welfare of animals as sentient beings.

The Bill does not, however, define sentience.

The *Action Plan for Animal Welfare* makes only a brief reference to zoos in the fifth of its five 'key strands':

5. We will increase protections for kept wild animals by ending the low welfare practice of keeping primates as pets, improving standards in zoos, and enhancing conservation

and states that

We plan to improve current requirements applying to zoos including in relation to their conservation work.

The Action Plan notes the government's willingness to extend the purview of the law in the future to cover invertebrate taxa if such action is supported by the scientific evidence:

[The government has] commissioned research into the sentience of decapod crustaceans and cephalopods, and, in light of the findings, we will consider further protections.

The Animal Welfare (Sentience) Act became law in the UK in 2022. It requires that the government must have regard to the ways in which any of its policies or their implementation 'might have an adverse effect on the welfare of animals as sentient beings' (s.2(2)). The animals to which the Act applies are any vertebrates other than *Homo sapiens*, any cephalopod mollusc and any decapod crustacean (s.5(1)). Section 5(2) of the Act makes provision for the Secretary of State to extent the definition of 'animal' to include other invertebrates by statutory instrument. The Act does not define sentience.

In 1992, Switzerland became the first country in the world to recognise the importance of protecting the dignity of animals when it amended its federal constitution (Bolliger, 2015). This grants a moral value to animals irrespective of their sentience. Protection for an animal's inherent worth includes protection from humiliation. Bolliger defines 'humiliation' as

any demeaning conduct towards an animal that does not consider its nature.

He argues that an act of humiliation can affect an individual or an entire species, and that it is not necessary for an animal to be aware that it is being humiliated. Dignity does not require consciousness of its presence in humans or in animals.

Bolliger lists a number of ways in which animals may be humiliated, including being trained to perform tricks or unnatural behaviours as a means of public entertainment. An act of humiliation may be construed when an animal is caused to perform a submissive behaviour designed to illustrate human dominion over dangerous or powerful animals such as elephants, large felids, bears or crocodiles, especially when trained to do this in a circus or zoo. Such acts are seen far less often in modern zoos than was formerly the case. However, Bolliger also includes the following in his list of humiliating acts:

acts of humiliation are present when humans pester or otherwise provoke animals in exhibitions that do not allow an animal to retreat from view.

The question of protecting the dignity of animals could have far-reaching implications for zoos if it were to be included in animal welfare legislation worldwide. Animals kept in zoos may be exposed to intimidating and threatening behaviour from visitors as they try to attract their attention; noisy teenagers, inconsiderate ground staff and construction workers and a host of human activities that intrude on their lives and from which they may not be able to escape.

During one of my many visits to zoos with undergraduate students studying zoo biology I witnessed large groups of noisy school children being allowed to enter the indoor accommodation of a group of bonobos (*Pan paniscus*) after one of the females had given birth during the previous night. There was great excitement among the apes and much interest in the young infant and the discarded placenta. From time to time the staff emptied the bonobo house to reduce the disturbance to the animals, but then reopened it shortly thereafter. This seemed to me to represent a total lack of consideration for the new mother and her offspring, and in my view the bonobo house should have been closed for at least a few days.

Zoos, understandably, want to publicise the births of animals to attract visitors, promote their conservation message and increase their revenue. But this should not be at the expense of the welfare and dignity of the animals. When Asian elephant (*Elephas maximus*) calves are born at Chester Zoo in England it has been the practice to close the elephant house for several days thereafter to protect the mother and her offspring from the attention of an excited and noisy public.

Although many zoos provide their animals with the opportunity to be out of sight of the public, a survey of visitors to Hamilton Zoo in New Zealand ($n = 359$) found that they attached more importance to the viewing of animals than to the possibility that animals might need 'private places' (Ryan and Saward, 2004).

It is conceivable that a wide acceptance of the importance of the concept of animal dignity in considerations of animal welfare could eventually result in the demise of zoos as we know them today. Many zoos already keep some of their animals off-show and away from the public gaze in what are essentially breeding units. The conservation breeding of endangered species does not require the existence of zoos, provided that the necessary animal accommodation, staff and funding exist elsewhere.

It should be noted that there are philosophical arguments against the extension of the concept of dignity to animals. Zuolo (2016) has argued that the three main approaches that might justify the application of dignity to animals – the species-based approach, moral individualism and the relational approach – do not provide an appropriate basis.

6.5 Are Elephant, Ape and Cetacean Exhibits the New 'Human Zoos'?

Public displays of 'exotic' people from different ethnic groups, often in naturalistic settings – ethnological expositions – were common in the nineteenth and twentieth centuries and have been described in detail in Blanchard et al. (2008a). In earlier times, in Ancient Egypt, black 'dwarves' were exhibited from the Sudanese territories. During the Roman Empire conquered 'barbarians' and 'savages' were paraded through the street as a reminder of Roman superiority (Blanchard et al., 2008b).

More recently, in 1874, Carl Hagenbeck – the German animal trader and owner of the Tierpark in Stellingen – exhibited a family of six Lapps (Sami) with their tents, weapons and sleds alongside around 30 reindeer (*Rangifer tarandus*) in Hamburg, Germany. His other exhibitions included Indians, Australian aborigines, Sinhalese, Patagonians, Mongolians, Maasai and Kalmyks. Hagenbeck staged his last ethnic show in 1932 (Thode-Arora, 2008).

In August 1877 the Jardin Zoologique d'Acclimation in Paris hosted a new exhibition of animals from Somalia and Sudan, including exotic species of cattle, camels, giraffes, elephants, ostriches and a 'dwarf rhinoceros'. These were accompanied by 14 'Nubians' and the attraction was a great success (Schneider, 2008). This was followed by other ethnographic exhibitions featuring Eskimos (Inuit) from Greenland, Lapps (Sami) and gauchos from Argentina.

In 1904 a Mbuti man from the Congo, Ota Benga, was exhibited at the Louisiana Purchase Exposition in St Louis, Missouri. Two years later he was taken to New York and exhibited at the Bronx Zoo, accompanied by a sign that read 'African Pygmy, "Ota Benga"' (Fig. 6.6). At the time, William Hornaday was the zoo director. He went on to help found the Smithsonian's National Zoo in Washington. On 9 September

Fig. 6.6 Ota Benga was taken to the Bronx Zoo, New York, in 1906. This photograph was taken *c.*1915–1916 (courtesy of the United States Library of Congress, reproduction number LC-DIG-ggbain-22741).

1906 the *New York Times* carried an article entitled 'Bushman shares a cage with Bronx Park apes' (*New York Times*, 1906).

A detailed account of many examples of ethnographic exhibitions that have been held in zoos and elsewhere has been provided by Blanchard et al. (2008a). It includes photographs of some of these exhibits and images of the posters used to advertise them.

The exhibition of people in this manner is unthinkable today, but nevertheless we still exhibit our close relatives – chimpanzees, gorillas, orangutans and other apes – in naturalistic settings in zoos. Elephants and cetaceans are not suited to a life confined in small enclosures with limited social contact with conspecifics, but zoos continue to keep these animals in spite of the fact that there is good evidence that captivity compromises their welfare.

6.6 Conclusion

This book is not the place for a detailed philosophical discussion of the value of the concepts of animal rights, animal sentience and animal dignity in protecting individual animals or species. Nevertheless, if the current interest in the protection of the environment and the biodiversity crisis extends into an increased concern for animal welfare and a more robust attempt at establishing rights for at least some species, zoos may be compelled by public opinion – and possibly changes to the law – to re-examine their living collections and some of their current practices. Human attitudes to the environment, animal sentience and animal welfare are changing. It is not inconceivable that in the near future pressure from the public, animal welfare NGOs and governments will result in zoos being forced to acknowledge that it is no longer acceptable to keep large, intelligent, socially complex animals in captivity.

7 The Contribution of Zoos to Zoology

7.1 Introduction

Much of the scientific research conducted on animals living in zoos and aquariums is concerned with improving their welfare and breeding success. In many cases, this means trying to ameliorate problems caused by captivity, for example abnormal behaviours, poor reproduction and low neonate survival. But there is a separate body of scientific work conducted in zoos that examines the fundamental biology of animals, especially their natural behaviour (including cognitive abilities), ecology and physiology.

Research on zoo-living animals can provide data that cannot be collected in the wild. Behringer et al. (2018) have noted that such studies promote validation and refinement of methods, and the data collected can serve as a benchmark for studies on wild conspecifics and assist in the development of novel research tools. Animal living in zoos are accessible, sample collection is relatively easy, environmental factors can be controlled (up to a point) and the life histories of, and genetic relationships between, the animals are often well known.

In addition to the basic zoological knowledge that we can gain from studying animals living in zoos, there also exists a body of knowledge concerning the history of zoos and animals living in them, some of which has been obtained from examining archaeological sites and animal remains.

7.2 Zooarchaeology

We cannot be sure when the first zoo came into existence; we can only know the earliest zoos that archaeologists have found, or think they have found. Archaeological interpretation of historical artefacts depends largely on inference. Animal remains are frequently found alongside those of humans and their buildings. Such animals are usually those that are domesticated – such as dogs, cattle and goats – or those that have been hunted for their meat and skins. Where such animal remains exist, their skeletons bear the marks made during the process of butchery. But some archaeological sites contain evidence of associations between human populations and wild animals that do not appear to have been killed for food and so must have been kept for some other purpose.

If a zoo is a collection of exotic animals, then the HK29A site at Predynastic Hierakonpolis in Upper Egypt appears to qualify; and if it does, then it is the oldest zoo found to date. Animals are frequently found associated with archaeological remains in Egypt. What makes HK29A special is that the types of animals found there do not conform to those commonly found elsewhere.

The city of Hierakonpolis began to take shape around 4000 BCE and its cemeteries in particular have yielded interesting faunal remains, particularly that of the elite members of society at HK6. The remains of over 100 animals buried whole have been found. Along with domestic animals – dogs, cattle, sheep and goats – wild animals have been discovered: aurochs (wild cattle), wild cat, swamp cat, hartebeest, wild donkey, Anubis baboon, hippopotamus and elephant (Linseele et al., 2009).

The Royal Menagerie at the Tower of London housed exotic animals from the 1200s to 1835. When the Menagerie closed, some of the remaining animals were used to found London Zoo in Regent's Park. The Menagerie is presently commemorated by the presence of wire sculptures of baboons, lions and elephants, and interpretive materials (Fig. 7.1). In 1937 the remains of some of the big cats kept at the Tower were excavated and then left unstudied for around 70 years until they were examined by O'Regan et al. (2006). These authors used radiocarbon dating to establish that the remains ranged in date from the thirteenth to the seventeenth centuries. Two lion skulls and fragments of a leopard were described, and the authors noted that one of the

Fig. 7.1 Baboon sculptures at the Tower of London. (A black and white version of this figure will appear in some formats. For the colour version, please refer to the plate section.)

lion skulls had a partially occluded foramen magnum. As this anomaly has also been recorded in captive cat skulls from the early twentieth century – whose provenance was unknown – they commented that the condition has a long history.

I have included a brief reference to the zoo at Hierakonpolis and the lions of the Royal Menagerie here because they are the subjects of relatively recent publications. This history of menageries and zoological gardens has been more than adequately covered by Johnson (1994), Hoage and Deiss (1996), Kisling (2000), Baratay and Hardouin-Fugier (2002) and others (see the Bibliography).

7.3   Zoos as Sources of Biological Knowledge

Zoos and menageries have been important sources of biological knowledge for centuries. Aristotle's *History of Animals* contains descriptions of the anatomy, behaviour and other aspects of the biology of animals that he can only have observed in the menageries that had been established in the city states of Greece by the fourth century BCE (Cresswell, 1883). As part of his account of the biology of elephants, Aristotle wrote:

The elephant has four teeth on each side, with which he grinds food, for he reduces his food very small, like meal. Besides these, he has two tusks: in the male these are large, and turned upwards; in the female they are small, and bent in the contrary direction. The elephant has teeth as soon as it is born; but the tusks are small, and therefore inconspicuous at first. It has so small a tongue within its mouth, that it is difficult to see it.

The travelling menageries of the Victorian era in England brought exotic animals to the masses and produced educational materials describing their animals (Fig. 7.2). Bostock and Wombwell's Royal Menagerie was a large menagerie that toured widely in Britain and abroad from 1805 until 1932. It housed a very wide range of animals, including elephants, camels, lions and tigers. It was so successful that it eventually became three menageries that toured independently and gave several royal command performances before Queen Victoria. The sale of the menagerie to London Zoo in 1932 probably marked the end of the large self-sufficient travelling menageries.

The behaviour of animals in zoos attracted the interest of scientists and keepers from the beginning of scientific zoos, and was becoming well established in the Victorian era. The work of the first superintendent of London Zoo, Abraham Lee Bartlett, was compiled and edited by his son Edward and published as *Wild Animals in Captivity: Being an Account of the Habits, Food, Management and Treatment of the Beasts and Birds at the 'Zoo' with Reminiscences and Anecdotes* (Bartlett, 1899). This contains accounts of many species, including Bartlett's observation on elephants:

My fondness for elephants led me to study them and pay particular attention to their habits and treatment in captivity. I found that the males when approaching maturity, or when about twenty years of age, required very careful management, for about this period, if well fed and in good condition, they become restless and somewhat uncertain in temper, and in many instances extremely dangerous to be approached. This condition generally would last four or five weeks,

Catalogue of the Collection. 7

The Bengal Tiger.

The Cheetah, or Hunting Leopard.

The Spotted or Laughing Hyæna.

420. The Wombat.

The Wombats are short, clumsy and thick-legged Marsupials that are natives of Australia, and live entirely on vegetable matter, such as grass and roots, which they dig up with their strong claws. They are harmless creatures, and their flesh is said to be palatable.

430. Young Specimen of the Boomah.

The largest of the kangaroo species from Australia. By their peculiar formation kangaroos neither run nor walk but move along by a series of bounds. They are pouch-bearing animals and are much hunted in their native country.

440. Black-headed Sheep, from Port Said.

450. Zebu, or Sacred Brahmin Bull of Hindostan.

An animal still worshipped by some of the Hindoo Tribes, but principally used as a beast of burden.

460. Axis Deer, from Ceylon.

The handsomest of the deer family.

470. The Thar

Is a Himalayan Wild Goat that inhabits forests from Kashmir to Bhutan. It differs from typical Goats in its short, thick, and much compressed horns, and is of a ferocious disposition.

480. Hog Deer, from China.

490. Giant Male Hippopotamus.

About 8 years of age, added to the collection at enormous cost. It hails from the White Nile, and is positively the first living specimen ever seen in Scotland.

Fig. 7.2 A page from Bostock and Wombwell's Menagerie catalogue 1917 (courtesy of Chetham's Library, Manchester).

and is well known to elephant keepers by the term 'must.' I heard of the deaths of many persons who had been killed from time to time by elephants while in this state. The first elephant that ever came immediately under my charge was the celebrated 'Jumbo.'

The 'father of zoo biology' Heini Hediger published his work *The Psychology and Behaviour of Animals in Zoos and Circuses* in 1969. Many other important books on zoos have been published since (see the Bibliography).

7.3.1   Taxonomy

Zoos have been an important source of animals for taxonomic research since their early beginnings. Naturalists and others reporting to the Zoological Society of London (ZSL) in its early days frequently published descriptions of new species and the results of collecting expeditions in the *Proceedings of the Zoological Society of London* (now the *Journal of Zoology*):

Gray, J. E. (1854). On a new species of rhinoceros. *Proceedings of the Zoological Society of London*, 22, 250–251.

Austem, E. E. (1896). Notes on a recent zoological expedition on the Lower Amazon. *Proceedings of the Zoological Society of London*, 64, 768–788.

Flower, S. S. (1899). Note on the proboscis monkey, *Nasalis larvatus*. *Proceedings of the Zoological Society of London*, 67, 785–787.

Sharpe, R. B., Mackinder, H. J., Saunders, E. and Camburn, C. (1900). On the birds collected during the Mackinder Expedition to Mount Kenya. *Proceedings of the Zoological Society of London*, 69, 596–609.

The new rhinoceros species to which Gray alluded in the paper listed above was the white rhinoceros (*Ceratotherium simum*).

The role of Italian zoos in mammalian systematic studies, conservation biology and museum collections has been discussed by Gippoliti and Kitchener (2007).

7.3.2   Anatomy

Captive animals, especially those from zoos, have been an important source of animals for dissection for millennia. Writing in the fourth century BCE, Aristotle provided many anatomical details of animals that were not native to his homeland in his *History of Animals* (Cresswell, 1883). He wrote of elephants:

the testicles are not external but internal and near the kidney.

London Zoo is widely regarded as the world's first scientific zoo, and some of the early dissections conducted on its deceased animal occupants were reported in newspapers and the ZSL's *Proceedings*:

Owen, Prof. (1854). On the anatomy of the great anteater (*Myrmecophaga jubata*). *Proceedings of the Zoological Society of London*, 22, 154–157.

Crisp, E. (1865). On some points relating to the anatomy and habits of the Bactrian camel (*Camelus bactrianus*), and on the presence of intestinal glands not before noticed. *Proceedings of the Zoological Society of London*, 33, 257–265.

Owen, Prof. 1865. On the morbid appearances observed in the dissection of the penguin (*Aptenodytes forsteri*). *Proceedings of the Zoological Society of London*, 33, 438–439.

When the Asian elephant *Jack* (Fig. 7.3) died at London Zoo in 1847 his body was dissected by Professor Richard Owen, an accomplished comparative anatomist and Hunterian Professor of Anatomy at the Royal College of Surgeons who invented the term 'dinosaur' and later became the first superintendent of the British Museum (Natural History) (Rees, 2021). Various portions of the body were then distributed to learned institutions for further study, including the Royal College of Surgeons, the Anatomical School of Oxford, King's College, London, and the Royal Veterinary College (*The Illustrated London News*, 1847b). Professor Owen regularly dissected animals that died at the zoo and was a prolific contributor to the *Proceedings of the Zoological Society*, where he published anatomical papers on a wide range of taxa including the cheetah, giraffe and orangutan (Owen 1834; 1839a; 1839b). In addition to presenting his findings at meetings of the ZSL, Owen also frequently chaired these meetings.

Zoos still provide dead and living specimens for anatomical study. The first description of a preputial gland for the genus *Nasau* was identified in a male coati (*N. nasau*) in a specimen from Edinburgh Zoo, Scotland (Shannon et al., 1995). An anatomical study of the elbow joint of a Bengal tiger (*Panthera t. tigris*) was made

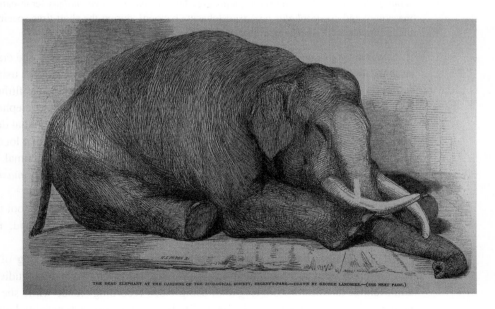

THE DEAD ELEPHANT AT THE GARDENS OF THE ZOOLOGICAL SOCIETY, REGENT'S-PARK.—DRAWN BY GEORGE LANDSEER.—(SEE NEXT PAGE.)

Fig. 7.3 *Jack* the elephant died in 1847 at London Zoo and was dissected by Prof. Richard Owen. (source: Death of the Zoological Society's elephant. *The Illustrated London News*, 19 June 1847, pp. 1–2).

using magnetic resonance imaging and gross dissection of a cadaver of a six-year-old female tiger obtained from Cocodrilos Park Zoo in the Canary Islands, Spain (Encinoso et al., 2019). The structure of the cranium of the Komodo dragon (*Varanus komodoensis*) was examined by Pérez et al. (2021) using two 17-year-old females born in captivity at *Reptilandia*, Las Palmas, Spain.

Kitchener (2020) has suggested that zoos and museums should work together to ensure that when zoo animals die their bodies are preserved for research into anatomy and functional morphology. He noted that small carnivorans are poorly represented in zoos, and their ecology and behaviours are poorly understood. Museum collections of zoo specimens allow the impacts of captivity (e.g. diet and activity levels) to be investigated.

7.3.3　Physiology

Studies of animals living in zoos give scientists opportunities to make important advances in comparative physiology. Early publications often simply reported observations. For example, Heape (1898), writing in the *British Medical Journal*, refers to observations made by Bartlett – the superintendent – at London Zoo that monkeys regularly menstruate and that the superintendent of Calcutta Zoo had confirmed this for animals in his zoo:

> The late Mr. Bartlett, superintendant of the Zoological Gardens in London, and Mr. Sayal, superintendant of the Zoological Gardens in Calcutta, both assure me that monkeys menstruate regularly in their establishments, and I myself have observed regular menstruation in *semnopithecus entellus*,[1] [northern plains grey langur] *macacus cynomolgus* [crab-eating macaque], and *cynocephalus porcarius* [chacma baboon], during the short time specimens of these animals were under my notice in Calcutta.

Some animals can be trained to participate in physiological studies. The energetics of walking in the African elephant (*Loxodonta africana*) has been studied using three young males at Zoo Atlanta, Georgia, who were trained to wear a loose-fitting mask enclosing the mouth and trunk (Langman et al., 1995). While the elephant was walking he was preceded by a motorised golf cart which carried a pump that delivered air to the face mask via a tube. The experiments led to the conclusion that locomotion in the African elephant has the lowest energy cost of any living land mammal. Similar experiments were conducted using two female Asian elephants (*Elephas maximus*) at Audubon Zoo, New Orleans (Langman et al., 2012) (Fig. 7.4).

The mechanics of walking in Asian elephants was investigated using foot pressure patterns at Whipsnade Zoo and Woburn Safari Park, both in England, utilising pressure platforms (Panagiotopoulou et al., 2012).

Studies of apes living in zoos can help to advance our understanding of human evolution. Bipedal locomotion in chimpanzees (*Pan troglodytes*) was studied in an adult female at Chester Zoo, England, using video recordings made while she walked across a force plate (Thorpe et al., 2004). Studies of chimpanzee bipedality have found

---

[1] The author omitted to capitalise the names of genera in this text.

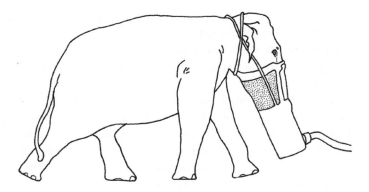

Fig. 7.4 The ability of zoo staff to train elephants to wear equipment and walk in a controlled manner has made it possible to study their oxygen consumption and thus the energetics of locomotion (drawing by Dr A. J. Woodward, first published in Rees (2021)).

it to be mechanically inefficient and an inappropriate model for the study of the early evolution of bipedalism in hominids.

A study of a herd of Asian elephants at Chester Zoo found that when the environmental temperature was above 13°C, dusting frequency was positively correlated with temperature (Rees, 2002). Although the purpose of this behaviour could not be established, providing access to sand or soil clearly offers an important opportunity for elephants to exhibit a natural behaviour.

The physiological and behavioural changes that occur during hibernation in the Japanese black bear (*Ursus thibetanus japonicus*) were studied in a specially designed exhibit at Ueno Zoological Gardens (Itoh et al., 2010).

The contributions of studies conducted in zoos to our understanding of the physiology and development of bonobos (*Pan paniscus*), chimpanzees (*P. troglodytes*) and other primates have been described by Behringer et al. (2018). Studies of food passage time have contributed to our understanding of the relationships of body size and morphology to ecology in gorillas (Remis, 2000).

Zoos provide opportunities for scientists to make important discoveries about the reproductive physiology of animals. Some of these are of great practical importance to improving breeding in zoo populations (see Chapter 11). However, others add significantly to our basic biological knowledge. Genetic fingerprinting was used by Watts et al. (2006) to determine that parthenogenetic offspring had been produced by two female Komodo dragons (*Varanus komodoensis*) at Chester Zoo. One of these females subsequently produced additional offspring sexually. This work illustrated the importance of reproductive plasticity to the survival of this species in the absence of males.

7.3.4 Ecology

An animal living in a captive environment outside its normal home range has a new ecology quite separate from its ecology when living wild. It has relationships with the biological and physical components of its captive environment. These can affect its wellbeing and are legitimate subjects for study by zoo scientists. However, captive animals retain many of the characteristics of their wild conspecifics and may provide useful opportunities for us to learn more about the ecology of some species in the wild.

The daily food consumption of an animal may be calculated from knowledge of its dung production rate and gross assimilation efficiency. The former can be estimated from field studies – by following an animal around and collecting its dung – but the latter requires a study that can only be performed in captivity. The dry weight of all food eaten over a fixed period of time must be measured along with the dry weight of all of the dung produced over the same period.

Using data collected from two 11-year-old African elephants (*Loxodonta africana*) kept at Knowsley Safari Park, England, I calculated the gross assimilation efficiency (GAE) of this species to be approximately 22.4 per cent. The use of this figure to calculate food consumption in wild African elephants resulted in estimates that were lower than those that had previously been calculated from a figure for GAE calculated by Benedict (1936) using an Asian circus elephant (*Elephas maximus*) named *Jap*, the only figure previously available for an elephant. Other studies concerning nutrition and feeding are considered in Chapter 13.

7.3.5   Behaviour

Zoos provide researchers with an opportunity to study the behaviour of animals that have rarely been studied in the wild. Clearly, some of this behaviour may be influenced by captivity, for example, feeding and ranging behaviour. Other behaviours, however, may be similar to those found in the wild, for example, courtship and mating. Understanding the natural behaviour of animals in the wild is essential if zoos are to provide suitable conditions for captive individuals.

Zoos can provide opportunities for biologists to study behaviours that are difficult to observe in the wild, such as behaviours that occur underwater. The submarine foraging behaviour of alcids (auks) has been studied in an exhibit at *SeaWorld San Diego* (Duffy et al., 1987). Aggressive interactions were common, as was competition for food. The study also examined diving duration and methods of underwater propulsion.

A study of parental behaviour in maned wolves (*Chrysocyon brachyurus*) at Belo Horzonte Zoo in Brazil suggested that only the male is directly involved in rearing the young, and the female is only concerned with their protection (Veado, 1997) (Fig. 7.5). It has been suggested that the presence of the male may be important in reducing aggression within the family group (Veado, 2005).

A study of the activity budgets and behavioural synchrony exhibited by 14 pairs of Patagonian maras (*Dolichotis patagonum*) housed at Córdoba Zoo, Argentina, was undertaken by Baechli et al. (2021) (Fig. 7.6). Between 08.00 and 18.00 each day the maras were found to spend most of their time resting (43 per cent), feeding (25 per cent)

Fig. 7.5 Maned wolf (*Chrysocyon brachyurus*). Studies of this species in a zoo environment have elucidated the role the male plays in rearing the young.

and alert (13 per cent), and both members of each pair engaged in the same behaviours 48 per cent of the time, compared with 57 per cent in the wild (Taber, 1987). The authors noted that this high level of synchrony is necessary for the maintenance of the pair and that a high level of alertness is important to survival in the wild. All of the pairs produced offspring during the study season and Baechli et al. considered that their study would provide useful information for future reintroduction programmes.

The two African elephants used in Rees (1982) to study their feeding ecology were also used to study the synchronisation of defecation in this species. Social facilitation is the process by which an animal in a group is induced to perform a behaviour after observing its performance by others. This often results in individuals within a group coordinating their behaviour so that they, for example, feed at the same time, with some degree of mutual stimulation. The elephants in this study were found to defecate within the same 15-minute period more frequently than would be expected by chance. This synchronisation may have a selective value, especially in elephant calves, making them less likely to stop to defecate when the herd is moving, and so protecting them from potential predators (Rees, 1983).

The social facilitation of mounting behaviour has been observed in a juvenile bull Asian elephant (*Elephas maximus*) in a zoo, and this may be an important element of the development of normal sexual behaviour in this species (Rees, 2004b).

Fig. 7.6 Patagonian mara (*Dolichotis patagonum*). Behaviour is highly synchronised in this species and this may be important to the survival of wild individuals.

7.3.6   Population Biology, Genetics and Disease

Whether a species expands or becomes extinct is largely determined by the difference between its birth rate and its death rate. Knowledge of demographic data is therefore essential for the formulation of effective conservation policies. A Demographic Species Knowledge Index that classifies the available information for 32,144 (97 per cent) of extant described species of mammals, birds, reptiles and amphibians has been produced by Conde et al. (2019). They located comprehensive data on birth and death rates for just 1.3 per cent of tetrapod species and no data at all for 65 per cent of threatened tetrapods. The authors demonstrated that the use of data from zoos and aquariums in the Species360 network could result in an almost eightfold gain in knowledge, especially in relation to the availability of life tables and population matrices. However, they found that the origin of the data available for 66 per cent of the species managed in zoos and aquariums was either unknown or not reported, and noted that this is a cause for concern because these data are widely used for conservation and comparative studies. This is a common difficulty encountered when examining historical data that were not originally collected with a view to subsequent scientific analysis and, in any event, even reliable data held for managed populations may not be directly relevant to wild populations.

The study of the genetics of animals held in zoological collections is critical to their long-term survival in captivity because the genetic relationships between individuals of the same species need to be understood to prevent inbreeding and to identify evolutionarily significant units (ESUs). In addition to studies concerned with conservation genetics (see Chapter 11), there is also a body of work concerned with the genetics and evolution of taxa which is of importance in extending our fundamental knowledge of these disciplines.

Cell lines from the Frozen Zoo established by San Diego Zoo, California, and Catoctin Wildlife Preserve and Zoo, Maryland, have contributed to a study of X chromosome evolution in the Cetartiodactyla – a superorder of mammals that combines the orders Cetacea and Artiodactyla, created since it was discovered by molecular analysis that the former evolved from ancestors belonging to the latter (Proskuryakova et al., 2017).

Monitor lizards have a high aerobic capacity and distinctive cardiovascular physiology similar to that of mammals. Lind et al. (2019) compared the genome of Komodo dragons (*Varanus komodoensis*) from Zoo Atlanta, Georgia, with that of related species and found evidence of positive selection in pathways related to cardiovascular homeostasis, haemostasis and energy metabolism. They also found species-specific expansions of a chemoreceptor gene family related to pheromone and kairomone sensing in both *V. komodoensis* and other lizard linkages. These findings taken together reveal the genetic underpinnings of the Komodo dragon's unique cardiovascular and sensory systems and suggest that alteration of haemostasis genes help individuals of the species evade the anticoagulant effects of their own saliva.

Cancers are common in some taxa but rare in others. Zoo collections provide a unique opportunity to study disease in a wide range of species, aided by recent advances in zoo record-keeping at an international level. Vincze et al. (2022) investigated cancer mortality risk (CMR) across 191 species of mammals (110,148 individuals) using data from Species360 and the Zoological Information Management System (ZIMS), and found substantial differences in cancer mortality rate across major orders. They showed that the phylogenetic distribution of CMR was associated with diet such that carnivorous mammals (especially those that feed on mammals) suffered the highest mortality rate. Rates were much lower in the Primates and Artiodactyla, the latter having the lowest CMR. Across species, CMR was largely independent of body mass and adult life expectancy. In some species death due to cancer accounted for more than 20–40 per cent of the managed adult populations, but in 47 of the 191 species studied the CMR was zero. The highest CMR (57.1 per cent) was observed in the kowari (*Dasyuroides byrnei*), a small carnivorous marsupial. This study identified ruminants (Artiodactyla) as a possible useful model for cancer research due to the very low cancer mortality rate in this group.

7.4 Cognition

Hopper (2017) has noted that cognitive research in zoos is increasing, with most studies involving primates, elephants and bears, but little work on birds, reptiles, insects and other taxa. However, Garcia-Pelegrin et al. (2022) noted that zoos are a relatively untapped resource for studying animal cognition. Animals living in zoos allow close observation and more experimental control than would be the case in the wild. The authors advocated the use of multi-zoo collaborations to increase sample sizes.

7.4.1   Self-Recognition

Self-recognition studies have examined primates, including gorillas at San Francisco Zoo and chimpanzees (*Pan troglodytes*) at Oakland Zoo, California, (Parker, 1994), lesser apes (Hylobatidae) at Perth Zoo, Western Australia, and Adelaide Zoo and Gorge Wildlife Park, South Australia (Suddendorf and Collier-Baker, 2009); *Cebus* monkeys (Collinge, 1989); bonobos (*P. paniscus*) (Walraven et al., 1995); an orangutan (*Pongo* sp.) at Jardin des Plantes, Paris (Robert, 1986); and capuchin monkeys (*Cebus apella*) (Paukner et al., 2004). They have also been conducted on other taxa, including elephants (*Elephas maximus*) at the Bronx Zoo, New York (Plotnik et al., 2006) bottlenose dolphins (*Tursiops truncatus*) in the New York Aquarium (Reiss and Marino, 2001) and grey wolves (*Canis lupus*) at Wolf Park, Indiana (Gatti et al., 2021).

7.4.2   Problem Solving and Tool Use

Animals living in zoos have been used to study problem solving and tool use, especially in primates and elephants. Over 40 years ago Lethmate (1979) described tool use in zoo-housed orangutans. Much more recently a study of tool use among six chimpanzees at Lincoln Park Zoo, Chicago, established that they could use two tools in the correct sequence to extract a food reward, as they do in the wild (Bernstein-Kurtycz et al., 2020). Tool use and problem solving by eight captive brown bears (*Ursus arctos*) was studied by Waroff et al. (2017) (Fig. 7.7).

Twelve Asian elephants housed at the Smithsonian's National Zoological Park (also known at the National Zoo, Washington, DC) and the Oklahoma City Zoo were used in an examination of their ability to use water as a tool in a 'floating object task' previously only tested in primates (Barrett and Benson-Amram, 2020). In this task the animal adds water to a tube in order to reach a floating food reward. One of the elephants solved the task on her own. Elephants at one of the zoos who observed a conspecific successfully complete the task showed increased interest in the task thereafter, demonstrating social learning.

7.4.3   Language and Communication

Zoo-living animals may be a source of vocal signals that can be used to test hypotheses concerning communication between animals in the wild. For example, the use of signature whistles in communication among wild bottlenose dolphins (*Tursiops truncatus*) has been studied by using playbacks of synthetic signature

Fig. 7.7 Bear puzzle box. A brown bear (*Ursus arctos*) attempting to remove food from a puzzle box as part of a research project conducted by Helen Chambers, University of Salford (source: Kathryn Page).

whistles modelled after those of captive dolphins from Zoo Duisburg, Germany and *The Seas* at Epcot, Florida (King and Janik, 2013).

Schneiderová et al. (2016) studied the vocal activity of lesser galagos (*Galago senegalensis* and *G. moholi*) housed at four zoos: Dierenpark Amersfoort in the Netherlands, and Prague Zoo, Ostrava Zoo and the Zoo and Botanical Garden Plzeň, all in the Czech Republic. They found that vocalisations fell into three categories – contact, attention/alarm, agonistic – and were made spontaneously and regularly. Species-specific differences were apparent and may be used to distinguish between them.

Some species are able to communicate information about available food types to conspecifics. Clay and Zuberbühler (2009) studied the calls made by 10 captive bonobos (*Pan paniscus*) at two different facilities. They found that the composition of call sequences produced by the animals was related to the type of food found by the caller and concluded that these sequences convey meaningful information to other group members which may guide their foraging behaviour.

In a similar experiment using bonobos at Twycross Zoo, England, experimenters broadcasted sequences of four calls produced by a familiar individual who was responding to the presence of either kiwi (a preferred food) or apples (a less popular food) (Clay and Zuberbühler, 2011). The bonobos attended to the entire sequence of calls (rather than individual calls) to make inferences about the type of food

Fig. 7.8 The vocalisations of the greater rhea (*Rhea americana*) have been studied at the zoo of the Federal University of Mato Grosso in Brazil.

encountered by the caller, thereby demonstrating the importance of call combinations in communication in this species. Twycross Zoo was the only institution in the UK that kept bonobos at the time.

Tanner and Byrne (1999) described the development of spontaneous gestural communication in a group of lowland gorillas (*Gorilla gorilla*) at San Francisco Zoo. They found that the gestural repertoires of individuals of the same age varied in type and quality. The repertoire of facial movements in chimpanzees (*P. troglodytes*) has been studied in the colony at Chester Zoo, England (Vick et al., 2007).

Vocalisations of the greater rhea (*Rhea americana*) were studied at the zoo of the Federal University of Mato Grosso in Brazil using passive acoustic monitoring and automated signal recognition software (Pérez-Granados and Schuchmann, 2020) (Fig. 7.8). The authors suggested that their work could make a contribution to the understanding of the evolution of vocalisation in ratites. Juvenile contact calls in crocodilians have been studied using animals from La Ferme aux Crocodile, Pierrelatte, France (Vergne et al., 2012) and the vocal repertoire of Asian elephants (*Elephas maximus*) was studied using domesticated elephants in Thailand and elephants at Oregon Zoo in Portland, Oregon (Glaeser et al., 2009).

7.4.4   Lateralisation

In relation to brain function asymmetry, lateralisation is the tendency for some neural functions or cognitive processes to be specialised to either the left or the right side of the brain. The existence of handedness has been widely studied in captive primates and other species and has implications for the evolution of language and cognition.

In a study of handedness in 22 captive gorillas (*Gorilla gorilla*), 10 individuals showed no preference in any of the behaviours performed as part of their daily routine, and the majority of the remainder showed a preference in just one of the behaviours (Harrison and Nystrom, 2010). Fagot and Vauclair (1988) studied 10 gorillas at Barcelona Zoo in Spain and found that when reaching for food three showed a right-hand preference, three showed a left-hand preference and four showed no hand preference. However, in a spatial task that involved precisely aligning two openings seven out of the eight gorillas tested exhibited a left-hand preference. This study highlighted the importance of the type of test used to determine handedness.

Other studies that have examined handedness have included chimpanzees (*Pan troglodytes*) (Forrester et al., 2012); orangutans (*Pongo pygmaeus*) (O'Malley and McGrew, 2006); siamangs (*Hylobates syndactylus*) (Redmond and Lamperez, 2004); Eastern grey kangaroos (*Macropus giganteus*) and red kangaroos (*Osphranter* [*Macropus*] *rufus*) (Giljov et al., 2015); and slow lorises (*Nycticebus* spp.) (Poindexter et al., 2018).

The relationship between handedness and target animacy has been examined in northern pig-tailed macaques (*Macaca leonina*) at Tianjin Zoo, China (Zhao et al., 2015) and laterality in trunk use in six African elephants (*Loxodonta africana*) has been studied at ZooParc de Beauval in France (Lefeuvre et al., 2022).

Foot preference for a string-pulling task in macaws (*Ara* spp.) living in zoos was studied by Regaiolli et al. (2021). Among the seven individuals tested, they found right-footed, left-footed and ambi-preferent birds.

7.4.5 Personality

Weinstein et al. (2008) defined personality as 'those characteristics of individuals that describe and account for consistent patterns of feeling, cognition and behaving'. Studies of animal personality often use behaviour measures, behaviour ratings or adjective ratings (e.g. Uher and Asendorpf, 2008).

Studies of animal personality began in the early part of the twentieth century (Pavlov, 1906), with mammals (especially primates) having been the most popular subjects, studies of chimpanzees beginning in the 1930s (Crawford, 1938; Hebb, 1949). When Gosling (2001) examined 187 personality studies conducted on 64 species he calculated that 84 per cent of these were conducted on mammals (29 per cent of the total involved primates), 8 per cent were on fish, 4 per cent on birds and the remaining 4 per cent were studies of reptiles, amphibians, arthropods and molluscs.

Personality has been studied in a number of felid species living in zoos, and often involves assessments made by keepers, for example snow leopards (*Uncia uncia*) (Gartner and Powell, 2012), cheetahs (*Acinonyx jubatus*) (Chadwick, 2014), tigers (*Panthera tigris*) (Bullock et al., 2021), Scottish wildcats (*Felis silvestris grampia*) (Gartner and Weiss, 2013) and Asiatic lions (*Panthera leo persica*) (Pastorino et al., 2017).

7.5   Behaviour Methods

Zoos can provide opportunities for developing and testing tools for studying animals, especially methods of collecting and analysing behaviour.

The types of sampling methods used in observational studies of social behaviour found in the literature of the time have been discussed by Altmann (1974) in an early paper which discussed the suitability of each for different types of investigation. Martin and Bateson (2007) published a short account of the methods used to study behaviour – now in its fourth edition, Prof. Patrick Bateson having been replaced following his death by his daughter Prof. Melissa Bateson (Bateson and Martin, 2021) – and I have drawn together some of the important methods used in zoos and other captive environments (Rees, 2015).

A number of researchers have developed indices to measure the strength of friendships in social species or the quality of their relationships with others. A friendship index between individual Celebes crested macaques (*Macaca nigra*) living in a zoo enclosure was calculated by Michelatta and Waller (2012) by assessing the extent of grooming and proximity between individuals in a dyad. During a study of male gorillas, a relationship quality index (RQI) was developed by Pullen (2009) based on the ratio of dominance to affiliative behaviours. She found that bachelor males exhibited a significantly lower RQI than males in breeding groups.

Nearest-neighbour calculations have been used in zoos to study reproductive behaviour in giraffes (Bercovitch et al., 2006), social structure in lemurs (Curtis and Zaramody, 1999), hand-reared and orphaned gorillas (McCann and Rothman, 1999; King et al., 2003) and chimpanzees (Clark, 2011b) and tigers (Miller and Kuhar, 2008). Marolf et al. (2007) used a maintenance of proximity index (MPI) (sometimes called the Hinde Index) to study the lemur species *Eulemur coronatus* and *E. rubriventer*, and found that males were responsible for maintaining partial proximity in the former and females in the latter.

Some useful studies undertaken in zoos have examined issues relating to sampling in behaviour studies or helped to develop new methods of studying behaviour. A 'behaviour discovery curve' was used by Jule et al. (2009) to predict the optimal observation time necessary to collect data when establishing an ethogram for captive red pandas (*Ailurus f. fulgens*) (Fig. 7.9). They produced a curve fitted to a logarithmic model that predicts the rate of new behaviours that will be observed in any given length of observation time.

The number of days over which an animal is studied may significantly affect the calculation of an activity budget. For example, when the frequency of stereotypic behaviour exhibited by an adult cow Asian elephant (*Elephas maximus*) kept in a zoo was calculated for 35 days it was around 0.09 (9 per cent). When the 35 days were

Fig. 7.9 Red panda (*Ailurus fulgens*).

treated as five separate seven-day studies the frequencies during these seven-day periods ranged from 0.003 (0.3 per cent) to 0.189 (18.9 per cent) (Rees, 2015).

7.6 The Integration of Zoo Exhibits and Research Facilities

Living Links to Human Evolution is an exhibit at Edinburgh Zoo, Scotland, which doubles as a research facility for studying cognitive evolution in primates and various aspects of their behaviour (Macdonald and Whiten, 2011) (Fig. 7.10). Projects are undertaken in cooperation with researchers from St Andrews University and also the universities of Sterling, Edinburgh and Abertay. The facility houses brown capuchin monkeys (*Sapajus apella*) and squirrel monkeys (*Saimiri sciureus*) together in two discrete 'identical' enclosures so that experiments can be conducted in one enclosure while the other is used as a control.

The work conducted in this facility includes studies of personality, cognition and social behaviour in primates – that is, work on their fundamental biology, studies related to animal welfare, and also work on human behaviour and public engagement with science.

The centre has been used for several studies concerned with the welfare of the monkeys (Leonardi et al., 2010; Buchanan-Smith, 2013) and the impacts on their social networks, activity and wellbeing of moving the two species to their new

Fig. 7.10 The *Living Links to Human Evolution* exhibit at Edinburgh Zoo, Scotland, allows keepers and researchers to interact with brown capuchin monkeys (*Sapajus* [*Cebus*] *apella*) (a) and squirrel monkeys (*Saimiri sciureus*) (b) in transparent cubicles (c).

environment (Dufour, 2011). The role of social anointing in improving the coverage of topically applied anti-parasite medicines in capuchin monkeys has been investigated in an unusual study by Bowler et al. (2015).

Other studies have examined the relationship between personality, social status and facial morphology in capuchin monkeys (Lefevre et al., 2014; Wilson et al., 2014) and compared personality structure in capuchins with that in chimpanzees (*Pan troglodytes*), orangutans (*Pongo* spp.) and rhesus macaques (*Macaca mulatta*) (Morton et al., 2013a).

Fig. 7.11 Research pods in the *Budongo Trail* chimpanzee (*Pan troglodytes*) exhibit at Edinburgh Zoo, Scotland. Chimpanzees voluntarily enter the pods to participate in research on cognition using computer touchscreens. Inset: Visitors can inspect the pods through windows.

Polgár et al. (2017) have examined individual differences in squirrel monkeys' reactions to visitors, research participation and personality ratings. The effects of individual cubicle research on the social interactions and individual behaviours of capuchin monkeys have been investigated by Ruby and Buchanan-Smith (2015). The need to take personality selection bias seriously in animal cognition research has been considered by Morton et al. (2013b) using capuchin monkeys as a case study.

A number of studies have used *Living Links* to examine the public's engagement with science (Bowler et al., 2012) and social learning in humans (Claidière et al., 2012, 2014; Whiten et al., 2016).

The *Budongo Trail* exhibit at Edinburgh Zoo, Scotland, contains a second primate research facility housing chimpanzees (*Pan troglodytes*) where individuals voluntarily enter research pods to participate in research on cognition (Fig. 7.11). The facility has been used to study lip-smacking in chimpanzees and speech-rhythm evolution (Pereira et al., 2020), vocal learning of functionally referential food grunts (Watson et al., 2015), social preferences (Clark, 2011b), tool use (Harrison and Whiten, 2018) and learning (Sonnweber et al., 2015). Chimpanzee intellect has been studied by examining their personality, performance and motivation with touchscreen tasks (Altschul et al., 2017).

Welfare studies have included investigations of self-directed behaviours and regurgitation and reingestion (Wallace et al., 2019) and the relationship between personality and welfare (Robinson et al., 2017). Herrelko et al. (2012) investigated the influence of chimpanzee personality on cognitive research, including the impact of this research on their welfare. In addition, studies have been made of the behaviour of chimpanzees when unfamiliar animals are introduced to each other (Herrelko et al., 2015) and the effect of playing music to chimpanzees (Wallace et al., 2017). Studies of public engagement have included an examination of the role of the exhibit in promoting positive attitudes towards chimpanzees and conservation in visitors (Craig and Vick, 2021) and an evaluation of public engagement activities to promote science (Whitehouse et al., 2014).

7.7 Developing and Testing Technology for Field Studies

Behavioural research has been constrained for many years by a lack of technology for collecting data and a lack of statistical and computational tools for its analysis. Animals living in zoo environments are useful proxies upon which to test methods and technologies before committing to using them with *in-situ* populations.

In order to quantify grazing behaviour in seven African elephants (*Loxodonta africana*) at Knowsley Safari Park, England, I filmed them on 16 mm monochrome Ilford Motion Picture film Mk.5 at 18 frames per second using a clockwork Bolex H16 reflex cine camera fitted with an Angenieux zoom type 20x12B lens (focal length 12–240 mm) mounted on a heavy wooden tripod from the back of a vehicle (Rees, 1977). Feeding rates (trunkfuls/min) calculated from the film were then used together with estimates of the size of individual trunkfuls of vegetation and the weights of similar samples collected by hand to estimate food consumption. Previous field studies of feeding in elephants had relied upon simple observation (e.g. McKay, 1973; Wyatt and Eltringham, 1974).

At that time, the samples taken were constrained by the cost of purchasing the film and paying for it to be processed. This was the technology of the day and the same work could be carried out now, some 45 years later, using a smartphone or digital camera at almost no cost.

Camera traps are now routinely used in natural habitats to identify, count and study the distribution and movements of animals, and to record behaviours. However, the type, location, height and orientation of the camera is critical to its successful capture of useful images and video recordings. Numbats (*Myrmecobius fasciatus*) kept in enclosures at Perth Zoo, Western Australia have been used to test various camera traps (Seidlitz et al., 2021). This study found that optimising camera trap height and model increased detection and individual identification rates for this species. Detection rates of known events were 89, 51 and 37 per cent for the three types of camera tested. As the height of elevated, downward-angled cameras increased, the number of suitable images depicting dorsal fur patterns captured also increased; the optimum height was 25 cm as this achieved the highest detection rate. The authors emphasised the importance of rigorous species-specific testing of cameras rather than reliance on generic set-up instructions.

Conservationists and field biologists are now using drones to map and monitor biodiversity (Wich and Koh, 2018). In safari parks, savannah animals are kept in large

Fig. 7.12 White rhinoceroses (*Ceratotherium simum*) at Knowsley Safari Park, England, have been used to test anti-poaching drone technology.

open paddocks simulating the habitats found in parts of Africa, through which visitors drive in their cars. Animals at Knowsley Safari, England (formerly Knowsley Safari Park), were used in the development of anti-poaching drone technology using a neural network to detect white rhinoceroses (*Ceratotherium simum*) and cars (important tools for poachers) from the air (Chalmers et al., 2021) (Fig. 7.12). The results demonstrated that rhinoceroses could be detected by the model used with 91 per cent sensitivity and cars with 100 per cent sensitivity.

Asian elephants (*Elephas maximus*) at Whipsnade Zoo and African elephants (*Loxodonta africana*) at Colchester Zoo (both in England) have been used to assist in the development of an early warning system to reduce human–elephant conflict. The system, developed as part of the Human–Elephant Alert Technologies (HEAT) Project, uses thermal cameras to identify the heat signature of elephants. Some 30,000 thermal images taken at Whipsnade Zoo have been used to 'teach' the system what an Asian elephant looks like (ZSL, 2021b).

New methods of monitoring biodiversity at a community scale will be important in the future conservation of ecosystems. The use of air sampling to collect and identify environmental DNA (eDNA) from vertebrates has been tested using the animal community living within Copenhagen Zoo, Denmark (Lynggaard et al., 2022). Air was filtered at three localities in the zoo and metabarcoding detected 49 vertebrate species from 26 orders and 37 families: 30 mammals, 13 birds, 4 fishes, 1 amphibian and 1 reptile. This included exotic species kept at the zoo, native species found in the zoo (including pests) and species used in animal feed. In a similar study, Clare et al. (2022)

collected air samples from Hamerton Zoo Park in Cambridgeshire, England, and identified DNA from 25 species of mammals and birds, including 17 terrestrial resident zoo species. They also identified food items from air sampled in enclosures and detected local native taxa. Airborne eDNA was detected several hundred metres from the source. These studies showed the potential of air as a source of eDNA for monitoring biodiversity in natural terrestrial habitats, including the detection of consumed prey following predation.

Acoustic monitoring has been widely used to detect the presence of animal species in the field (e.g. Acevedo and Villanueva-Rivera, 2006). It may also assist in distinguishing between individual animals in free-living populations of vocally active species when making population estimates using capture–mark–recapture techniques. Sadhukhan et al. (2021) recorded howls from Indian wolves (*Canis lupus pallipes*) kept at Jaipur Zoo, India, and from free-ranging wolves in Maharashtra. They trained a supervised agglomerative nesting hierarchical clustering (AGNES) model using 49 howls from five wolves and achieved 98 per cent accuracy in identifying individuals. When the model was used to analyse 20 novel howls from four different individuals it achieved 75 per cent accuracy in classifying these howls to individual animals. The authors concluded that this method provides a non-invasive means of reducing bias in population estimates obtained using capture–mark–recapture methods and could be adapted for use with other species that produce individually distinctive vocalisations.

7.8   Conclusion

From their earliest days, zoos have made an enormous contribution to zoology. They have given scientists access to exotic species from all over the world and given them the opportunity to study their anatomy, behaviour, reproduction and many other aspects of their biology. They have also provided insights into human evolution, disease transmission, animal personality, tool use and language. Some modern zoos have integrated research facilities into the design of their enclosures while others have been used to test new technology for studying animals in the wild. Developments such as these and the increased collaboration between universities and zoos will ensure that zoos continue to make a significant contribution to our understanding of animal biology.

8   Animals and Their Enclosures

8.1   Introduction

Zoo enclosures are inevitably a compromise between what is technically possible, the space and financial resources available, the biological requirements of the animals, the safety requirements for zoo staff and the requirement to provide visitors with an enjoyable experience. This chapter is concerned with enclosure design, the utilisation of the usable space by the animals and the animals' interactions with visitors.

8.2   Enclosure Design

Early zoos consisted largely of open field enclosures for paddock animals and barren cages for carnivores, primates and other animals perceived to be dangerous and capable of escape. Modern naturalistic enclosures often provide stimulating habitats for the animals while making them less visible to the visiting public. Zoo designers and managers must strive to provide good welfare for animals while still making them accessible to visitors, and this is likely to result in design that provides a compromise that maintains visitor interest.

A study of 458 zoos found that zoos keeping large numbers of many species (especially larger species) tend to attract high visitor numbers (Mooney et al., 2020). This is problematic for zoos because keeping large groups of large mammals requires a great deal of space, and all zoos have a limit to the amount of space available.

In 1924 Heinrich and Lorenz Hagenbeck – sons of Carl Hagenbeck, the famous animal trader and founder of the revolutionary Tierpark Hagenbeck – were granted a patent for 'new and useful Improvements in Animal Inclosures' by the United States Patent Office (Serial No. 686,235). Although the ideas patented were novel at the time, they now comprise some of the basic principles used in the design of modern zoo exhibits. The Hagenbecks proposed the use of moats and 'invisible' barriers to separate visitors from animals and to allow incompatible species to appear to cohabit the same enclosure:

This invention relates to improvements in inclosures for animals, birds, reptiles and the like, an object being to provide means whereby animals, birds and reptiles may be confined without the use of bars, wires or similar view obstructing restraining means, the invention being especially adapted for zoos and other places of animal exhibition.

Another object of the invention is the provision of an inclosure of this character wherein the different animals, birds and reptiles are effectively separated in a manner to minimize the prominence of the means of separation, while the landscape may be arranged to simulate the character of the country where the particular animals, birds and so forth are usually found, so that they apparently remain in their natural habitat.

Another object of the invention is the provision of means for separating and enclosing the animals in such manner that a panoramic view may be obtained of a part of or the entire exhibition, while the separating and enclosing means remain practically hidden.

Longitudinal sections through hypothetical enclosures illustrating the designs proposed by the Hagenbecks were included in their patent, showing moats filled with water, dry moats and depressed fences (Fig. 8.1)

Some modern zoo exhibits are architecturally and structurally complex and have been the subject of technical publications. The challenge of designing the mesh geometry for two spherical domes forming the glass roof covering the two pools of the hippopotamus house at Berlin Zoo has been described by Schlaich and Schober (1997) (Fig. 8.2). The pools are circular, one about 21 m in diameter for the pygmy hippopotamuses (*Choeropsis liberiensis*) and the other about 29 m in diameter for the larger common or river hippopotamuses (*Hippopotamus amphibius*). The new house also has underwater viewing windows. The engineering challenge was to create a mesh geometry that would allow the joining of two spherical domes with a smooth transition between them using flat glass panels.

The history and architecture of some of the best-known elephant houses built in Europe have been described by Haywood (2014), from the Elephant Stables designed by Decimus Burton and constructed in London Zoo in 1831 to the postmodernist creation of Sir Norman Foster in Copenhagen Zoo, Denmark, covered by glazed domes (Table 8.1). Sir Hugh Casson's Elephant and Rhino Pavilion was built in the 1960s, but, while the building won great architectural acclaim, it no longer houses pachyderms and the Zoological Society of London's (ZSL) elephants were sent to its other facility (Whipsnade Zoo) in 2001 (Fig. 8.3).

Although some zoos have invested heavily in new animal houses and enclosures, others have retained historically important buildings and made relatively few changes to them. At Dudley Zoo in England there are 12 important Lubetkin's modernist buildings that are protected due to their architectural importance, but which are totally inappropriate for the species they originally housed, including a polar bear pit, an elephant house and a small pool that was once home to a killer whale (*Orcinus orca*) called *Cuddles*.

At London Zoo, Decimus Burton's giraffe house has remained substantially unchanged since it was opened in 1836. Unfortunately it was damaged by bombing during the Second World War (1940) and repaired much later, in 1960–1963. The structure is protected as a Grade II listed building and the repaired structure remains largely true to the original design (Fig. 8.4).

An excellent account of the history of zoological gardens in the West, which includes many important photographs, has been published by Baratay and Hardouin-Fugier (2002).

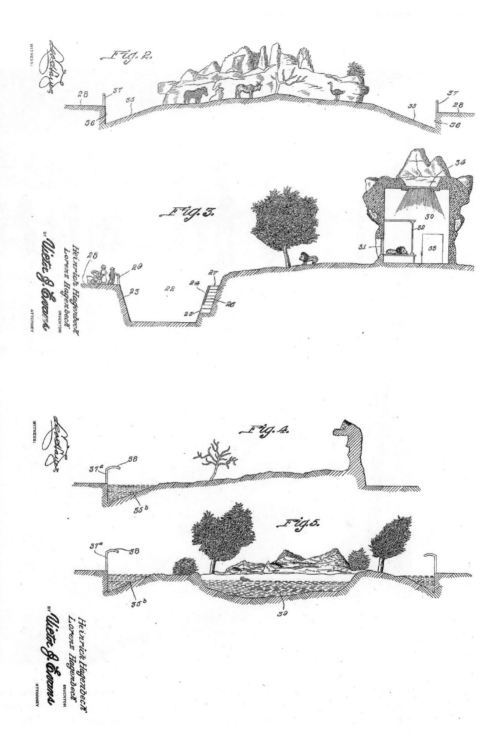

Fig. 8.1 In 1924 Heinrich and Lorenz Hagenbeck were granted a patent for 'new and useful Improvements in Animal Inclosures' by the United States Patent Office (Serial No. 686,235). These drawings formed part of the patent application (source: US Patent Office).

Fig. 8.2 The hippopotamus house at Berlin Zoo. The building contains underwater viewing windows and complex roof geometry. (a) The viewing window for the group of Nile hippopotamuses (*Hippopotamus amphibius*). (b) A small child watching a pygmy hippopotamus (*Choeropsis liberiensis*).

Table 8.1 Examples of architecturally important elephant houses constructed in Europe.

Zoo	Year of construction	Building	Architect	Architectural style
London, England	1831	Elephant Stables	Decimus Burton	Hybridised blend of imperial and subaltern indigenous architectures
London, England	1867	Elephant and Rhinoceros House	Anthony Salvin Jr	English Gothic Revival
Berlin, Germany	1873	Elephant House	Ende und Böckmann	Indian pagoda
Whipsnade, England	1935	Elephant House	Berthold Lubetkin/Tecton	Modernist
Dudley, England	1935–1937	Elephant House	Berthold Lubetkin/Tecton	Modernist
London, England	1962–1965	Elephant and Rhino Pavilion	Sir Hugh Casson	Brutalist
Copenhagen, Denmark	2008	Elephant House	Sir Norman Foster	Postmodernist

Source: based on information in Haywood, 2014 and first published in Rees, 2021.

Fig. 8.3 The Casson Elephant and Rhinoceros Pavilion at London was constructed between 1962 and 1965.

(a)

1885

(b)

2008

Fig. 8.4 The giraffe house at London Zoo has remained largely unchanged in appearance since it was constructed in 1836 (a) (source: An Amateur Photographer at the Zoo. *The Graphic*, 5 September 1885, p. 271). The original building was damaged by bombing during the Second World War (1940) and substantially rebuilt in 1960–1963 (b).

8.3 Animal Escapes

Escapes of animals from zoos occasionally attract the attention of the media, especially if the animals are considered dangerous. However, there has been limited scholarly discussion about such escapes.

The escape of animals from zoos in the United States has been discussed by Kim (2008) from the point of view of risk management. The author expressed the opinion that voluntary industry containment standards and the ineffective federal oversight of zoos have dampened progress made by zoos in safeguarding visitors from animal attacks. However, the facts that accidents will happen was recognised.

Fàbregas et al. (2010) examined a random sample of 1,568 animal enclosures in 63 Spanish zoos to assess their security. They concluded that 221 enclosures (distributed across 47 zoos) were not secure against animal escapes based on their assessment of the suitability of the physical barrier and the possibility that animals could be released by members of the public.

In a study of Australian zoos, no relationship was found between bird escapes and the type of holding facility (e.g. aviary versus free-range/open pond) (Cassey and Hogg, 2015).

8.4 How Large Should an Enclosure Be?

Very few animal species kept in captivity can be provided with the same amount of space they would utilise in the wild. Comparisons between the size of zoo enclosures and the home ranges or territories of animals in the wild are meaningless. Many species exhibit plasticity in their ecological requirements, depending upon the availability of resources such as food, places to shelter or breed and mates. Predatory species will increase their territory size when food is in short supply and decrease it when food is abundant. Similarly, herbivores will forage widely – and even migrate – when vegetation is sparse, but do not need to do this when food plants are plentiful. Animals do not require space for its own sake, but sufficient space to satisfy their biological needs such as food, sheltering places and the provision of mates. For some – perhaps most – species a small heterogeneous enclosure is probably more likely to fulfil their psychological needs than is a larger, barren, homogeneous space.

Husbandry guidelines and manuals provide minimum or recommended sizes for enclosures. These are often presented as minimum dimensions with a suggestion that larger enclosures would be preferable. It has been suggested that 75 per cent of the enclosure's vertical space should be available for small cats, but this appears to be a completely arbitrary figure (Mellen, 1997).

Many of the space requirements specified for particular species appear to have little or no scientific basis. In Wales, the recommended size for accommodation for domestic rabbits is that it should be sufficiently long for a rabbit to be able to take at least three hops from one end to the other (*Code of Practice for the Welfare of*

Rabbits (2009 No. 44)). But why not two hops, or ten? In some jurisdictions minimum requirements are established by law for certain farm animals. For example, in Member States of the European Union, Council Directive 1999/74 EC of 19 July 1999 laying down minimum standards for the protection of laying hens defines two systems: enriched cages where laying hens have at least 750 cm^2 of cage area per hen; and alternative systems where the stocking density does not exceed nine laying hens per 1 m^2 usable area, with at least one nest for every seven hens, plus adequate perches.

A number of organisations publish minimum space requirements for a variety of taxa, including the Association of Zoos and Aquariums (AZA), the Central Zoo Authority (CZA), the International Wildlife Rehabilitation Council (IWRC) and the National Wildlife Rehabilitators Association (NWRA).

Hussain et al. (2015) compared the sizes of enclosures for 95 species at Lahore Zoological Garden in Pakistan with international norms as specified by documents produced by AZA, CZA, IWRC and NWRA (Table 8.2). Overall, they determined that only about 46 per cent of enclosures met international standards.

The importance of designing bat enclosures so that they allow the animals to maintain a capacity for sustained flight has been discussed by Bell et al. (2019). They found that Rodrigues fruit bats (*Pteropus rodricensis*) flew more often and in more complex paths – involving turns – than Livingstone's fruit bats (*P. livingstonii*), a larger species, kept in the same enclosure. Livingstone's fruit bats' flights were more likely to end in a crash-landing than those of the Rodrigues fruit bats. These differences suggest that the enclosure was not large enough for the former to display a full range of flight behaviour.

A review of the psychological priorities of the Amur tiger (*Panthera tigris altaica*) as a model species for assessing the enclosure requirements of wide-ranging species living in zoos has been provided by Veasey (2020) based on an assessment conducted by an expert panel. He concluded that the impacts of habitat compression on the welfare of captive carnivores may be the result of a reduction in cognitive opportunities that covary with habitat size rather than the reduction in habitat size *per se*.

8.5   Usable Space

Arboreal species can make better use of a small enclosure than can animals that cannot climb, provided suitable structures are provided. Obviously, the amount of space available to an arboreal animal will depend upon the complexity of the climbing structures. Mesh fencing, although unattractive, can significantly increase the usable volume of a cage for animals such as monkeys and parrots. It is unlikely that an enclosure could be designed where 100 per cent of the volume could be used, except perhaps for some small birds and, of course, small fish kept in an aquarium.

Browning and Maple (2019) note that increased enclosure space can improve animal welfare for a number of reasons (Fig. 8.5) and that traditional methods of accessing this often underestimate the space available.

Table 8.2 Enclosures that met the minimum size standards at Lahore Zoological Garden.

Taxon	No. species/ enclosures	No. enclosures meeting minimum size standard	Percentage of enclosures meeting minimum size standard
Mammals (carnivores)	11	7	63.6
Mammals (herbivores)	16	8	50.0
Mammals (primates)	6	4	66.7
Reptiles	10	3	30.0
Birds	45	15	33.3
Running birds (ratites)	3	3	100.0
Water birds	4	4	100.0
Total	95	44	46.3

Source: based on Hussain et al., 2015.

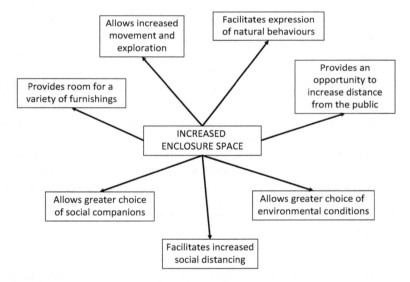

Fig. 8.5 The benefits of an increase in enclosure space.

Space measurements for an exhibit often specify floor area, height (for arboreal species) and water surface area (or volume) for species that utilise water. Some provide very little detail in relation to dimensions. For example, the EAZA Hornbill Management and Husbandry Guidelines (Galama et al., 2002) state the following in relation to enclosure size for hornbills:

4.6 Dimensions of the enclosure

Elongated enclosures are recommended for large hornbill species, as they need to exercise their wings. The minimal width of the enclosure should be 4 times the wingspan of the species housed in it, enabling the birds to easily pass each other in flight ... A minimal enclosure

Table 8.3 Main pool minimum space requirements for the ringed seal (*Phoca hispida*).

Average species length (m)	Group size	Land area (m^2)	Additional land area per extra animal (m^2)	Pool area (m^2)	Additional pool area per extra animal (m^2)	Minimum volume (m^3)
1.6	1–6	18	3	72	12	153

Table 8.4 A generalised list of enclosure features that may be important in determining the usable space of an enclosure.

Component	Description
Floor space	The basic floor space of an enclosure is a large part of the usable space for that exhibit. For strongly terrestrial animals, such as kangaroos, this might still be the primary measure of usable space. For arboreal species it will play less of a role. For raised or uneven surfaces, the surface area will be higher than the simple enclosure dimensions.
Elevated platforms	The surface area of elevated platform spaces.
Rocks	Perhaps a type of elevated platform, the sitting and climbing surfaces of rocks count for those animals that can use them.
Arboreal pathways	The length of ropes/logs/other pathways between elevated spaces.
Climbing structures	The height (and possibly diameter/circumference) of climbing poles/trees.
Cage sides	For many primates and birds, the mesh of cage sides is usable space to move around on.
Air volume	For flying (or leaping) animals, the total air volume of the enclosure could function as usable space for locomotion.
Water volume	For aquatic animals, the volume of ponds and pools would count as usable space.

Source: based on Browning and Maple, 2019.

height of 3.0 m is recommended to allow the hornbills to perch above the public and keepers. As forest hornbills in the wild are often found high in, or above, the canopy, it is assumed that the higher the enclosure (and its furnishings) the more suitable the enclosure is.

Enclosure requirements outlined in the *EAZA and EAAM* [European Association for Aquatic Mammals] *Best Practice Guidelines for Otariidae and Phocidae* for ringed seals (*Phoca hispida*) refer to the minimum land area necessary along with the minimum pool area and volume for a main pool and, separately, for a secondary pool (Gili et al., 2018) (Table 8.3).

The additional land area required per extra animal was calculated as πr^2, where r is half the length of the animal and therefore represents the size of a circle where the length of the animal is the diameter.

Browning and Maple (2019) have listed the enclosure elements other than space and volume that may be important to animals (Table 8.4).

It is common to find enclosures in zoos where, although the inhabitants could potentially use all of the available volume, this is not possible because much of it is made inaccessible by poor design. For example, voluminous indoor enclosures for orangutans (*Pongo* spp.) often have high ceilings, but the apes can only reach the highest points in a few places and by a limited number of routes along climbing frames or tree branches. However, such enclosures must always be a compromise between the needs of the animals (a dense 'forest' of structures through which to climb), the requirements of keepers (to clean enclosures efficiently), and the desires of the visitors (to see the animals unobscured by enclosure furniture).

8.6　Enclosure Shape

The shape of an enclosure is important for many species because, among other things, it determines the perimeter length per unit area, and it may affect the likelihood that individuals will see and interact with each other. For some species, it may be necessary to construct refuges for subordinate animals or females. Where aggression is likely to occur, circular enclosures or houses remove the possibility of individuals being trapped in corners.

Coe et al. (2009) have noted that linear zoo habitats provide the maximum length of perimeter for bachelor male gorillas to patrol compared with non-linear enclosures. This is simply a matter of geometry. The shortest possible perimeter is created by providing a circular enclosure.

Studies of the social behaviour of animals in enclosures often use an association index to calculate the degree of association exhibited by dyads (associations of two animals) (White et al., 2003; Chadwick et al., 2013; Rees, 2021). Many zoo enclosures are relatively small so individuals may be recorded as 'associating' when in fact they are near each other as a result of random movements. The extent to which these random meetings distort association data may be affected by both the size and shape of the enclosure.

Chadwick et al. (2015) used Monte Carlo simulations to study the effect of chance encounters on the calculation of association indices for animals kept in enclosures, using data collected on male cheetahs (*Acinonyx jubatus soemmeringii*) at Chester Zoo, England (see Fig. 11.3). We produced a method that provides a robust estimate of the probability of a chance encounter in a square of any area and showed that this produced acceptable estimates of the probability of chance encounters in regular shapes (rectangles with sides of various proportions) and the shapes of six actual zoo enclosures. Applying this correction to calculated association indices controls for differences in enclosure size and shape, thereby allowing association indices between dyads housed in different enclosures to be compared. We found, not surprisingly, that applying a correction was particularly important in small enclosures, where chance encounters are more likely to occur. In large enclosures the correction was very small so had little effect.

8.7   The Effect of Expanding an Enclosure on Behaviour

When animals are introduced to a new enclosure, or an existing enclosure is enlarged, they may exhibit individual variation in their responses to the new space. When a herd of eight Asian elephants (*Elephas maximus*) were given access to a new bull pen for the first time, two entered immediately but, although whole-day recordings were made on average every six days, it was over 20 weeks before the entire herd was recorded within it (Rees, 2000).

A study of the effect of enclosure expansion on the activity budgets of Eastern black-and-white colobus monkeys (*Colobus guereza*) at Adelaide Zoo, South Australia, found that the addition of aerial walkways increased the behavioural repertoire of the animals, increasing locomotion, feeding and social behaviours (Robbins and Sheridan, 2021). An overall increase in activity was attributed to the larger area available and the increased environmental complexity. The changes to the enclosure had the effect of more closely aligning activity budgets with those of their wild counterparts.

Herrelko et al. (2015) examined the effect of making an increasing amount of space available to chimpanzees (*Pan troglodytes*) on their behaviour when two groups (n = 22) were introduced to each other at the *Budongo Trail* exhibit at Edinburgh Zoo in Scotland. During this process the animals experienced a variety of enclosure restrictions and changes to group composition. The researchers found that as the number of separate rooms made available was increased, yawning and arousal-related scratching decreased. However, only yawning decreased as the available area increased. The results of this study suggested that during introductions the number of accessible areas available to chimpanzees is more important than the total area available. The authors suggest the welfare impacts of reintroductions may be reduced by using modular enclosures.

8.8   Rotational Exhibits

The zoo architect Jon Coe claims to have first introduced 'animal rotation' displays in zoos in 1995 (Coe, 1995). He describes the concept as essentially a 'time share' arrangement for zoo animals, whereby animals of different species may occupy the same area at different times, forming a 'consecutive mixed-species display' (Coe, 2004). This method of management of enclosure space effectively expands the space available to each species. The idea is not new and a similar type of animal management was practised at least 20 years earlier in at least one safari park in England.

At Knowsley Safari Park, near Liverpool in England, in the 1970s, groups of African large mammal species were routinely displayed together in the same enclosure and the mix of species was changed from time to time. In 1976 the park kept a herd of seven African elephants (*Loxodonta africana*). They spent the night in the elephant house, but each morning they were walked through the park to one of the drive-through reserves. They spent the day in this reserve – accompanied by a keeper in a Land Rover – and returned to the elephant house in the late afternoon. They used three different reserves in all, and also had access to woodland behind the elephant house some mornings before the park opened.

Which reserves the elephants used depended partly on the state of the ground and the vegetation, so the rotation was intended to protect the land rather than to provide variety for the animals; farmers have been practising such rotations for many years to protect their fields from overgrazing and trampling. Nevertheless, some days the elephants were kept in a reserve on their own, with a pride of lions (*Panthera leo*) in the next reserve (that they could see and smell through a chain link fence); some days they were kept with a herd of Cape buffalo (*Syncerus caffer*) and had access to a large pond; and some days they were kept with giraffes (*Giraffa camelopardalis*), a pair of hippopotamuses (*Hippopotamus amphibius*), blue wildebeest (*Connochaetes tarinus*) and several other species of African ungulates. I know this because on some days in 1976 I was the keeper in the Land Rover.

Rotational exhibits can provide considerable enrichment for their occupants and increased interest for zoo visitors. Lukas et al. (2003) examined the effects of relative novelty on captive gorilla behaviour when two groups were regularly alternated between complex naturalistic enclosures at Zoo Atlanta, Georgia. They found that gorillas exhibited increased exhibit use and were more visible to visitors during a four-day novel phase compared with subsequent days. While in their 'Away' exhibit the gorillas increased locomotion, social distance, use of grass areas and visibility to visitors compared with their 'Home' exhibit.

8.9 Enclosure Use

8.9.1 Introduction

Studies of enclosure use by animals have recently been reviewed by Brereton (2020a). He identified the various assessment techniques available as including zone occupancy, spread of participation indices (traditional and modified), and electivity index, supplemented with measures of behavioural diversity and stereotypy prevalence. Brereton noted that there is a bias towards mammals in enclosure-use studies, but acknowledged that some initial studies of birds and fishes have been made.

I have discussed the calculation and usefulness of various measures of enclosure use elsewhere (Rees, 2015).

8.9.2 Recording Locations and Movements

Approximate locations of animals in enclosures can be indicated on a simple map (e.g. cheetahs (*Acinonyx jubatus soemmeringii*): Chadwick et al., 2013) or by dividing the enclosure into simple zones (e.g. Indian leopards (*Panthera pardus*): Mallapur et al., 2002; lion-tailed macaques (*Macaca silenus*): Mallapur et al., 2005;

Asian elephants (*Elephas maximus*): Rees, 2021). However, determining precise locations is not possible without the use of special equipment such as GPS devices (e.g. African elephants (*Loxodonta africana*) at Disney's Animal Kingdom, Florida: Leighty et al., 2010) or movement tags (MTags) (e.g. dolphins (*Tursiops*: Lauderdale et al., 2021) It is important to consider the behavioural effects of GPS collars on animals before embarking on a long-term study. When two red pandas (*Ailurus fulgens*) at Rotterdam Zoo in the Netherlands were fitted with collars they exhibited increased head shaking and suffered from abrasions caused by the devices (Van De Bunte et al., 2021).

An accelerometer is an electromechanical device that measures acceleration forces and step count and may therefore be used to detect the movement of an animal in its enclosure. The movements of African elephants at the San Diego Zoo Wild Animal Park were monitored using a body-worn GPS unit and an accelerometer (Rothwell et al., 2011). The GPS unit measured actual distance travelled, but when data from this was compared with that collected by the accelerometer, the latter was found to overestimate step count and consequently walking distance. The authors presented a linear regression equation that would allow estimates of daily walking distance to be calculated for elephants fitted with an accelerometer only.

8.9.3   Space Utilisation

For many species, the quality of the space within an enclosure may be much more important than the quantity. A small but elaborate enclosure may provide much more opportunity for an animal to engage in normal behaviour than a large homogeneous space.

When confined, animals have very little control over what happens to them. Increasing the complexity of an enclosure can increase an animal's available choices and help it to cope with novel or stressful situations. The provision of pen dividers, retreat areas or vegetation can improve many indicators of animal welfare by creating opportunities for concealment from other animals or from people.

The effect of enclosure design on space utilisation has been examined in singly-housed Indian leopards (*Panthera pardus*) by Mallapur et al. (2002). They studied leopards in four zoos in southern India and found that they used the edge or edge zone of the enclosure for stereotypic pacing, the rear for resting and the remainder of the enclosure for other activities. Leopards housed in on-exhibit enclosures that were structurally enriched (e.g. with sleeping platforms, logs and trees) showed higher levels of activity than those housed in barren enclosures.

Rose and Robert (2013) used the spread of participation index to examine enclosure use by a herd of eight sitatunga (*Tragelaphus spekii*) and found that significantly enhanced behavioural repertoires occurred in the biologically relevant 'natural' zones of the enclosure (long grasses, reeds and shallow water) compared with the less relevant (open, short-grassed) areas. Three behaviours (standing, sitting/ruminating and eating) showed significant differences in performance between natural and artificial zones and sitatunga spent more time in the natural areas. They concluded that enclosure design based on 'facets of natural ecology' is important for the expression of 'wild-type' behaviour in this species.

Mallapur et al. (2005) studied the enclosure space used by 47 captive lion-tailed macaques (*Macaca silenus*) housed in 13 zoos across India, and found that use of the part of the enclosure closest to the visitor area (the edge zone) correlated with the exhibition of abnormal behaviours, food-related behaviours and social interactions. Macaques housed in barren enclosures used the edge zone to a significantly greater extent than did those housed in complex exhibits. Autogrooming, social interactions and food-related behaviours significantly correlated with the use of the enriched zone.

Many animals living in zoos use the available space in their enclosures unequally. This may be because resources (e.g. food, water, sheltering places) are concentrated in particular places, they avoid or favour particular microclimates, keep away from areas where they are disturbed by visitors or are intimidated by the presence of potential predators in adjacent exhibits.

The effects of spatial restriction on the behaviour of rabbits (*Oryctolagus cuniculus*) were studied by Dixon et al. (2010). They found that rabbits were more active and interacted more with environmental resources in larger pens compared with smaller pens. The authors suggested that smaller hutches can compromise rabbit welfare because larger pens provide behavioural opportunities that are restricted in smaller pens.

When Hogan et al. (1988) compared the behaviour of Przewalski's horses (*Equus ferus przewalskii*) in enclosures of two different sizes they found the frequency of pacing, aggression and mutual grooming was higher in the smaller enclosure (17 × 30 m grassless pen) than in the larger enclosure (3.4 ha pasture).

Studies of the ways animals use space and resources in their enclosures can help to inform future enclosure design.

In a four-year study of the habitat use and structural preferences of Western lowland gorillas (*Gorilla g. gorilla*) at Zoo Atlanta, the quality of the available space was found to be more important than the quantity (Stoinski et al., 2001). The animals spent 50 per cent of their time using less than 15 per cent of the exhibits, preferring areas near structures, especially the holding building. Age, sex and rearing history did not appear to affect preferences, but social factors appeared to be important. Preference for spending time near the holding building (indoor accommodation) has also been observed in other species. For example, Asian elephants at Chester Zoo, England, spent on average 22.6 per cent of their time throughout the day in the area near the entrance to the elephant house, and this rose to 35 per cent in the period immediately prior to returning inside for their afternoon feed (Rees, 2021) (Fig. 8.6).

A study of chimpanzees (*Pan troglodytes*) and gorillas (*Gorilla gorilla*) found that in an older hardscape environment both species avoided open spaces and positioned themselves near mesh borders and in corners more than would be expected by chance (Ross et al., 2009). When they were transferred to a new naturalistic enclosure – designed using preference data collected from the old enclosure – both species altered their patterns of enclosure use. The chimpanzees used the environmental elements at rates similar to the proportions in which they were available. The preferences of

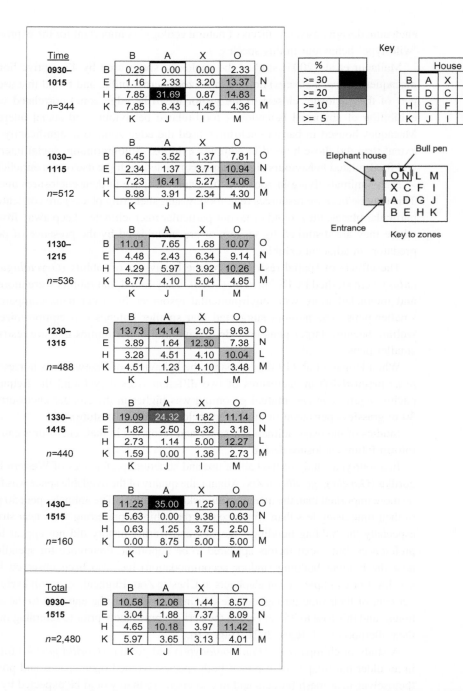

Fig. 8.6 Use of enclosure zones by eight Asian elephants (*Elephas maximus*) at Chester Zoo in England (first published in Rees (2021) and reproduced by permission of Elsevier). Numbers represent the percentage of all recordings of elephants made in each zone. The elephants were normally released into their outdoor enclosure at 10.00 and returned to the elephant house at 16.00, when they were fed. Towards the end of the day they spent most of their time near the main door to the building, where the five adults exhibited a variety of stereotypic behaviours.

gorillas for mesh barriers and doorways disappeared, but they maintained their preference for corners. Ross and Lukas (2006) found that gorillas spent more time than expected near mesh, corners and doorways, and less time than expected in open areas (Fig. 8.7).

In a study of the effects of visitors on little penguins (*Eudyptula minor*) at Taronga Zoo, Australia, almost 60 per cent were recorded in a corner of their exhibit that was out of sight of visitors (Chiew et al., 2020).

Electivity indices were used by Hunter et al. (2014) to study space use by two African wild dogs (*Lycaon pictus*) living in two naturalistic, outdoor enclosures at the San Diego Zoo, California, and the Bronx Zoo, New York. They identified underutilised and overutilised features of the enclosures and suggested that future designs should focus on providing the overutilised features, and not include the types of features that were underutilised.

Design can be used to manipulate the behaviour of animals so that it reflects that observed in the wild, for example by encouraging arboreal animals to spend time in the trees. The use of the four vertical levels (skylights, upper canopy, lower canopy and flooded floor) of an indoor enclosure by three adolescent orangutans (*Pongo pygmaeus*) was studied by Hebert and Bard (2000). The enclosure contained four large moulded trees and interwoven vines. They found that the apes preferred to spend time in the upper canopy – which contained tree limbs where they could rest – and next favoured the lower canopy and skylights, especially those skylights where they could not be seen by the public. The flooded floor was designed to keep the animals off the ground and in the trees, and was used just 1 per cent of the time.

Goff et al. (1994) examined mother–offspring use of vertical and horizontal space in chimpanzees and found that mother–offspring pairs spent most of their time using levels above the floor and spent the majority of time in areas near horizontal bench substrates 1.2–2.5 m above the floor.

A study of four wild-caught tarsiers (*Tarsius bancanus*) maintained in large enclosures found that they preferred mid-level heights and upright, small-diameter substrates (Roberts and Cunningham, 1986). The sleeping sites of Philippine tarsiers (*Tarsius syrichta*) kept in a semi-captive environment in the Philippines preferred dense, low-level vegetation in secondary forest, using perching sites on average 2 m above the ground (Jachowski and Pizzaras, 2005).

The differential use of space by eight bird species (3 vultures, 2 storks, 2 ducks and 1 goose) in a free-flight aviary was examined by Leuck (1977). She found that the use of space was correlated with food location, the presence of shade or similarity to substrates utilised in the wild. Space utilisation was strongly influenced by environmental structure, and social interactions within and between species may also have affected space use. Only the vultures made appreciable use of heights within the aviary of 1.5 m above ground level.

A study of a walk-through mixed-species free-flight aviary containing 24 bird species at Lowry Park Zoo in Florida found that range sizes decreased during periods of high visitor numbers, and there was increased use of vegetation cover and movement away from the visitor path (Blanchett et al., 2020).

Fig. 8.7 (a) Chimpanzees (*Pan troglodytes*) and (b) bonobos (*P. paniscus*) often spend time near buildings or the entrance to their indoor accommodation and avoid being out in the open. (A black and white version of this figure will appear in some formats. For the colour version, please refer to the plate section.)

The nature of the access to vertical space may affect its usage in some taxa. For example, individually housed Round Island geckos (*Phelsuma guentheri*) kept in vivaria heavily used locations containing furnishings or hides but avoided vertical glass walls, so this may not be an appropriate substrate (Wheeler and Fa, 1995). Enclosure size (0.03 m^3 and 0.22 m^3) did not affect activity cycles (nocturnal with crepuscular peaks) and, although thermogradients were not actively used, the geckos oriented towards natural light when given the opportunity. Provision of an alternative focal heat source to UV light would provide warmth without light.

Adding complexity to enclosures is enriching for corn snakes (*Pantherophis guttatus*). Given a choice, they prefer to occupy an enriched rather than a standard enclosure, and this choice benefits their behaviour and welfare (Hoehfurtner et al., 2021).

8.9.4 Enclosure Use and Animal Welfare

Care should be taken when locating accommodation for a particular species in a zoo to avoid creating an environment where the occupants suffer negative welfare by virtue of the location of their enclosure. Some species may experience stress by virtue of the proximity of their enclosure to that of a potential predator or competitor. Leopard cats (*Prionailurus bengalensis*) stressed by the presence of large cats (*Panthera* spp.) experienced lower cortisol levels and exhibited less pacing when provided with hiding places (Carlstead et al., 1993).

Stereotypic route-tracing consists of, for example, repeatedly walking the same route around an enclosure or swimming the same route around a tank. It is observed in some animals kept in zoos (e.g. canids, felids, bears, pandas, elephants) and also in farm animals (e.g. horses (McGreevy et al., 1995)). Part of the route often includes some of the boundary of the enclosure and may involve walking along a fence line or repeatedly pacing back and forth along a short length of fencing (Figs. 8.8 and 8.9). In some cases an animal may place its feet in the same place on each circuit of its enclosure or swim in the same pattern around a pool. A polar bear (*Ursus maritimus*) at Detroit Zoo in Michigan repeatedly swam in a circuit around its pool, standing in the same place on the underwater tunnel used by visitors and taking a breath of air before submerging again, swimming across the pool to look through the underwater window to a seal pool (presumably for prey) before returning to the top of the tunnel (Fig. 8.10).

Route-tracing by carnivores living in zoos has attracted considerable attention. Few frequent zoo visitors can have failed to notice big cats pacing around the periphery of their enclosure or walking repeatedly back and forth along a fixed length of fencing (Fig. 8.11). Clubb and Mason (2007) suggested that the welfare of many carnivores kept in zoos is compromised as a result of inadequate enclosure sizes. They claim that the brown bear (*Ursus arctos*) and snow leopard (*Uncia uncia*) adapt well to captivity, but the polar bear and clouded leopard (*Neofelis nebulosa*) are hard to breed and tend to develop abnormal behaviours.

Fig. 8.8 This spectacled or Andean bear (*Tremarctos ornatus*) (inset) has left evidence in the soil of its route-tracing behaviour.

Fig. 8.9 Giant panda (*Ailuropoda melanoleuca*) route-tracing in an arc between a viewing window and the door to the indoor accommodation.

Fig. 8.10 This polar bear (*Ursus maritimus*) at Detroit Zoo, Michigan, was performing stereotypic route-tracing whereby it repeatedly moved in 'circles', standing on the top of an underwater tunnel, diving beneath the surface and swimming to an underwater window to a seal pool, and then returning to the tunnel and surfacing for air. (A black and white version of this figure will appear in some formats. For the colour version, please refer to the plate section.)

Kroshko et al. (2016) examined data sets for 23 species of carnivores and found that those with a naturally large home range and a long chase distance during hunts exhibited more severe route-tracing behaviours. The authors suggested that being wide-ranging and a pursuit predator may be inflexible 'behavioural needs' which cannot be relinquished even when homeostatic needs (e.g. nutritional requirements) are met. Interestingly, natural travel distance and range size did not predict infant mortality in this study, but had been suggested as risk factors in previous research on carnivores by Clubb and Mason (2007), two of the authors of this work. In this earlier paper the authors concluded unequivocally that 'naturally wide-ranging lifestyles also predicted relatively high captive infant mortality rates', repeating a claim made in an earlier paper published in *Nature* that received considerable attention from the media and anti-zoo organisations: 'Natural home-range size (HR) predicted captive-infant mortality' and 'infant mortality in the wild was unrelated to range size' (Clubb and Mason, 2003). The methodology used by Clubb and Mason in this earlier work was

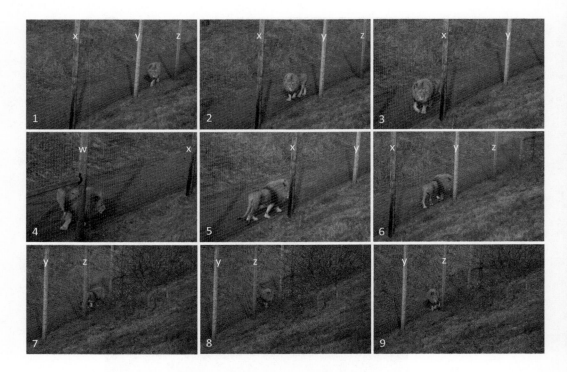

Fig. 8.11 This male African lion (*Panthera leo*) is performing a common stereotypic route-tracing behaviour observed in felids: pacing along a section of fencing. The fence posts are labelled to indicate the relative position of the lion in each image.

criticised by zoo professionals at the time, who insisted that large carnivores in zoos were not routinely suffering as a result of poor enclosures (Randerson, 2003). Importantly, the data set used by Kroshko et al. (2016) was nearly double the size of the original data set used by Clubb and Mason and they imposed stricter quality controls on these data – for example, the minimum sample size included.

Elephants living in zoos commonly exhibit a range of stereotypic behaviours, including pacing back and forth and route-tracing. An adult female Asian elephant (*Elephas maximus*) at Chester Zoo in England repeatedly walked in a large circle, always in the same location and always anti-clockwise (Rees, 2021). She had brought this stereotypic pattern with her from Whispnade Zoo, where she and the other elephants were kept at night in individual small and barren circular indoor pens in an elephant house designed by Berthold Lubetkin. The behaviour had developed when she paced around the edge of this pen and at Chester Zoo she performed this circular route-tracing in a curved corner of her outdoor enclosure (effectively representing one-quarter of the periphery of her old accommodation at Whipsnade) (Fig. 8.12).

In a study of the factors that influenced the characteristics of locomotor stereotypies in elephants, 77 elephants (*Loxodonta* 5.31; *Elephas* 8.33) were videotaped at

Fig. 8.12 A female Asian elephant (*Elephas maximus*) walking repeatedly in an anti-clockwise circle. The direction never changed and had developed in her former circular indoor accommodation at Whipsnade Zoo. Based on a figure first published in Rees (2021).

39 North American zoos (Greco et al., 2017). Analysis of the recordings revealed that most of the elephants that performed locomotor stereotypies also performed whole-body stereotypies. The risk of having locomotion present in their stereotypic repertoire increased with the amount of time housed separately and in those elephants who were members of several social groups. The authors of the study concluded that locomotor stereotypies might be prevented by enhancing elephants' social environment and increasing the spatial complexity of their enclosures. Regrettably, too many zoos are still keeping elephants in small groups where the opportunity to increase their social contacts is severely restricted (Rees, 2021).

Miller et al. (2021) found that dolphins exhibited a low rate of route-tracing behaviour in a study of 47 individuals (*Tursiops truncatus* and *T. aduncus*) held at 25 facilities distributed around the world. Few animal management factors or habitat characteristics were related to this behaviour. There was a positive relationship between behavioural diversity and the number of habitats accessible and a significant inverse relationship with the maximum depth of the habitat.

There have been relatively few studies of route-tracing in captive birds. In an investigation of the factors causing stereotypies in birds Keiper (1969) found that route-tracing was reduced in a very large aviary cage and when a swinging perch arrangement was present. Wild-caught birds exhibited a greater number of route traces compared with laboratory-reared birds. Keiper concluded that route-tracing was associated with the physical restrictions imposed on movement by the cage. Garner et al. (2003) studied experimentally caged blue tits (*Parus caeruleus*) and marsh tits (*P. palustris*) and found that stereotypic route-tracing in these birds correlated with general behavioural disinhibition.

Injury in arboreal species may restrict their use of vertical space. However, a study of a female chimpanzee (*Pan troglodytes*) living in a zoo following the surgical amputation of her right forelimb found that her injury made little difference to her daily activities (Ang et al., 2017). After surgery she exhibited no change in her use of vertical space. There was a decrease in the frequency of locomotion on the ground, but the overall distance she travelled significantly increased. The authors noted that their study illustrated that, given an adequate environment in which to remain active, chimpanzees are able to adapt to significant anatomical changes.

Enclosure utilisation and activity budgets of disabled Malayan sun bears (*Helarctos malayanus*) were studied by Lewis et al. (2017). They observed 12 adult sun bears and found that amputees used enclosures less evenly, and groomed and paced less than non-amputees. Bears that were partially sighted neither used enclosures differently nor behaved differently from controls.

8.9.5 Overcrowding

Animals kept in zoo enclosures and cages inevitably have little space available to them compared with their wild conspecifics. This can result in unnatural behaviours – for example, because subordinate individuals cannot completely escape from dominant individuals – and may lead to unnaturally high levels of association between individuals simply because they are confined within a small space.

A zoo-living adult cow Asian elephant (*Elephas maximus*) has been observed exhibiting unusual submissive behaviours – bowing the head and lying prostrate – possibly as a result of being forced to occupy the same relatively small enclosure as a bull with whom she had an agonistic relationship (Rees, 2004c).

The maintenance of a social hierarchy within a group of animals assumes that each of the individuals knows its rank in relation to all of the others. If an unnaturally large group is kept in a confined space, social behaviour may be disrupted because there are too many individuals for each to recognise all the group members (D'Eath and Keeling, 2003).

Abnormally high frequencies of aggressive interactions may occur when animal densities are high. For example, submission and appeasement gestures increased in captive rhesus macaques (*Macaca mulatta*) kept in crowded conditions (de Waal 1996).

The influences of enclosure size and animal density on Père David's deer (*Elaphurus davidianus*) at Dafeng Nature Reserve in China were studied by Li et al. (2007) (Fig. 8.13). They compared behaviour and faecal cortisol levels – an indicator of stress – in a group of stags in a large enclosure at low density (200 ha; 0.66 deer/ha) with stags kept in a small display pen with high density (0.75 ha; 25.33 deer/ha). The frequency of conflict behaviour was higher in the small display pen than in the large enclosure, and the stags in the small pen exhibited higher cortisol concentrations. In a second experiment, faecal cortisol levels in a group of 12 stags were measured before and after they were moved from a 100 ha enclosure to a small pen of 0.5 ha. No difference was found between the cortisol levels on sampling days, but the mean level was significantly higher on the day after transfer than on the day before.

Fig. 8.13 Père David's deer (*Elaphurus davidianus*) at Knowsley Safari. The species was extirpated in the wild and successfully captive-bred in Europe. Captive-bred individuals have been reintroduced into the wild in China.

8.9.6   Social Systems and Enclosure Use

The effect of dominance rank on enclosure use may vary within a species, depending upon the age of the individuals. A study of four adult Western lowland gorillas (*Gorilla g. gorilla*) housed together at the Cincinnati Zoo in Ohio showed that they limited their space use to particular enclosure sections, but there was no consistent relationship between the degree to which an animal limited its space use and dominance rank (Hedeen, 1983). However, in two groups of juvenile gorillas at the same zoo, dominant individuals had higher spread of participation index values than did subordinate cage-mates (Hedeen, 1982).

In their study of the use of exhibit space and resources by a herd of five adult female African elephants (*Loxodonta africana*) at Disney's Animal Kingdom in Florida, Leighty et al. (2010) found preliminary evidence that position in the dominance hierarchy affected the percentage of space occupied and contributed to access to resources. Collar-mounted GPS recording devices indicated that dominant animals used a higher percentage of the available space than did subordinates, and the latter avoided narrow or enclosed regions of the enclosure. Dominant females occupied the watering hole more than subordinates, but the mud wallow was used equally by all animals.

Aggression between adult male and non-adult male spider monkeys (*Ateles*) is common in zoos and sometimes has fatal consequences. Davis et al. (2009) surveyed

26 zoos and found that these monkeys are predominantly kept in small social groups consisting of one adult male housed with two females and their young. They suggested that the management of this species in captivity should reflect the wild condition of male philopatry and female dispersal. In addition, they suggested that enclosures should be designed to allow individuals to segregate themselves from others, simulating the fission that occurs in the wild, to avoid aggression.

Forthman and Bakeman (1992) studied the enclosure use of four adult sloth bears (*Ursus ursinus*) at Los Angeles Zoo, California, and found that this was conditional on variations in both the physical and social environments. The two females were most social when exhibited with a familiar male, and social behaviour and non-social activity decreased when an unfamiliar male was present.

8.10   Multi-species Exhibits

Some species are kept in multi-species exhibits in zoos and aquariums. This can result in a better educational experience for visitors than single-species exhibits if the cohabiting species occur together naturally (i.e. are part of the same biological community), or a poorer educational experience if they do not. For example, Knowsley Safari Park in England has exhibited Asian antelopes, nilgai (*Boselaphus tragocamelus*), in the same enclosure as red-necked wallabies (*Macropus rufogriseus*) from Australia, and in the past in other zoos large felids such as lions and tigers were often housed together, resulting in hybrids (ligers and tigons).

Both welfare losses and gains may result from allowing different species to occupy the same space. For example, although the presence of other species may be enriching from a behavioural point of view, it may provide an opportunity for the spread of disease. Only a single incidence of cross contamination of salmonella occurred between African elephants (*Loxodonta africana*) and Hamadryas baboons (*Papio hamadryas hamadryas*) in a mixed-species exhibit at Safari Beekse Bergen in the Netherlands (Deleu et al., 2003). However, an epizootic of amoebiasis (*Entamoeba* sp.) resulted in the death of 21 juvenile tortoises in a mixed exhibit of tortoises at a zoo in Florida (Hollamby et al., 2000).

Popp (1984) examined interspecific aggression in mixed-species ungulate exhibits at the Audubon Zoological Gardens in New Orleans, Louisiana, and found that the total rate of aggression was highest between distantly related species and increases in aggression were triggered by mating activity, births and the introduction of a new animal to an exhibit.

Interspecies aggressive interactions among African ungulates kept in mixed-species groups at Dvůr Králové Zoo in the Czech Republic over 20 years resulted in deaths more frequently when they occurred between closely related species than when they occurred between more distantly related species (Hanzlíková et al., 2014). This may have been because of the very similar resource requirements of the closely related species. However, when only fights between adults were considered, more aggressive interactions occurred between taxonomically more distantly related species.

Andersen (1992) found a negative correlation between exhibit size and frequency of interspecific interactions in a multi-species exhibit of plains zebra (*Equus burchelli*) and eland (*Taurotragus oryx*) in three Danish zoos. The author recommended that multi-species exhibits should contain ample resources (e.g. food and water) and sufficient space at feeding sites.

Few agonistic interactions were observed between red-bellied tamarins (*Saguinus labiatus*) and saddleback tamarins (*S. fuscicollis*) kept in stable associations at Belfast Zoo, Northern Ireland (Hardie, 1997). Most aggressive interactions involved disputes over food, with *S. labiatus* typically displacing *S. fuscicollis* merely by approaching. Some interspecies grooming and huddling was observed. The species are sympatric in the wild and Hardie suggested that there may be some benefit in exhibiting tamarin species together.

A successful multi-species exhibit of Goeldi's monkeys (*Callimico goeldii*) and pygmy marmosets (*Callithrix pygmaea*) at Edinburgh Zoo in Scotland has been described by Dalton and Buchanan-Smith (2005). In the wild, the species are sympatric and the two species utilised the available space in their enclosure differently from each other but in a similar way to that observed in the wild. This separation may have contributed to their peaceful coexistence.

Law et al. (2021) studied ring-tailed lemurs (*Lemur catta*) kept with other lemur species in 10 polyspecific exhibits in zoos in the UK. Agonistic interspecific interactions were slightly more frequent overall than affiliative interactions in these groups, but not significantly so. The frequency and duration of agonistic and affiliative interactions varied among exhibits. In single-sex exhibits and those containing individuals less than one year old increased frequencies of affiliative interactions were recorded, whereas larger groups and those without infants exhibited a reduced frequency of agonistic events.

It is clear that, while aggression between species may be problematic in some multi-species exhibits, others contain species that are able to coexist peaceably. An understanding of the factors affecting the frequency of aggressive interactions in multi-species groups is important in informing management decisions about housing and group composition. Intraspecies aggression in zoo exhibits is discussed in Section 9.12.

8.11 Substratum and Welfare

The effect of the substratum on the welfare of animals living in zoos is not a new concern. It was suggested by Ball (1886) that the use of boards for flooring in the lion cages at Dublin Zoo in Ireland instead of tiles contributed to their success in breeding lions, although he presented no evidence for this. He reported that London Zoo had recently boarded cages in their lion house as a result of Dublin's success.

An analysis of the information returned by the 62 zoos accredited by AZA that responded to a survey concerning biofloors used in great ape enclosures found that 45 (72.6 per cent) had indoor exhibit spaces that could be viewed by visitors, but only 13 exhibits (28.9 per cent of these) provided a biofloor (Leinwand et al., 2021). The key motivation reported for installing biofloors was animal welfare. Those facilities that did not have a biofloor most often cited facility constraints as the reason. As enclosure maintenance (pest control and cleaning procedures) varied little between floor types, the authors concluded that biofloors promote positive welfare without compromising husbandry efforts.

Early studies of the importance of substratum were conducted on primates kept in laboratories. For example, caged rhesus macaques (*Macaca mulatta*) were shown to benefit from the foraging opportunities afforded by artificial turf (Bayne et al., 1992) and the addition of woodchip bedding to outdoor chimpanzee enclosures reduced the frequency of abnormal behaviours (Brent, 1992).

Pair-housed female squirrel monkeys (*Saimiri sciureus*) were provided with treat-enhanced artificial turf foraging boards in an attempt to reduce abnormal behaviours and aggression, and to alter their time budget so that it approximated that observed in the wild (Fekete et al., 2000). During the first 30 minutes of the two-hour daily enrichment period, board-related behaviours occupied 36.3 per cent of the activity budget, inactivity declined by 35.3 per cent and locomotion increased by 3.8 per cent. These changes represented an adjustment of the time budget so that it was more like that observed in the wild, but were absent after 1.5 hours. The foraging opportunity afforded by the turf boards had no effect on either the levels of aggression or frequency of stereotypic behaviour (consisting of pacing, head-swinging and tail-chewing).

The authors of a study of substrate preferences in six Asian elephants (*Elephas maximus*) at Oregon Zoo concluded that rubber flooring in the elephant house may provide a more comfortable surface than concrete for locomotion and standing-resting behaviour (Meller et al., 2007). When given a choice, all eight Asian elephants at Dublin Zoo in Ireland preferred to sleep on sand rather than concrete (Walsh, 2017).

8.12   Shade, Temperature and Air Quality

There has been little research on the importance of providing shade in animal enclosures (Fig. 8.14). Langman et al. (1996) made a thermal assessment of a sea lion enclosure at the Audubon Zoo in Louisiana and concluded that the thermal properties of the materials used in zoo enclosures are an important determinant of the animals' heat loads, and should be considered when exhibits are designed. The original exhibit contained simulated rocks made from gunite (a type of concrete) that reflected 41 per cent of shortwave radiation, but after renovation the same area reflected only 8 per cent of shortwave radiation. Short wave heat load was reduced by darkening the gunite surfaces.

The thermal microclimate provided by a shade structure for African elephants (*Loxodonta africana*) at Zoo Atlanta in Georgia was quantified by Langman et al. (2003) by comparing the ambient conditions inside a Stevenson screen with the ambient conditions in shaded and non-shaded areas of the exhibit. They concluded

Fig. 8.14 Sun shades are important in enclosures where large trees are not present. (a) Knowsley Safari, England. (b) Detroit Zoo, Michigan.

that shade structures alone may not provide adequate protection from radiant heat for captive species.

Analysis of 33 enclosures at Lincoln Park Zoo (Chicago, Illinois) indicated that shade availability varied widely with the percentage of shaded space, ranging from 85 per cent to 22 per cent across enclosures during the summer months, with the least shade available at midday (Wark et al., 2020).

The Sichuan takin (*Budorcas taxicolor tibetana*) is a cold-adapted species native to the mountainous regions of China and Tibet (Fig. 8.15). Wark et al., (2020) installed a shade structure (a cloth sail) in an enclosure at Lincoln Park Zoo used by three takin and evaluated its use. The animals showed a preference for shaded areas near walls, but also used the sail, which increased the total available shade by 10 per cent. As temperature changed there was a gradual transition in shade use by the takin as they switched from sun-bathing behaviours under colder temperatures to shade-seeking behaviours under higher temperatures. The authors suggested that cloth sails are a cost-effective method of increasing shade in zoo enclosures and emphasised the importance of providing animals with an appropriate thermal environment to enhance their welfare.

Colobus monkeys (*Colobus guereza*) living on a concrete island with limited shade spent more time resting in the shade than in the sun when temperatures exceeded 23°C (Wark et al., 2014). At these temperatures, the monkeys spent more time in social contact and performed more self-directed behaviours. When they were given access to an indoor area and temperatures were higher than 23°C, the monkeys spent more than half of their time indoors compared with just 10 per cent of their time indoors at lower temperatures.

Fig. 8.15 Sichuan takin (*Budorcas taxicolor tibetana*). Cloth sails have been used effectively to provide shade for takins during hot weather.

Asian elephants (*Elephas maximus*) with little access to shade in their outdoor enclosure at Chester Zoo in north-west England spent more time dusting as ambient temperature increased, but the reason for this was unclear (Rees, 2002). In the wild, Asian elephants are forest-dwellers and spend the heat of the day under the shade of the forest canopy (Sitompul et al., 2013). The use of shade trees by wild Asian elephants has been documented by Ramakrishnan et al. (2014). Whipsnade Zoo provides large artificial shades for its Asian elephants.

Western lowland gorillas (*Gorilla g. gorilla*) at Zoo Atlanta, Georgia, spent more time away from their holding building and other structures when temperatures were cold and more time near them when they were hot (Stoinski et al., 2001).

In modern indoor zoo environments the quality of the air is often controlled by a sophisticated heating, ventilation and air-conditioning (HVAC) system. Although designed to provide a suitable environment for animals, sometimes such systems can contribute to health problems.

Wilson et al. (2006) investigated a single death and poor health among five alligators (*Alligator mississippiensis*) in a holding facility in a zoo in the south-eastern United States. They analysed indoor air and water from the facility and the operation of the HVAC system. Compared with two control sites that were also similarly tested, the alligator facility and its HVAC system were extensively contaminated with a range of fungi and there were higher levels of fungal spores in the air. Zoo records revealed pulmonary disorders in three alligators, one of which died as a result, and a general malaise was observed in the other two. There was no morbidity or death associated with the control sites. The authors of the study concluded that environmental mould contamination was responsible for the health issues experienced by the alligators.

8.13 Visibility and Visitor Effects

With respect to enclosure design, in many cases the interests of animals and visitors conflict. People visit zoos to see animals; animals need to have places where they can escape from the gaze and noise of visitors (Fig. 8.16). The visibility of free-ranging quokkas (*Setonix brachyurus*) in a walk-through exhibit at Melbourne Zoo, Australia, was reduced when visitors were present (Learmonth et al., 2018). However, visitor interest was not compromised at a naturalistic exhibit in a Chinese zoo studied by Davey (2006b) when he compared visitor behaviours at naturalistic and barren exhibits. He concluded that the needs of animals and visitors can be balanced in exhibit designs.

Some animals rest in predictable but secluded places within their enclosures. Signs were erected at a naturalistic tiger enclosure in Zoo Atlanta, Georgia, indicating the likely locations of the animals, but these did not increase the number of visitors who found the animals and very few visitors reported using the signs (Bashaw and Maple, 2001).

Rotational exhibits consist of two or more adjacent enclosures between which animals are 'rotated' on different days to increase the total amount of space to which they have access. Regularly alternating gorillas between two naturalistic exhibits at Zoo Atlanta increased their activity levels and use of exhibit space, but did not always increase their visibility due to the constraints of the exhibit design (Lukas et al., 2003).

The presence of visitors at an enclosure may have a negative, positive or neutral effect on animals. It may provoke fear, provide stimulation or have no discernible effect. Some zoos obscure enclosure windows to reduce the visibility of visitors (Fig. 8.17).

The welfare of Western lowland gorillas (*Gorilla g. gorilla*) can be improved by using camouflage net barriers to reduce the animals' sight of visitors (Blaney and Wells, 2004). In the presence of the net they showed lower levels of aggression towards conspecifics and a reduction in stereotypic behaviours compared with the period prior to the installation of the net. Furthermore, visitors' perceptions of the gorillas were affected by the camouflage net; when it was present they considered the gorillas to look more exciting and less aggressive than when it was absent.

O'Malley et al. (2021) examined the impact of 360° visitor viewing access to Western lowland gorillas at the Brookfield Zoo, Illinois, by comparing aspects of their behaviour and physiology before and after viewing access to the perimeter of the enclosure was reduced by 70 per cent. When access was restricted, they recorded a reduction in solitary grooming and two measures of the adrenal system compared with levels when the public had 360° access. However, other behaviours remained constant across conditions. The authors concluded that 360° access may have a negative impact on gorillas. However, studies of the effects of visitors on different species of great apes have produced inconsistent results. When half of the viewing window of an orangutan (*Pongo pygmaeus abelii*) enclosure was covered, the animals preferred to position themselves in front of the uncovered section, suggesting that they were visually stimulated by the presence of visitors (Bloomfield et al., 2015).

Fig. 8.16 (a) A tiger (*Panthera tigris*) emerging from thick undergrowth at Edinburgh Zoo, Scotland. (b) Two tigers in an open area of their enclosure at Knowsley Safari, England. The extent to which animals are visible to visitors is in part determined by enclosure design. Thick vegetation may provide large predators with a naturalistic environment, but often at the expense of visibility. If enclosures are too open, their inhabitants are unable to escape the gaze of the public. (A black and white version of this figure will appear in some formats. For the colour version, please refer to the plate section.)

Fig. 8.17 The viewing window to the dorcas gazelle (*Gazella dorcas*) enclosure at Barcelona Zoo in Spain is obscured by silhouettes of coloured leaves to reduce the effects of the presence of visitors. Inset: dorcas gazelle. (A black and white version of this figure will appear in some formats. For the colour version, please refer to the plate section.)

A study of little penguins (*Eudyptula minor*) at Taronga Zoo, Australia, examined the effect on their behaviour of covering one out of four viewing area windows (Chiew et al., 2020). The window covered was the main viewing window and this had the effect of reducing the proportion of visitors present at this window by 85 per cent, thereby reducing threatening visitor behaviours at this window such as loud vocalisations, sudden movements and touching the window. In the area near the main window the proportion of visible penguins increased by some 25 per cent, the proportion preening in the water increased by about 180 per cent and the vigilant proportion decreased by about 70 per cent. This study provides evidence that the presence of visitors may be fear-provoking in this species, but it was not clear which aspects of visitor contact were responsible.

A study of two corvid species – crows (*Corvus corone*) and ravens (*C. corax*) – housed in outdoor aviaries at the Cumberland Wildlife Park, Upper Austria, showed that individuals that had been housed in the park for several years still showed behavioural responses to the close proximity of visitors to their enclosure, such as increased preening in crows and increased vocalisations in ravens (Wascher et al., 2021). However, the authors did not observe increased locomotion, stress-related behaviours or any other indications of reduced welfare.

A study of five species of flamingos – Caribbean (*Phoenicopterus ruber*), Chilean (*P. chilensis*), greater (*P. roseus*), lesser (*Phoeniconaias minor*) and Andean

(*Phoenicoparrus andinus*) – at the Wildfowl and Wetland Trust's Slimbridge Wetland Centre in England found that all five species exhibited preferred areas of occupancy within their enclosures, but no evidence of a visitor effect was detected (Rose et al., 2018). The presence of visitors did not cause a change in activity patterns or a reduction in enclosure use. Diurnal activity patterns were similar to those published for wild flamingos, with greater activity occurring later in the day.

The location of an enclosure in a zoo may affect the number of visitors it receives. This is an important factor to consider when deciding where to house species whose behaviour is strongly influenced by the presence of visitors. Golden-bellied mangabey (*Cercocebus chrysogaster*) cages were located at three slightly different locations within Sacramento Zoo, California, and the number of visitors to each was counted by Mitchell et al. (1990). More people visited the cage nearest the entrance/exit of the zoo than the cage away from the exit but on the main path, and this second cage was visited more often than the third cage, which was slightly off the main walkway. These results suggest that species sensitive to disturbance should not be placed near the zoo entrance or on the main path around the zoo.

Staff at Disney's Animal Kingdom in Florida have developed a system for the purpose of monitoring and improving the visibility of animals on display (Kuhar et al., 2010). The system involves the collection of visibility data, using visibility criteria to which managers are held accountable, and a process for planning to improve animal visibility without compromising animal welfare.

8.14 Using Signage to Control Visitor Behaviour

Exposure to noise from visitors may affect the behaviour and welfare of some species (see Section 9.13). Dancer and Burn (2019) examined the effect of signage requesting quietness at the enclosures of Sulawesi crested macaques (*Macaca nigra*) on visitor noise and found that visitor noise levels were slightly reduced when simple signs requesting visitors to be quiet (control signs) were present compared with noise levels in the absence of signs. When signs were displayed that incorporated salient 'watching' human eyes they were no more effective than the control signs.

Public feeding of animals in zoos was once common, but modern zoos prohibit this. The use of signs to control public feeding in the Birmingham Zoo, Alabama, was studied by Bitgood et al. (1988b). They studied the effectiveness of three types of 'do-not-feed' signs at the monkey island in the zoo: an instructional sign saying 'Do not feed'; an instructional sign with an explanation ('These animals are on a special diet'); and a sign that compared the public feeding of the animals with feeding a child an inappropriate diet. A baseline feeding rate was established when no signage was present. The sign that was most effective in reducing public feeding was the instructional sign that explained that the animals received a special diet.

8.15 Conclusion

In the best modern zoos, animal enclosures have evolved from small barren cages and paddocks to sophisticated, enriched, mixed-species, naturalistic and immersive environments. Aquariums have moved on from keeping aquatic species in small tanks to creating elaborate underwater experiences for their visitors. Developments in enclosure design have improved the lives of animals living in zoos and enriched the visitor experience. These developments have been driven by scientific studies of the effects of the presence of visitors on animals, their substrate preferences, the way animals utilise the space available to them and a myriad of other environmental factors that affect animal welfare.

9 Animal Welfare

9.1 Introduction

Animal welfare science is a relatively new area of academic study. It began with concern about the effects of captivity on farm and laboratory animals but then encompassed animals in zoos and wild animals. Apart from the moral imperative to treat animals well, it is essential that those kept in zoos experience good welfare so that they thrive and reproduce. Nevertheless, zoos inevitably place many constraints on the behaviour of animals and some develop abnormal behaviours as a result. This chapter considers some of the welfare issues experienced by animals living in zoos and the attempts that have been made to address them.

9.2 Historical Perspectives

The first professor of animal welfare in the world was Donald Broom, who was appointed to the Chair in Animal Welfare in the Department of Veterinary Medicine of the University of Cambridge in 1986. To put this in its historical context, consider that the first professor of zoology and comparative anatomy at Cambridge (Alfred Newton) was appointed 120 years earlier in 1866 (just seven years after the publication of Darwin's *Origin of Species* in 1859). The University of Glasgow appointed Lockhart Muirhead as its first professor of natural history (which included zoology) in 1807, although this position did not become the Chair in Zoology until 1902.

The journal *Animal Welfare* was first published in February 1992. The journal published its first zoo-based study in November of the same year: a study of the influence of a puzzle feeder on the activity and behaviour patterns of orangutans, gorillas and chimpanzees at London Zoo (Gilloux et al., 1992). Over a decade earlier the first issue of *Zoo Biology* published a paper on the effects of spatial crowding on social behaviour in a chimpanzee colony at Arnhem Zoo (Nieuwenhuijsen and de Waal, 1982).

Legislators were much more expeditious in recognising the need to protect animals from cruel treatment, initially focusing their attention on farm animals when Martin's Act (An Act to Prevent the Improper Treatment of Cattle 1822) was passed by the UK Parliament. This was the first national legislation anywhere in the world to punish animal cruelty.

Early academic discussions of animal welfare were largely ethical in nature and conducted by philosophers. They focused primarily on the conditions in which animals were kept on farms and in scientific laboratories. Peter Singer's influential book *Animal Liberation* was first published in 1975, but, although it argued that non-human animals should be afforded equal moral consideration, he did not discuss the plight of animals kept in zoos even though many at that time were kept in less than optimal conditions and animals were routinely being taken from the wild (Singer 1975).

Zoo animal welfare came of age as an academic discipline – an applied science – with the publication of *Zoo Animal Welfare* by Terry Maple and Bonnie Perdue (Maple and Perdue, 2013) and the convening of a conference entitled *From Good Care to Great Welfare* at the Center for Animal Welfare at Detroit Zoo, Michigan, in 2011. At this conference zoo scientists and administrators sat alongside delegates from animal welfare organisations and academics to discuss common interests (Kagan 2013).

It is tempting to believe that those who managed early Victorian zoos knew little about the proper management of their animals. Although many were kept in barren cages, some consideration was given to their ecological requirements. The *Proceeding of the Zoological Society of London* recorded a meeting held on 28 December 1830 at which a letter from Mr J. C. Cox Esq. FLS was read concerning 'the subject of preserving a proper temperature for exotic animals' (Yarrell, 1830). He also referred to the need to maintain a high humidity for species from the forests of Ceylon (now Sri Lanka) and suggested that 'this may readily be produced by watering the flues used for heating the houses in which they are kept'. He went on to discuss the importance of good ventilation:

For the general regulation of the admission of cold air a convenient plan is to have a leaden or iron weight balanced in a vessel filled with mercury, attached to a sliding sash, which will thus rise or fall in proportion to the height of the mercury.

In fact, many early beliefs about the temperature requirements of animals from tropical countries when kept in zoos located in temperate latitudes have since been found to be erroneous.

Ball (1886) published a history of lion breeding in the Gardens of the Royal Zoological Society of Ireland (Dublin Zoo) between 1857 and 1885 when he was Director of the Science and Art Museum in Dublin and Honorary Secretary of the Society. He described 131 cubs from 34 litters, and believed this to be the best breeding record for lions of any contemporaneous zoo. The greatest age attained was 16 years and juvenile mortality was about 22 per cent.

Improvements in animal accommodation at London Zoo in the nineteenth century were well received by the press (Fig. 9.1). The new 'Lion House' was recognised as a great improvement on the old facility, as indicated by an article in *The Illustrated London News* of 1 April 1876:

The new building covers a much greater space than was formerly allotted to the same kinds of animals under the central terrace ... There are fourteen compartments for the beasts, very much larger than they occupied before, generally about twice as large, besides a back parlour,

THE NEW LION-HOUSE, ZOOLOGICAL SOCIETY'S GARDENS.

Fig. 9.1 New Lion House, London Zoo, in 1876. At this time it housed tigers, lions, Indian and African leopards, pumas and a jaguar. (Source: *The Illustrated London News*, 1 April 1876, p. 325).

or 'growlery,' for each of them, to which the keepers have a private approach through the corridor behind ... The place is well warmed by a series of hot-air pipes. (*The Illustrated London News*, 1876)

Two years previously, in 1874, the same publication commented favourably on the large carnivore accommodation at the Berlin Zoological Gardens:

a zoological garden, which ... is not excelled by any in Europe, and where the animals enjoy ample space, air, and light. The larger carnivora are all provided with double cages, connected by a sliding panel of iron, the smaller of which looks, inside, a well-ventilated, handsome building, and serves for a sleeping-place; the other being roofed with thick glass, and closed in with strong iron bars – although sufficiently spacious to afford the animals a good run – and ornamented with rockwork. (*The Illustrated London News*, 1874)

The article also noted the zoo's 'vast aviaries' and the fact that some species were allowed to 'roam over the extensive grounds at their own sweet will'.

The animal collections of modern zoos are biased towards the charismatic megafauna, particularly large mammals. These taxa attract visitors and at least some species are of conservation concern. However, charismatic species – such as elephants, great apes and cetaceans – attract more welfare-related concern from animal activist groups and the media than do non-charismatic taxa, and the former are also the subject of more research on their welfare than are the latter (Hosey et al., 2020a).

9.3   What Is Welfare?

Broom (1986) defined the welfare of an individual animal as 'its state as regards its attempts to cope with its environment' and Broom (1991) emphasises that 'Welfare is a characteristic of an animal, not something given to it, and can be measured using an array of indicators'. Welfare – or wellbeing – is about feelings such as 'suffering' or 'contentment' that cannot be measured directly but may be inferred (Mason and Veasey, 2010a).

A great deal of current animal welfare legislation is based on the concept of the Five Freedoms established in relation to farm animals in the UK as a result of the Brambell Report (HMSO, 1965). These are freedom from hunger and thirst; discomfort; pain, injury or disease; fear and distress; and freedom to express normal behaviour. More recently the 'five domains' model has been introduced which, rather than emphasising the removal of negative experiences, focuses on providing positive experiences: nutrition; environment; health; behaviour; and mental state (Mellor and Reid, 1994).

9.4   How May Welfare Be Measured?

Physical welfare may be assessed by examining the body condition of an individual and comparing it with that of its conspecifics, by looking for abnormalities in an animal's gait and other motor functions, and in its physiology.

Measuring psychological welfare is much more problematic. Keeping individuals in a social group may be beneficial for some individuals but very stressful for others. Individuals may experience chronic stress without exhibiting indicative behaviours.

It may be possible to measure welfare indirectly by examining the conditions under which animals are kept. For example, if they have access to clean water, an appropriate diet, access to shade and enrichment devices, an appropriate substrate on which to walk and lie down, and live in a cohesive social group, they are more likely to experience good welfare than if the opposite is true. However, using indirect measures of welfare may be misleading because although in theory an enriched environment is more likely to provide opportunities for better welfare than would a barren environment, the mere presence of an enrichment device does not ensure that the animals' lives are enriched.

The various welfare indices used to measure the wellbeing of animals have been reviewed by Mason and Veasey (2010b) – behavioural and cognitive responses, physiological responses and the potential negative effects of chronic stress on reproduction and health – and they have discussed the extent to which these are understood for elephants.

Perhaps the most basic aspect of animal behaviour that we can study in a zoo is how an animal spends its day. We can achieve this by constructing an activity budget that shows how much time it spends feeding, resting, walking and engaging in other behaviours. This is important for a number of reasons. For rare or illusive species we may know little about their behaviour in the wild, and studying their activities in captivity is the next best thing. Measuring the amount of time spent on particular behaviours may indicate whether or not an animal is experiencing good welfare and give keepers the information necessary to inform husbandry adjustments. If captive individuals are to be used for reintroduction programmes, determining their activity budgets may help scientists to decide whether or not they are behaviourally equipped to be released. For many species, baseline data on activity in zoos is not available, and where it is available comparisons between animals kept at different institutions or between animals kept in zoos and wild populations may be difficult due to methodological inconsistencies.

The use of bioacoustic information to assess animal welfare in zoos has been discussed by Schneider and Dierkes (2021). They used specialist software – Localize Animal Sound Events Reliably (LASER) – to localise animal sounds using cross correlation of the time differences of the arrival of sounds at two microphones in separate locations. Such sounds may provide information about the individual's sex, subspecies, reproductive state, social status, physiological state and welfare. Much of the study used recorded sounds, but it also included tests under real conditions using sounds made by giant otters (*Pteronura brasiliensis*) at Dortmund Zoo in Germany. The authors suggest that this software could be useful in monitoring animal welfare in zoos.

9.4.1 Measuring Stress

Experiencing stress of various types is part of normal life for animals living in the wild. The body usually copes with this by adapting and returning to its normal state. When it is unable to do this, the animal becomes distressed (National Research Council, 2008).

The threats of physical harm, restraint, infection, hunger and thirst are all stressors, but some stressors may not be unpleasant, for example exercise and sexual activity. The coping mechanisms elicited by stressors include behavioural reactions, the activation of the sympathetic nervous system and adrenal medulla, the secretion of hormones (such as glucocorticoids and prolactin) and the mobilisation of the immune system. A stress response may involve one or more of these responses. None by itself is sufficient to denote stress, and the absence of these responses does not necessarily indicate the absence of stress (Moberg, 2000). Many of the biochemical variables commonly used as measures of animal welfare reach extreme levels during play and mating, when the animals are unlikely to be experiencing impaired welfare (Knowles et al., 2014).

Considerable effort has been expended in the development of non-invasive methods of measuring stress in animals. Narayan et al. (2013) have developed a non-invasive method of measuring physiological stress from faecal samples using a faecal cortisol metabolite (FCM) enzyme immunoassay. The measurement of faecal glucocorticoids to assess stress in parrotfishes (Scaridae) kept in aquariums has been described by Turner et al. (2003), and the use of blood sugar level as an indicator of stress in the freshwater fish *Labeo capensis* has been discussed by Hattingh (1977). When 'resting samples' are taken to determine baseline levels, it is important to

Fig. 9.2 Spotted hyenas (*Crocuta crocuta*), Yorkshire Wildlife Park, England.

appreciate that stress indicators (e.g. concentration of serum aldosterone, plasma ACTH and cortisol) may exhibit diurnal variation (Schmitt et al., 2010).

One way of measuring the effect of captive conditions on animals is to compare aspects of their physiology with that of wild conspecifics. Captivity affects immunological responses in some terrestrial and marine mammals, but the extent to which this can be generalised to other taxa is unknown. Studies comparing the immune systems of wild individuals and those kept in zoos are rare. A study of spotted hyenas (*Crocuta crocuta*) found that concentrations of certain immunoglobulins were significantly lower in captive than in wild individuals, but no difference was found in the overall function of the constitutive innate immunity (Flies et al., 2015) (Fig. 9.2). Fair et al. (2017) demonstrated that various immune markers were expressed more in wild free-ranging common bottlenose dolphins (*Tursiops truncatus*) than in captive individuals. However, major acute phase proteins were variously expressed (Cray et al., 2013).

These studies suggest that the elevated immune response in free-ranging animals is the result of their increased exposure to pathogens in their natural environments. However, when concentrations of immune markers were compared in blood samples taken from free-living and captive plains zebra (*Equus quagga*) and mountain zebra (*E. zebra*) by Seeber et al. (2020) they concluded that captivity and lactation may influence immune functions in zebra mares.

9.4.2 Using Keepers' Knowledge to Assess Welfare

Increasing use is being made of the knowledge of keepers and other caregivers in assessing the welfare of animals living in zoos.

Animal welfare monitoring using keeper ratings as an assessment tool based on species-specific Welfare Score Sheets has been discussed by Whitham and Wielebnowski (2009). Úbeda et al. (2021) used questionnaires to study the welfare of 26 captive killer whales (*Orcinus orca*). Each animal was rated by an average of 12.5 raters and the mean inter-rater reliability for the subjective wellbeing and welfare questionnaires was high. In a study of the welfare issues affecting elephants living in zoos, Chadwick et al. (2017) consulted representatives of 15 elephant-holding facilities in the UK, along with other elephant experts.

An online questionnaire was used by Riggio et al. (2020) to collect information about zookeepers' perceptions of the welfare of canids in zoos ($n = 116$). Keepers who perceived that their zoo provided canids with access to 'Brambell's Five Freedoms' experienced higher job satisfaction than those that did not. Female keepers attached more importance to these freedoms than did male keepers, and the authors suggested that zoos should pay more attention to the psychological aspects of canid welfare.

9.4.3 Longevity as a Measure of Welfare

The term 'longevity' refers to the length of the life of an animal. This can only be determined after its death and will reflect, among other things, the environment in which it lived, including its husbandry, diet and exposure to stress.

A number of studies have inferred poor welfare from statistics on the longevity of animals living in zoos. Such studies can only be used to draw inferences about the past living conditions of animals, especially in long-lived species that may have been transferred between zoos – and even between circuses and zoos – many times during their lives. If an elephant dies in a new elephant exhibit at the age of 35 years, this may be a result of poor welfare experienced over the first 33 years of her life (perhaps in a circus or a zoo in a country where conditions in zoos are poorly regulated) rather than an effect of the welfare she experienced in the last two years before her death.

Clubb et al. (2008) analysed data from 4,500 elephants, and concluded that those living in zoos had half the median life span of conspecifics in protected populations in elephant range states. However, this analysis used survivorship curves that reflected the experiences of elephants in the past and cannot take into account the effect of the improved conditions that now exist in many modern zoos. Longevity may be affected by a number of factors in zoos, including injuries sustained from exhibits (Leong et al., 2004), poor adaptation to captivity or to a zoo's climate (Karstad and Sileo, 1971; Gozalo and Montoya, 1991) and increased perinatal mortality caused by inbreeding (Wielebnowski, 1996). It may also be affected by high levels of obesity in some species (Taylor and Poole, 1998) and the efficient spread of disease in the close quarters of captivity, within and between species in a zoo environment (de Wit, 1995; Ward et al., 2003).

Data on more than 50 species of mammals were examined by Tidière et al. (2016) to test the theory that animals live longer in zoos than in the wild. They found this to be true for 84 per cent of the species studied. However, long-lived species with low reproductive rates and low mortality in the wild benefited less than short-lived species with high reproductive rates and high mortality, partly because there are fewer opportunities for conditions in captivity to reduce mortality in long-lived species.

The database of the International Species Information System (ISIS) was used by Kohler et al. (2006) to construct life tables for 51 species held in zoos, mostly mammals. They examined 35,229 individual animals present in zoos between the beginning of 1998 and the end of 2003, including apes, small primates, carnivores, hoofstock, kangaroos, crocodilians, ratites and raptors. Female survivorship significantly exceeded that of males above age five years in most of the groups studied, but there was no significant difference between the mortality of wild-born and captive-born animals.

Some species live much longer in zoos than in the wild and others that did not previously do well in zoos now have a much greater life span due to improvements in husbandry. Wild Burchell's (plains) zebra (*Equus quagga*) have a mean expectation of life of about nine years, reflecting the high juvenile mortality rate of perhaps 50 per cent. In captivity this species may live to the age of 40 years (Nowak, 1999). Kitchener (2004) noted that 32 per cent of captive polar bears (*Ursus maritimus*) were more than 20 years old, compared with only 3 per cent in a well-studied wild population in Hudson Bay, Canada (Ramsey and Stirling, 1988). In the 1960s, gorillas in zoos had a maximum longevity of 33.4 years (Jones, 1962), but this had increased to 54 years 30 years later (Nowak, 1999). The longevity of Goeldi's monkeys (*Callimico goeldii*) over the same period increased from just 2.3 to 17.9 years.

9.5 The Opportunity to Exhibit Most Normal Behaviour

It is not possible for most zoos to provide all of the conditions required for most of their species to perform all of the behaviours that they would perform in the wild. Migratory species cannot be allowed to migrate; most predators cannot be allowed to kill their own prey. Some natural behaviours, however, should be easier to accommodate, and allowing animals to live in a naturalistic social group should be one of them.

Comparing the behaviour of animals kept in zoos with that of their wild conspecifics may be useful in highlighting potential welfare issues in some circumstances. In a comparative study of the agonistic behaviour of male Eastern grey kangaroos (*Macropus giganteus*) in captivity and in the wild, Höhn et al. (2000) found that agonistic behaviour was more frequent in the zoo situation, except for ritualised fights.

Activity budgets have been used by Veasey et al. (1996) as indicators of welfare in giraffes (*Giraffa camelopardalis*) by comparing 19 individuals kept in four zoos with those living wild in Hwange National Park, Zimbabwe. Although they found differences between the behaviour of wild and zoo-kept giraffes, Hutchins (2006) has cautioned against assuming that differences between the behaviour of wild and captive individuals of the same species is a consequence of poor welfare. In many species, behaviour varies considerably in the wild depending upon prevailing environmental conditions. The time spent on any particular activity may vary considerably between individuals of the same sex and individuals of different sexes. For example, between

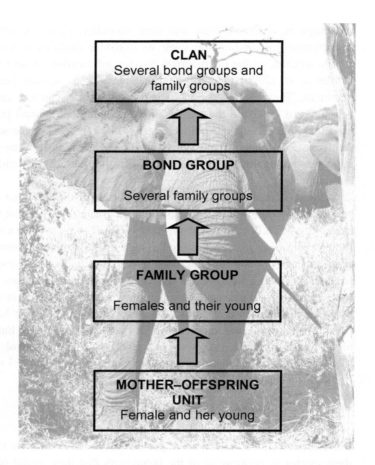

CLAN
Several bond groups and
family groups

BOND GROUP
Several family groups

FAMILY GROUP
Females and their young

**MOTHER–OFFSPRING
UNIT**
Female and her young

Fig. 9.3 Social structure of wild African elephants (*Loxodonta africana*) (source: first published in Rees (2013). Reproduced with permission from Wiley-Blackwell.)

10.00 and 14.00 an adult bull Asian elephant (*Elephas maximus*) at Chester Zoo in England spent on average 41.4 per cent of his time feeding, whereas the equivalent figure for the five adult cows was between 27.4 and 36.9 per cent (Rees, 2009a).

Keeping naturalistic social groups of some taxa in zoos is challenging because of their size or complex social structure. In the wild, African elephants (*Loxodonta africana*) have a complex social structure whereby mother–offspring units (a female and her young) form family groups. These combine to form bond groups and several bond groups and family groups combine to form a clan (Fig. 9.3). Zoos cannot hope to replicate this degree of social complexity and many have failed to maintain even stable family groups due to inadequate accommodation and the demands of cooperative breeding programmes (Rees, 2021).

In some cases it is possible to use animals kept in zoos in naturalistic social groups to examine the effects of housing the same species in unnatural groups in laboratory conditions. For example, when the development of play behaviours in chimpanzees (*Pan troglodytes*) reared in peer-only groups was compared with play development in a socially complex naturalistic group, little difference was detected (van Hooff et al., 1996).

9.6 Effects of Social Structure on Welfare

The presence and the absence of conspecifics can affect the welfare of animals kept in zoos. In social animals stress can result from unstable dominance hierarchies, high densities, intraspecific competition for resources and changes in group dynamics.

Methods of minimising social stress in captive bottlenose dolphins (*Tursiops aduncus*) have been discussed by Walpes and Gales (2002). They demonstrated a correlation between quantitative behavioural indices and physiological measures of stress and health. The authors presented evidence that social instability and the resultant aggression played a part in illness and mortality in dolphins and recommended that management should include the regular quantitative assessment of behaviour and associations and the maintenance of appropriate age and sex groupings.

A study of 14 Bornean orangutans (*Pongo pygmaeus*) living in a housing system at Apenheul Primate Park, the Netherlands, that facilitated a simulated fission–fusion social system similar to that found in the wild suggested that this system reduces the group size effect that leads to social stress in these animals (Amrein et al., 2014).

Alberts (1994) noted that in captive conditions, where dispersal is not possible, dominance hierarchies among male lizards often emerge in species that would otherwise be territorial. This may create social stress and prevent males of low status from reproducing. Alberts discussed the ramifications of this for exhibit design, the choice of animals kept together, the provision of resources (in space and time) and the orientation of enclosures within breeding facilities.

The loss or addition of an individual to a zoo-housed group has the potential to cause stress. Jain et al. (2021) found that the loss of her only companion was a significant source of stress for a female giraffe (*Giraffa camelopardalis*) at Lincoln Park Zoo, Illinois. Faecal hormone analysis was used to evaluate the introduction of a new adult female sable antelope (*Hippotragus niger*) to a resident individual at Lincoln Park Zoo (Loeding et al., 2011). Faecal glucocorticoid metabolites (FGM) were higher before than during or after the introduction and little aggression was observed between the two animals. The individual identified by keepers as the more dominant animal was the new antelope, and she had a higher mean FGM and faecal androgen metabolites (FAM) than the resident animal. Both animals were reproductively active throughout the year and the authors concluded that faecal hormone analysis can provide useful information to minimise the risk of aggression, injury and stress during introductions.

Altering social groups can have beneficial effects on some species. Bottlenose dolphins (*Tursiops*) that were split into groups and reunited or rotated between subgroups exhibited higher behavioural diversity than those managed in the same group (Miller et al., 2021). In these animals, their management was found to be more important than habitat size in ensuring good welfare.

The use of social network analysis (SNA) as a tool for the management of populations of social animals in zoos has been described by Rose and Croft (2015). They emphasise its value in helping to design social groupings that minimise social stress.

9.7 Stress and Reproduction

Social stressors may cause chronic stress in rhinoceroses and this may have an impact on the sustainability of zoo populations (Carlstead and Brown, 2005). In black rhinoceroses (*Diceros bicornis*) higher mean corticoid concentrations were found in zoos where the animals were more exposed to the public around the perimeter of their enclosures than in zoos where they were less exposed. In this species higher variability in corticoid secretion was correlated with higher fighting rates between breeding partners and higher mortality rates. Less fighting and lower corticoid variability were seen in pairs that were introduced when females were in oestrus than in pairs that were kept together every day. Noncycling white rhinoceroses (*Ceratotherium simum*) exhibited more corticoid variability than cycling individuals and the former showed higher rates of stereotypic pacing. High corticoid variability appeared to be an indicator of chronic or 'bad' stress. In this species, lower mean corticoid levels were found in animals that rated higher on a 'friendliness to keeper' scale.

Corticosterone was found to inhibit courtship behaviour in male rough-skinned newts (*Taricha granulosa*) when it was injected intraperitoneally in sexually active individuals, suggesting that exposure to stressful stimuli may inhibit sexual behaviours in amphibians (Moore and Miller, 1984). However, male red-sided garter snakes (*Thamnophis sirtalis parietalis*) subjected to four hours of capture stress exhibited no suppression of mating behaviour even though there was a significant increase in plasma corticosterone and a significant decrease in testosterone (Moore et al., 2000).

A study of the cause of the failure of the 15 Geoffroy's tamarins (*Saguinus geoffroyi*) at Cleveland Metroparks Zoo in Ohio to rear offspring successfully found no signs of stress in the animals, based on physiological indicators (Kuhar et al., 2003). However, those tamarins kept in the colony-housed condition exhibited high levels of aggressive and territorial behaviour compared with those in the non-colony housing, indicating social unrest which could have been contributing to poor reproductive success.

9.8 Maternal Deprivation

The need to move animals from their natal group for breeding purposes – often to another zoo – is one of the compromises zoos must impose on the social structures of many of their animals. Young animals are often separated from their parents in zoos much earlier than would be the case in the wild. Parenting behaviour and the effects of hand-rearing are important areas of research, especially in relation to the breeding of rare species (e.g. Bloomsmith et al., 2003; Kreger et al., 2004).

A study of the effect of early experience on adult copulatory behaviour in zoo-born chimpanzees found that the animals least likely to copulate as adults were those that had been hand-reared in the total absence of conspecifics (King and Mellen, 1994). A study of 117 gorillas living in 17 zoos showed that wild-caught and captive-born, hand-reared gorillas exhibited a higher incidence of regurgitation and reingestion behaviours than that of mother-reared infants (Gould and Bres, 1986).

The link between early separation from the mother and the development of stereotypic behaviour has been discussed by Latham and Mason (2008). Maternal deprivation may occur due to the removal of young from the mother earlier than would occur in the wild or as a result of inadequate maternal care by an inexperienced mother or the existence of a restrictive environment that limits normal maternal behaviours.

Rearing history may affect the behaviour of animals later in life. Hand-reared sloth bears (*Ursus ursinus*) have been found to show significantly higher frequencies of stereotypic and self-directed behaviours than did mother-reared animals (Forthman and Bakeman, 1992).

Some of the effects of a failure to allow for adequate maternal care in mammals living in zoos, and various management strategies that can be used to promote adequate and appropriate care behaviours, have been discussed by Baker (1994).

9.9 Personality and Welfare

The welfare of individual animals is inextricably linked to their personalities. A study of chimpanzees at Edinburgh Zoo, Scotland, found that individuals who experienced positive welfare were happier, extraverted and emotionally stable (Robinson et al., 2017).

The welfare of captive killer whales (*Orcinus orca*) was assessed using a multi-trait questionnaire by Úbeda et al. (2021). They found that personality was related to welfare and subjective wellbeing in this species and concluded that questionnaires are a reliable and valid tool for assessing killer whale welfare.

The coping styles of fish have been discussed by Castanheira et al. (2017) in relation to differences in their physiological and behavioural responses to stressors. Although this study was concerned with farmed fish, they nevertheless considered the effects on disease susceptibility, health and welfare.

Tetley and O'Hara (2012) have reviewed the use of ratings of animal personality as a tool for improving the breeding, management and welfare of mammals living in zoos. They found that zoo animal personality is commonly assessed using observer ratings of personality traits determined by people who know the animals. The authors concluded that zookeepers were able to rate these traits reliably, that these ratings were valid and related to behaviour, and they advocated the integration of validated personality assessment questionnaires into existing management practices in zoos to inform decisions on captive breeding and animal welfare.

9.10 Condition Scoring

The physical condition of an animal's body may be represented by a single number called a body condition score. Condition scoring is widely used with domestic animals (cattle: Edmonson et al., 1989; horses: Dugdale et al., 2012; sheep: Pollott and Kilkenny, 1976; pigs: Chikwanha, 2007; donkeys and mules: Burden, 2012; laying hens: Gregory and Robins, 1998; dogs: Dorsten and Cooper, 2004; cats: Bjornvad et al., 2011). A health condition profile has been developed for salmon (Novotny and Beeman, 1990).

Condition scoring is increasingly being used to assess the welfare of animals living in zoos. Condition scores are species-specific and independent of weight, and should take into account the individual's age, individual differences (e.g. natural variation in size) and seasonal variation. Methodology varies between and within species, but generally visual assessments of muscle and fat reserves are made by individually scoring different parts of the body with reference to drawings or photographs and descriptions. Changes in body condition score may indicate a health or nutritional problem.

Condition scores have been devised for a number of taxa found in zoos, including rhinoceroses (greater one-horned rhinoceros (*Rhinoceros unicornis*): Heidegger et al., 2016: black rhinoceroses (*Diceros bicornis*): Reuter and Adcock, 1998), primates (rhesus macaques (*Macaca mulatta*): Summers et al., 2012), ungulates (African buffalo (*Syncerus caffer caffer*): Ezenwa et al., 2009: Bornean banteng (*Bos javanicus lowi*): Prosser et al., 2016; red deer (*Cervus elaphus*): Audigé et al., 1998), birds (Magellanic penguins (*Spheniscus magellanicus*): Clements and Sanchez, 2015) and reptiles (African side-neck turtles (Pelomedusidae): Rawski and Józefiak, 2014) (Fig. 9.4). The use of body condition scoring and a body condition index in corn snakes (*Pantherophis guttatus*) have been discussed by Gimmel et al. (2021).

It may be useful to compare condition scores of animals living wild and those of the same species living in zoos. An image-based body condition score for giraffes (*Giraffa camelopardalis*) has been used by Clavadetscher et al. (2021) to compare the condition of 232 zoo-housed and 532 free-ranging individuals. Using this method the authors concluded that zoo-housed giraffes were less constrained by dietary resources than their free-ranging counterparts, and they attributed this to improved diets in European zoos.

More than one condition scoring system exists for some taxa. Wemmer et al. (2006) developed a method for measuring the body condition of Asian elephants (*Elephas maximus*) that used visual assessment to assign numerical scores to six different regions of the body, which were then totalled to give a numerical index ranging from 0 to 11. Other systems have been devised for elephants by Mikota (2006), Fernando et al. (2009), Morfeld et al. (2014) and Wijeyamohan et al. (2015).

Fig. 9.4 (a) Black rhinoceros (*Diceros bicornis*) and (b) banteng (*Bos javanicus*) are two of a number of species for which condition scoring systems are available. (A black and white version of this figure will appear in some formats. For the colour version, please refer to the plate section.)

Some welfare assessments consider coat and tail condition (e.g. Berg et al., 2009 for ring-tailed lemur (*Lemur catta*)). Dorsal hair loss occurs in some captive primates as a result of excessive hair-pulling or over-grooming by cage-mates, and is considered to be associated with stress. Honess et al. (2005) developed an alopecia scoring system to measure hair loss in captive rhesus macaques (*Macaca mulatta*) as a method of welfare assessment.

Some individuals of some species may become obese in zoos. This topic is discussed in Section 13.3.4.

9.11 Abnormal Behaviours

9.11.1 Introduction

Normal behaviour helps animals to maintain homeostasis by allowing them to control and modify their environments. Animals living in zoos and aquariums display a wide range of abnormal behaviours.

9.11.2 The Range of Abnormal Behaviours

The abnormal behaviours observed in animals in zoos and aquariums fall into several categories: repetitive behaviours not observed in the wild; alterations in the frequency of natural behaviours (e.g. time spent active); failure to develop and exhibit natural behaviours (e.g. maternal behaviour); increases in aggression (including infanticide); abnormal sexual behaviour; and self-mutilation. Abnormal repetitive behaviours (ARBs) may be divided into stereotypies and impulsive/compulsive behaviours (Garner, 2005). Stereotypic route-tracing is common in captive animals and consists of repeatedly travelling along the same route, for example a path around the periphery of an enclosure. This type of behaviour is discussed in Section 8.9.4.

The extent to which chimpanzees (*Pan troglodytes*) living in zoos exhibit abnormal behaviour – that is, behaviour not typical of their wild counterparts – has been examined by Birkett and Newton-Fisher (2011). They examined 40 socially-housed chimpanzees living in six zoos located in the United States and the UK. They found that, despite enrichment efforts, all of the individuals exhibited some abnormal behaviour (Table 9.1) and individual differences could not be explained by sex, age, rearing history or prior housing conditions. Although most of the behaviour observed was 'normal', the authors concluded that abnormal behaviour is endemic in the zoo chimpanzee population and there is an urgent need to understand how the chimpanzee mind copes with captivity.

Although Birkett and Newton-Fisher (2011) identified 37 abnormal behaviours in their study of 40 chimpanzees, a study of 51 bonobos (*Pan paniscus*) living in six zoos identified just 13 abnormal behaviours (Laméris et al., 2021). Wild-born bonobos exhibited a greater diversity of abnormal behaviours compared with those that were mother-reared. Separate abnormal behaviours were influenced by sex, rearing history

Table 9.1 Percentage of chimpanzees ($n = 40$) exhibiting abnormal behaviours held in social groups in six zoos in the United States and the UK.

Abnormal behaviour	%	Abnormal behaviour	%	Abnormal behaviour	%
Eat faeces	83	Jerk	13	Spit	5
Pluck hair	58	Toss head	13	Twirl	5
Rock	53	Bang self against surface	10	Twitch body-part	5
Groom stereotypically	50	Configure lips	10	Bite–hit–lick combination	3
Pat genitals	38	Display to human	10	Floating limb	3
Poke anus	30	Stimulate self stylised	10	Groom stereotypically with tool	3
Regurgitate	30	Bounce	8	Move hand stereotypically	3
Fumble nipple	23	Clasp self	8	Pinch self	3
Pluck hair other	20	Drink urine	8	Poke eye	3
Bit self	20	Eat faeces other	8	Rub hands	3
Hit self	18	Incest	5	Touch urine stream	3
Clap	15	Pace	5	Walk on object	3
Manipulate faeces	15				

Source: based on Birkett and Newton-Fisher, 2011.

and personality. Laméris et al. noted that not all abnormal behaviours are indicators of poor welfare and some behaviours classified as 'abnormal' in captivity have been observed in the wild, such as coprophagy.

9.11.3 Coprophagy

Coprophagy – the eating of faeces – is a normal element of feeding behaviour in some species (e.g. rodents and rabbits) but considered abnormal in others. It was recorded in 83 per cent of the 40 zoo chimpanzees studied by Birkett and Newton-Fisher (2011) and was the most frequently recorded abnormal behaviour observed in these animals. Coprophagy has also been recorded in captive gorillas (*Gorilla gorilla*; Ackers and Schildkraut, 1985) (Fig. 9.5), African elephants (*Loxodonta africana*) (observed by the author during the measurement of assimilation efficiency (Rees, 1982)), brown capuchin monkeys (*Cebus appella*; Prates and Bicca-Marques, 2005) and bonobos (*Pan paniscus*; Laméris et al., 2021). Boyd (1988) found that copography in captive Przewalski's horses (*Equus ferus przewalskii*) was eliminated by the presence of forage *ad libitum* and this also greatly reduced the amount of pacing observed.

Although coprophagy is considered abnormal in zoos, it occurs in the wild in some species not normally considered coprophagous, for example, mountain gorillas (*G. beringei beringei*) (Harcourt and Stewart, 1978) and chimpanzees (*P. troglodytes*) (Krief et al., 2004). Leggett (2004) observed desert-dwelling African elephants (*Loxodonta africana*) in Namibia eating their own faeces and engaging in two other rare thermoregulatory behaviours that might be classified as abnormal if recorded in captivity: cooling their bodies with water extracted from their pharyngeal pouch and urinating on sand before throwing it onto their bodies.

Fig. 9.5 Coprophagy in a male Western lowland gorilla (*Gorilla g. gorilla*).

9.11.4 Self-Injurious Behaviours

Self-injurious behaviours (SIBs; self-mutilation) observed in animals living in zoos include feather-plucking, over-preening, hair-pulling and self-biting.

Self-injurious behaviours observed among primates in British and Irish zoos have been examined using a questionnaire survey by Hosey and Skyner (2007). Of 42 zoos that kept primates, responses were received from 35 (83 per cent), with SIBs reported in 16 species: 14 incidences in males and 10 in females. Based on the descriptions provided by the zoos, the authors divided SIBs into two groups: self-biting and hair-pulling. Self-biting was recorded in 2.0 per cent of individuals and hair-pulling in 1.8 per cent. Hosey and Skyner concluded that the incidence of SIBs in British and Irish zoos was low.

Bollen and Novak (2000) reported abnormal behaviour in 497 individuals from 68 primate species in North American zoos. This represented 14 per cent of all individuals, but this study is not comparable with that of Hosey and Skyner because it included other abnormal behaviour in addition to SIBs.

Feather-plucking is common in captive birds, especially parrots. This phenomenon has been described, for example, in Lear's macaw (*Anodorhyncus leari*) by Azevedo et al. (2016) and in golden parakeets (*Guaruba guarouba*) by Clyvia et al. (2015).

9.11.5 Stereotypic Behaviours

Stereotypic behaviours are morphologically similar patterns or sequences of behaviour performed repeatedly with no obvious function (Ödberg, 1978). They often appear when an animal is under stress, and the measurement of stereotypic behaviour is considered to be one of the best validated measures of animal welfare (Maple and Perdue, 2013). Reducing or eliminating this behaviour is generally considered to be associated with improved welfare, although some have argued that engaging in stereotypic behaviour may be a coping strategy in some individuals and that they may experience better welfare than those individuals with no coping strategy (Rushen, 1993; Wechsler, 1995). Shyne (2006) suggested that:

Stereotypies often arise when a captive animal has prolonged exposure to an ecologically relevant problem that it is incapable of solving within its enclosure.

Stereotypic behaviours include bar-biting (Fig. 9.6), tongue-playing, neck-twisting, head-bobbing, excessive grooming, swaying (Fig. 9.7) and repetitive pacing. Stereotypic behaviour has been widely reported in animals kept in zoos (e.g. Rees, 2009a), but also in farm and companion animals (e.g. birds (van Hoek and Ten Cate 1998)).

Oral stereotypies are common in giraffids, and in captive giraffes (*Giraffa camelopardalis*) may be the result of their inability to complete tongue manipulation, a highly motivated feeding behaviour pattern (Fernandez et al., 2008) (Fig. 9.6). Many species (e.g. lions, rhinoceroses, giant pandas, elephants) can be observed exhibiting stereotypic pacing towards, away from and back towards the door to their indoor accommodation at the end of the day prior to being fed indoors, and at other times. This is likely to be the result of the thwarting of highly motivated food acquisition behaviours. Stereotypic pacing is common before feeding time (e.g. Rees, 2002; 2009), but Carlstead and Seidensticker (1991) noted that pacing in a black bear was

Fig. 9.6 Giraffids are prone to developing oral stereotypies in zoos. Okapi (*Okapia johnstoni*) (a) and giraffe (*Giraffa camelopardalis*) (b). (A black and white version of this figure will appear in some formats. For the colour version, please refer to the plate section.)

Fig. 9.7 Greater one-horned rhinoceros (*Rhinoceros unicornis*) exhibiting stereotypic behaviour at the door to its indoor accommodation. Note the position of the head as the animal sways from side to side. (A black and white version of this figure will appear in some formats. For the colour version, please refer to the plate section.)

performed mainly around feeding time from August to November but mainly after feeding during the period May–July.

Asian elephants (*Elephas maximus*) at Chester Zoo, England, were observed exhibiting various types of stereotypic behaviour near the door of the elephant house prior to being allowed inside, where their main meal of the day was provided (Rees, 2009a). Feeding was the dominant activity at the beginning of the day (10.00) – because the animals were provided with food in their outdoor enclosure – but this was progressively replaced by stereotypic behaviour until this was the dominant activity by around 15.00. The proportion of time these five adult female Asian elephants spent exhibiting stereotypic behaviour was found to be negatively correlated with the proportion of time they spent feeding. This opens up the possibility that increasing feeding opportunities for elephants living in zoos could reduce the time spent engaged in stereotypic behaviour. The same elephants exhibited a negative correlation between the frequency of stereotypic behaviour and the maximum daily temperature, presumably in response to physiological stress (Rees, 2004a).

Stereotypic behaviour may be under-reported from zoos because some take the view that publicising that they have behavioural problems in some of their animals may attract the unwanted attention of animal rights and anti-zoo activists.

For most species the aetiology of ARBs in captive birds is little understood and has been the subject of relatively little research. What is known about these behaviours in birds has been discussed by Mellor et al. (2018).

Great care should be exercised in interpreting reports of abnormal behaviours observed in zoos for two reasons. First, studies often involve small numbers of animals at a single zoo (e.g. Azevedo et al., 2016). Second, 'abnormal' is difficult to define and the definition depends largely on the range of behaviours that have been recorded in the wild for any particular species. The manner in which the incidence of a particular behaviour is reported may affect the reader's perception of the situation. An animal welfare researcher may report that 25 per cent of a group of chimpanzees exhibited stereotypic behaviour, while the same data could be described by a researcher who is not primarily concerned with welfare by reporting that 75 per cent exhibited normal behaviour (or at least did not exhibit stereotypic behaviour). In any event, reporting the number of animals that exhibit abnormal behaviour tells us nothing about its frequency in any particular individual.

9.11.6 Alleviation of Stereotypic and Other Abnormal Behaviours

Stereotypic behaviours are common in animals living in captive environments, including laboratories, farms and zoos, and they are often very difficult to extinguish. However, studies have demonstrated that for some individuals of some species it is possible to alleviate some stereotypies.

A 12-year-old male brown bear (*Ursus arctos*) was rescued by the Karacabey Bear Sanctuary in Turkey after living with villagers as a pet for 10 years. After initially stereotypic pacing throughout the day this behaviour was extinguished following treatment with fluoxetine (an antidepressant) and relocation to a naturalistic enclosure that provided him with additional space and novel stimulation (Yalcin and Aytug, 2007). Fluoxetine was also used to successfully treat stereotypic pacing in an adult female polar bear (*Ursus maritimus*) who had been born in a zoo (Poulsen et al., 1996).

Fluoxetine has been shown to be effective in treating stereotypic behaviour other than pacing. Hugo et al. (2003) have demonstrated that it significantly reduces saluting, somersaulting, weaving and head tossing in vervet monkeys (*Chlorocebus pygerythrus*).

Training using positive reinforcement may reduce stereotypic behaviours in some species. Female rhesus macaques (*Macaca mulatta*) with a history of stereotypy (e.g. pacing, bouncing and somersaulting) were divided into an experimental group ($n = 6$) and a control group ($n = 5$) (Coleman and Maier, 2010). The animals in the experimental group were trained to touch a target and accept venipuncture; those in the control group were not so trained. The majority of the monkeys ($n = 4$) engaged in less stereotypic behaviour after training compared with baseline frequencies. The authors of the study concluded that training may be an effective means of reducing stereotypic behaviour in some individuals.

A study of stereotypic pacing in a single male American black bear (*Ursus americanus*) throughout the year found that this could be reduced slightly by placing bear odours in the enclosure and if small items of food were hidden in the fall (autumn), pacing was completely replaced by foraging (Carlstead and Seidensticker, 1991). The authors concluded that pacing was the result of the inability to perform mate-seeking behaviour predominating in the late spring and foraging behaviour in the late summer and fall.

A study of the effect of environmental enrichment on the frequency of abnormal behaviours (pacing and feather-plucking) in a pair of Lear's macaws (*Anodorhyncus leari*) at Belo Horizonte Zoo, Brazil, found that enrichment caused a significant decrease in abnormal behaviours in the male (Azevedo et al., 2016). The enrichment items consisted of bamboo forest, coconuts, grapes, pumpkins filled with hazelnuts, hazelnuts wrapped in banana tree leaves, corn on the cob, parrot sticks and cardboard boxes filled with grass and coconut. Only one item was offered each day.

9.12   Intraspecific Aggression

Intraspecific aggression occurs in the wild and in captive environments. However, in their natural habitats individuals may be driven out of social groups or leave of their own volition, thereby avoiding the worst consequences of aggression. It is another matter in captive environments, where aggressive individuals may need to be separated from others and occasionally culled to protect other members of the social group.

A study of 119 Japanese macaques (*Macaca fuscata*) living in 10 zoos recorded 1,007 wounds over a period of 24 months (Cronin et al., 2020). Most wounds occurred on the face, and females were more likely to be injured than males. This sex difference was more pronounced during the breeding season. On average, individuals received 4.67 (SEM = ± 0.55) wounds per year and 77.3 per cent of the population received at least one wound, although most wounds were superficial and did not require veterinary treatment. Macaques living in larger social groups on average received more wounds than those in smaller groups.

Intraspecific killings are occasionally reported from wild populations. Farhadinia et al. (2018) reported four cases of intraspecific killing among wild Persian leopards (*Panthera pardus*) in Iran, and Toni (2017) provided the first published record of a single incident of intraspecific killing in Eastern grey kangaroos (*Macropus giganteus*).

Such killings also occur in zoos. Nath and Chakraborty (2013) reported death due to traumatic injury and stress in a rhesus macaque (*Macaca mulatta*) resulting from a fight with a conspecific in the enclosure at Assam State Zoo, India. Deaths in zoos caused by intraspecific aggression have also been reported for collared peccary (*Dicotyles tajacu*; Lochmiller and Grant 1982), Eurasian lynx (*Lynx lynx*; Heaver and Waters, 2019); spider monkeys (*Ateles*; Davis et al., 2009) and African elephants (*Loxodonta africana*: BBC, 2021).

In lion tamarins (*Leontopithecus*), aggressive encounters within social groups are severe and resulted in an individual's death in 20.5 per cent of such interactions reported in a global survey (Inglett et al., 1989). Lamglait (2020) found that trauma was the most common cause of death among adult male springboks (*Antidorcas marsupialis*) at the Réserve Africaine de Sigean, France, mainly due to aggression during the breeding season. Slow lorises (*Nycticebus*) are venomous and bites inflicted by conspecifics are the major source of death in captive lorises (Nekaris et al., 2020).

An analysis of pink pigeons (*Nesoenas mayeri*) in European collections (1977–2018) found that the most common cause of mortality in adults was trauma (43.1 per cent) and that death resulting from intraspecific aggression was significantly higher in males than in females (Shopland et al., 2020). Intraspecific killing was a major cause of death among black storks (*Ciconia nigra*) in collections in Europe (King, 1994). Death as a result of trauma caused by conspecifics was recorded in subadult ostriches (*Struthio camelus*) and emus (*Dromaius novaehollandiae*) at the Réserve Africaine de Sigean in records from 1974–2015 (Lamglait, 2018).

There is evidence that the presence of visitors causes an increase in aggression in some primates. A study of 15 primate species found that they were more aggressive, more active and less affiliative in the presence of visitors (Chamove et al., 1988). This was particularly the case in arboreal species, especially smaller species. These effects were reduced by 50 per cent by lowering the height of the visitors by making them crouch down.

Hosey et al. (2016) studied chimpanzees (*Pan troglodytes*) and ring-tailed lemurs (*Lemur catta*) in three zoos and found no evidence that high visitor numbers resulted in increased woundings in these species.

The chemical control of aggression in animals living in zoos is discussed in Section 11.3.3.

9.13   Noise and Welfare

Animals living in zoos may be exposed to anthropogenic noise from visitors, construction work and, in some cases, funfair rides.

A four-year study of one female and one male giant panda (*Ailuropoda melanoleuca*) found that ambient noise, although it affected behaviour and resulted in increased excretion of glucocorticoids, caused no substantive detrimental effects on either wellbeing or reproduction (Owen et al., 2004). However, the female appeared especially sensitive to noise during oestrus and lactation.

A study of the effects of ambient noise on the maternal behaviour of a Bornean sun bear (*Helarctos malayanus euryspilus*) in a zoo predicted a negative effect but, in fact, the mother spent significantly more time with her cub on noisy days (Owen et al., 2013). She spent significantly less time feeding on noisy days and there was some evidence that the cub was more vocal on these days. The response of this species to ambient noise suggests it may increase energetic costs in the postpartum period.

Sulawesi crested macaques (*Macaca nigra*) housed in five zoos exhibited increased activity and vigilance with increases in visitor numbers and louder noise (Dancer and Burn, 2019) (Fig. 9.8).

An experiment conducted on the orangutans (*Pongo pygmaeus*) at Chester Zoo in England, in which visitor groups were told to behave either quietly or noisily (to

Fig. 9.8 In Sulawesi crested macaques (*Macaca nigra*) activity and vigilance increase with visitor numbers and louder noise.

eliminate the effect of group size), found that when confronted with noisy groups all animals looked more at the visitors and infants approached and held onto adults more (Birke, 2002).

The effect of visitors on the behaviour of six ebony langurs (*Trachypithecus auratus*) was studied using a 'visitor impact score' which combined measurements of visitor numbers, visitor activity and noise level (Roth and Cords, 2020). Higher impact scores predicted greater expression of displacement activities, affiliative behaviours (which may help to reduce stress) and aggression. Langurs also spent more time sleeping on high impact score days, possibly indicating learned helplessness.

Cronin et al. (2018) examined the effect of an annual event involving loud jets flying overhead on the mood of Japanese macaques (*Macaca fuscata*), chimpanzees (*Pan troglodytes*) and Western lowland gorillas (*Gorilla g. gorilla*), and found that the macaques underwent detectable affective changes in response to the loud event but the apes did not.

The potential impacts of future construction noise on selected zoo animals (alligator, emu, giraffe and elephant) were examined using recorded noise (Jakob-Hoff et al., 2019). The purpose of the study was to detect any aversive responses so that keepers could take mitigating action during the construction period. The emus, giraffes and elephants showed increases in behaviour associated with stress or agitation when exposed to loud sounds and may prefer quieter areas of their enclosures at these times.

A study of two giraffes at Lincoln Park Zoo, Illinois, found that disturbances caused by construction of a new exhibit near the giraffe enclosure were a source of

Fig. 9.9 Snow leopards (*Uncia uncia*) at Basel Zoo, Switzerland, responded to construction noise by spending more time resting and withdrawing to remote parts of their exhibit.

stress but affected one of the individuals more than the other (Jain et al., 2021). Sulser et al. (2008) found that snow leopards (*Uncia uncia*) at Basel Zoo, Switzerland, responded to construction noise by spending more time resting and withdrawing to remote parts of their exhibit (Fig. 9.9).

The use of animal monitoring (keepers' daily assessments), sound measurement and noise reduction in the management of zoo animals exposed to noise has been described by Orban et al. (2017). The work included an analysis of the efficacy of several sound-reducing barriers in a study of a single female giant anteater (*Myrmecophaga tridactyla*) (Fig. 9.10).

A study of the effects of visitor-generated noise on the welfare of 12 mammal species housed at Belo Horizonte Zoo, Brazil, was conducted by Quadros et al. (2014). No overall effect of noise on behaviour was found, but half of the individual animals increased vigilance behaviour as sound levels increased and about one-third increased their movements. The authors concluded that zoo visitors had a negative effect on individual animals, especially when noise levels exceeded 70 dB(A), the recommended limit for human wellbeing.

Hashmi and Sullivan (2020) examined the effect of visitor numbers and noise on primates: six gorillas (*Gorilla g. gorilla*), five orangutans (*Pongo pygmaeus*) and four gibbons (*Hylobates pileatus*). They found a range of visitor effects on these animals, from no effect to detrimental, with individual and species differences in reactions to the public. Both noise levels and visitor numbers had significant positive and negative effects on stereotypic behaviours, locomotion, inactivity and feeding behaviours. The authors of the study noted that increases in stereotypic and clinging behaviours and a decrease in activity suggested negative impacts on the welfare of

Fig. 9.10 Giant anteater (*Myrmecophaga tridactyla*).

these primates. The mixed results of the study suggested that differences in husbandry and personality moderated visitor effects and highlighted the need for off-show areas.

The behaviour of white-handed gibbons (*Hylobates lar*) in two Canadian zoos was affected by noise level and group size, particularly with respect to communicative behaviours (looking at visitors and open-mouth display) and locomotive behaviours (bipedal walking, brachiating and hanging) (Cooke and Schillaci, 2007). Behavioural changes associated with the presence of visitors, or high numbers thereof, have also been recorded in cotton-top tamarins (*Saguinus oedipus*) (Chamove et al., 1988) (Fig. 9.11).

There has been much recent concern about the effects of noise on marine mammals (e.g. Houser et al., 2020; Stevens et al., 2021). Studies of captive animals have contributed to our understanding of these effects. The possible effects of noise from drilling platforms on the behaviour and physiology of four captive belugas (*Delphinapterus leucas*) were examined using recordings from a semi-submersible drilling platform (SEDCO 708) (Thomas et al., 1990). No short-term behavioural or physiological effects of drilling noise were detected.

When seahorses (*Hippocampus erectus*) were experimentally exposed to loud noise in aquarium tanks, they exhibited stress responses at behavioural and physiological

Fig. 9.11 The behaviour of cotton-top tamarins (*Saguinus oedipus*), and many other primates, is affected by the presence of visitors.

levels and experienced costs to growth, condition and immune status (Anderson et al., 2011).

Studies of the effect of noise on animals in zoos generally involve small numbers of individual animals, mostly of mammalian species, and the measurement of a limited range of behaviours and physiological measurements of stress. There is individual variation in the response to visitor noise within and between species so there is no clear-cut effect on animal welfare. Furthermore, there is some evidence that visitor effects may be overestimated.

Few of the studies of the effects of visitors on animal behaviour have examined the effect of time of day and weather as variables. A study of ring-tailed lemurs (*Lemur catta*) in a walk-through exhibit at West Midland Safari Park, England, found that time, weather and visitor variables interacted in complex ways, but time and weather exerted the strongest influence on behaviour (Goodenough et al., 2019). Analysis indicated that visitors accounted for about 20 per cent of behavioural variation but this fell to about 6–8 per cent when the effects of time and weather were included. The results of this study suggest that visitor effects in zoos may be overestimated because behaviour may be mainly driven by co-variation with time and weather.

Hosey et al. (2020b) examined whether higher numbers of visitors at weekends had any effect on parturition in 16 mammalian species (artiodactyls, perrisodactyls, carnivores and primates) at four British zoos. They found that births were randomly distributed throughout the week and concluded that the presence of elevated visitor numbers at weekends were not sufficiently stressful to cause delayed parturition.

9.14   Transportation and Welfare

Animals are routinely moved between zoos as part of cooperative breeding programmes by land, air and sea, sometimes in simple wooden crates. In the 1960s an elephant transported between India and the UK might have been lifted onto a cargo ship by suspending it from a crane using a simple cradle placed between its front and rear legs. On arrival at the London Docks elephants were sometimes walked to London Zoo under police escort (Knight, 1967).

Modern-day animal transportation puts great emphasis on training large animals to enter transportation crates and reducing transport stress to a minimum. It was not always so. A drawing in *The Graphic* in 1872 showed the transfer of a 'hairy rhinoceros' (Sumatran rhinoceros (*Dicerorhinus sumatrensis*)) from her transportation crate – 'travelling den' – to her cage at London Zoo, in which teams of men can be seen pulling on ropes fixed to the chains attached to her feet (*The Graphic*, 1872) (Fig. 9.12).

TRANSFERRING THE HAIRY RHINOCEROS FROM HER TRAVELLING DEN TO HER CAGE

Fig. 9.12 Transferring the hairy rhinoceros (Sumatran rhinoceros (*Dicerorhinus sumatrensis*)) to her cage at London Zoo in 1872 (source: *The Graphic*, 2 March 1872).

When London Zoo sold *Jumbo* to the American showman P. T. Barnum, the elephant had to be crated for shipment across the Atlantic Ocean to the United States. On 25 February 1882 London newspapers printed drawings depicting *Jumbo*'s life at the zoo and attempts to persuade him into a wooden crate while bearing chains around the top of his trunk and back, and around his legs (*The Graphic*, 1882b; *The Illustrated London News*, 1882). After being unable to persuade – or force – the elephant into the crate on Saturday 18 February, or to walk him through the streets to the docks the next day, it was decided to leave 'The Box' at the entrance to the elephant house for a fortnight so that he would be forced to pass through it and in so doing become accustomed to it (Fig. 9.13a).

Transportation can cause stress in animals. This may be related to capture and handling procedures, loading onto vehicles and the animals' experiences during vehicle movement, such as exposure to noise, vibration, overcrowding and inter-actions with other animals. Transport stress has been widely studied in farm animals, especially cattle, sheep and pigs (Knowles et al., 2014). A number of studies have examined zoo animal transportation, especially in relation to identifying means of alleviating psychological and physiological stress.

Dembiec et al. (2004) simulated the experience of movement between zoos by transporting five tigers (*Panthera tigris*) in a small transfer cage for 30 minutes and then returning them to their original enclosure. Two of the tigers had previous experience of the procedure and the other three were naïve. The study demon-strated that even short-term transportation procedures can cause significant increases in immune-reactive (IR) cortisol concentration in tigers. Levels remained elevated longer in naïve individuals (9–12 days) than in the experienced tigers (3–6 days). The naïve tigers also exhibited a greater intensity of behavioural indicators of stress after transportation (e.g. faster pacing). The authors concluded that prior exposure to elements of the transport procedure may lead to a degree of habituation and therefore reduced stress. However, habituation does not always take place. Cortisol levels in cattle did not decrease with experience when they were subjected to repeated truck journeys during which they fell down (Fell and Shutt, 1986).

Circus elephants were once routinely transported by road and rail in the United States. The environmental conditions pertaining inside rail cars and trailers and their effects on elephants while they were being transported were examined by Toscano et al. (2001). They found no evidence of hypothermia or hyperthermia, even in extreme climatic conditions, and that ammonia and carbon dioxide levels were always below detectable levels. The authors concluded that the elephants' welfare was not compromised provided that care was taken in their handling.

Modern methods of elephant transport require the use of specialised vehicles in which the internal climate is controlled by a heating, ventilation and air-conditioning (HVAC) system, and the animals are monitored by keepers and veterinary staff using CCTV as they travel (Fig. 9.13b).

The transportation of marine mammals poses particular challenges for zoos. Dolphins are frequently transported on mattresses because there is a risk of damage

Fig. 9.13 (a) Staff at London Zoo attempting to put the famous African elephant *Jumbo* into a wooden crate in 1882 so that he could be shipped to the United States (source: *The Illustrated London News*, 25 February 1882, p. 200). (b) A truck carrying an elephant crate connected to a heating, ventilation and air-conditioning (HVAC) system that controls the environment. The vehicle is owned and operated by Stephen Fritz of Arizona (courtesy of Stephen Fritz).

to internal organs, especially the lungs, if their weight is not supported. The use of high-performance mattresses – similar to those used in human nursing care to prevent pressure ulcers – provides cardiopulmonary benefits to Indo-Pacific bottlenose dolphins (*Tursiops aduncus*), which had lower breathing rates, lower heart rates and higher exhaled CO_2 concentrations compared with using standard mattresses (Suzuki et al., 2008).

Few studies have examined the effect of transport stress on birds. A controlled study of the effect of transport, restraint and common clinical procedures on an established flock of 18 Hispaniolan Amazon parrots (*Amazona ventralis*) found that some significant haematological changes may arise as a result of routine handling and transportation (McRee et al., 2018). In the treatment group, white blood cell, heterophil and eosinophil counts increased over time, as did the heterophil to lymphocyte ratio, but no such changes were observed in the control group. The treatment group exhibited a mean corticosterone level that was approximately 60 per cent higher than the control group.

Wild-caught chukar partridges (*Alectoris chukar*) were exposed to the stress of capture and long-distance transportation. In the first few days of captivity the birds lost weight, had lowered haematocrit values and demonstrated changes in corticosterone concentrations. The process of introducing these birds into captivity put them into a temporary state of chronic stress from which they began to recover after nine days (Dickens et al., 2009).

A study of the effects of handling and transportation of tortoises (*Testudo hermanni*) found that circulating cortisol concentrations were 286 per cent higher after handling and transport, and homeostasis was not restored four weeks later (Fazio et al., 2014). Simulated transport of Eastern blue-tongued lizards (*Tiliqua scincoides*) found that cold temperatures are potentially noxious for these animals because they reduce activity and increase escape attempts (Mancera et al., 2014).

The stress of capture and transportation produces a marked increase in blood sugar in the freshwater fish *Labeo capensis* (Hattingh, 1977). Blood sugar concentration increased as a result of increases in ammonium ion concentration and decreases in the oxygen saturation of water.

The challenges to animal welfare during the transportation of wild mammals have been reviewed by Pohlin et al. (2021). They examined the literature in relation to the Five Domains model of welfare (nutrition, environment, health, behaviour and mental state) and concluded that wild mammal transport is associated with challenges relating to ensuring positive welfare in all five domains (Fig. 9.14). They noted that the negative physiological responses to transport can be mitigated by the administration of tranquilisers. The range of studies that had examined the factors influencing the physiological responses of mammals to transportation at the time of this analysis are summarised in Fig. 9.15. However, Pohlin et al. (2021) identified a need to undertake further studies of the effects of transportation on animals,

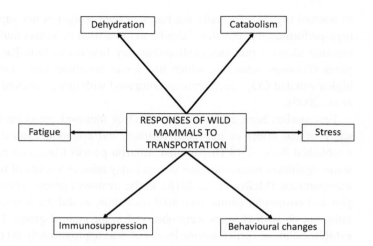

Fig. 9.14 Responses of wild mammals to transportation.

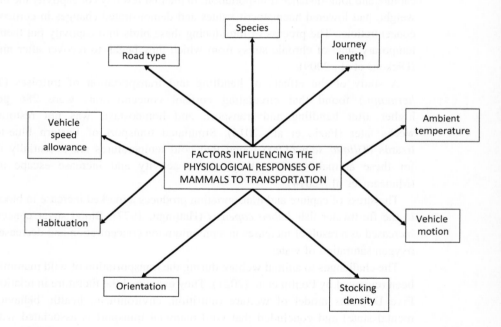

Fig. 9.15 Factors influencing the physiological responses of mammals to transportation.

encompassing species-specific and situation-specific physiological responses in order to ensure animal welfare

Transit tetany may occur in ruminants and equids as a result of the stress caused by travelling long distances, mainly as a consequence of hypocalcaemia. In extreme cases this may result in coma and death.

9.15 Stress Caused by Handling and Capture Myopathy

Fear responses caused by rough, unskilful or inconsistent handling by people can interfere with the basic endocrine processes that control reproduction and growth in farm animals and can affect milk yield, behaviour and heart rate in cows (Hemsworth and Coleman, 1998; Rushen et al., 1999). Some animals kept in zoos, such as spider monkeys (*Ateles* spp.) and woolly monkeys (*Lagothrix* spp.) are known to be particularly susceptible to the stress associated with restraint and handling (ILAR, 1998).

Being restrained or washed may cause stress. Magellanic penguins (*Spheniscus magellanicus*) that were rescued and washed following an oil spill exhibited elevated levels of corticosterone compared with those that were not oiled, or oiled but not washed (Fowler et al., 1995). A study of three captive adult beluga whales (*Delphinapterus leucas*) found that the levels of all stress-related hormones were significantly elevated during out-of-water physical examinations but not during wading-contact sessions (Schmitt et al., 2010).

During a reintroduction project, whooping cranes (*Grus americana*) showed an 8–34-fold increase in faecal steroid concentrations above baseline during shipment to a release site. This level returned to baseline within one week (Hartup et al., 2005).

Capture myopathy, also known as exertional myopathy or exertional rhabdomyolysis, is a muscle disease that results from the stress associated with capture in a range of bird and mammal taxa, especially wild ungulates. It is caused by the accumulation of lactic acid in muscles, resulting from oxygen debt. Signs include muscle damage and stiffness, depression and shock. In extreme cases death may occur, sometimes weeks later.

Cases of capture myopathy have been reported in a red-necked wallaby (*Macropus rufogriseus*; Kim et al., 2010) (Fig. 9.16); lion (*Panthera leo*; Özkan et al., 2018); Himalayan ibex (*Capra sibirica*; Zahid et al., 2018); dama gazelle (*Gazella dama*; Wallace and Bush, 1987); and a slow loris (*Nycticebus*; Nath and Chakraborty, 2013).

Deaths from capture myopathy at the National Zoological Park (also called the National Zoo), Washington, DC, from 1975 to 1985 have been reported by Wallace et al. (1985). This study only considered bovids, cervids and equids. Ten cases were identified, resulting in seven deaths: five after immobilisation – mostly after anaesthesia with xylazine and etorphine – and two after improper transport (although in these cases the anaesthetic history was unknown). During the 10-year period, the overall incidence of death from capture myopathy after immobilisation was 0.25 per cent.

Some reports of capture myopathy from zoos have discussed treatment and rehabilitation – for example, in a lesser flamingo (*Phoeniconaias minor*; McEntire and Sanchez, 2017) and a greater rhea (*Rhea americana*; Smith et al., 2005).

The increasing use of translocation as a conservation tool has highlighted the threat of capture myopathy to threatened species. The current state of knowledge about the condition, potential treatments and preventive measures, and future directions for research have been discussed by Breed et al. (2019).

Fig. 9.16 The red-necked wallaby (*Notamacropus rufogriseus*) is one of many species that may suffer from capture myopathy if roughly handled.

9.16 Invertebrate Welfare

Although invertebrate taxa constitute most of the animal species on Earth, their welfare has been largely overlooked. This has been justified by the assumption that invertebrates do not experience pain or stress, lack the capacity for higher-order cognitive functions and because the public has a negative view of these animals, often regarding them as pests. However, recent research has suggested that some invertebrates exhibit individual differences in behaviour similar to the personalities observed in vertebrates, some taxa have cognitive abilities comparable to those of some vertebrates, and at least some species are capable of experiencing both pain and stress (Horvath et al., 2013). Furthermore, recent studies have shown that sea slugs, bees, crayfish, snails, crabs, flies and ants all display cognitive, behavioural and/or physiological phenomena indicating internal states similar to what we would consider emotions in vertebrates (Mendl et al., 2011; Perry and Baciadonna, 2017).

9.17 Conclusion

The welfare of animals living in zoos and aquariums varies between species, individuals of the same species, husbandry conditions and zoos. It can be affected by an animal's personality, its early life experiences and its social environment, and any adverse effects on an individual's behaviour may be carried with it as it is moved from one institution to another. Great improvements have been made in the husbandry of many of the taxa kept in zoos, and zoo professionals have access to more information than ever to assist them in improving the lives of the animals for whom they care. It is imperative that zoos invest in the welfare of their animals because they are morally and, in most jurisdictions, legally required to do so. As public attitudes to animals change, the survival of zoos in their current form will depend in no small part on whether or not the public believes that animals living in zoos experience good welfare.

10 Enrichment and Training

10.1 Introduction

When animals are kept in barren environments in captivity they are liable to develop abnormal behaviours. Some of these are repetitive and some involve self-injury. They are not exclusively observed in animals living in zoos, but also in farm animals, animals kept in laboratories and in some companion animals. Providing complex and diverse environments helps to prevent or reduce the occurrence of these behaviours in zoos and, in recent decades, experiments on environmental enrichment have contributed to positive animal welfare in zoos. Alongside these developments, advances have been made in the training of animals and our understanding of the part that this may play in their welfare. This chapter examines some of the research on environmental enrichment and training that has been conducted in zoos, and includes some work that has been done in other captive environments.

10.2 What Is Enrichment?

Definitions of enrichment vary among authors. Rosenzweig et al. (1978) defined an enriched environment as 'a combination of complex inanimate and social stimulation'. Environmental enrichment has been described by Shepherdson (1998) as an animal husbandry principle that seeks to enhance the quality of captive animal care by identifying and providing the environmental stimuli necessary for optimal psychological and physiological wellbeing. More recently the Behavior Advisory Group of the Association of Zoos and Aquariums (AZA) has defined environmental enrichment as:

a process for improving or enhancing zoo animal environments and care within the context of their inhabitant's behavioral biology and natural history. It is a dynamic process in which changes to structures and husbandry practices are made with the goal of increasing behavioral choices to animals and drawing out their species appropriate behaviors and abilities, thus enhancing animal welfare. (AZA, 2009)

The term 'behavioural enrichment' is frequently used synonymously with environmental enrichment (e.g. Markowitz, 1982). However, the latter term is preferable as enrichment may confer non-behavioural benefits such as improved reproductive success.

The provision of environmental enrichment and the goal-oriented training of zoo animals are collectively termed 'behavioural husbandry' (Melfi, 2013).

The purpose of enrichment should be to encourage natural behaviours rather than simply to provide an animal with something to do. If an animal that usually exhibits abnormal repetitive behaviour or inactivity replaces this by repetitively operating a device that rewards it with food – such as a puzzle box – little has been gained. Hare et al. (2007) have described a group of one male and four female lions (*Panthera leo*) that were so entertained by novel burlap enrichment hung in their enclosure that it held their interest throughout the day and they would not enter their night quarters.

10.3 Historical Perspectives

Enrichment is not a new concept. The importance of providing climbing structures for some species was recognised by staff at London Zoo towards the end of the nineteenth century, as recorded in an article about the new Lion House (which also held other felids) in *The Illustrated London News* on 1 April 1876:

A piece of suitable timber, such as part of the trunk of a tree, with one or two branches, has been considerately placed in each of the dens for the exercise of climbing, of which the leopards are especially fond. (*The Illustrated London News*, 1876).

The publication *The Shape of Enrichment* was first published in 1992 and the First International Conference on Environmental Enrichment was held in Portland, Oregon, in 1993. Its proceedings were the basis of a book entitled *Second Nature: Environmental Enrichment for Captive Animals* (Shepherdson et al., 1998). Robert Young's *Environmental Enrichment for Captive Animals* was first published in 2003 (Young, 2003) and in 2011 Hal Markowitz published *Enriching Animal Lives* (Markowitz, 2011). The goals of enrichment are summarised in Fig. 10.1, based on Young (2003).

In the middle of the last century, Hediger (1950) recognised the inadequacy of most zoo environments. By the 1960s, seals (*Halichoerus* spp.) at London Zoo were being fed by an automatic device that carried fish around the edge of their pool (Morris, 1960). In the United States, Hal Markowitz pioneered the 'behavioural engineering' approach to enrichment at Portland Zoo (now Oregon Zoo) in the 1970s using operant conditioning techniques. He devised a food dispenser for white-handed gibbons (*Hylobates lar*) which dispensed food when they brachiated high in their enclosure (Markowitz, 1982).

Environmental enrichment is required by law in some jurisdictions or as a condition of accreditation by a zoo organisation because of increased concern for animal welfare. In the European Union the Zoos Directive 1999 (Art. 3) requires zoos to accommodate their animals:

under conditions which aim to satisfy the biological and conservation requirements of the individual species, inter alia, by providing species specific enrichment of the enclosures;

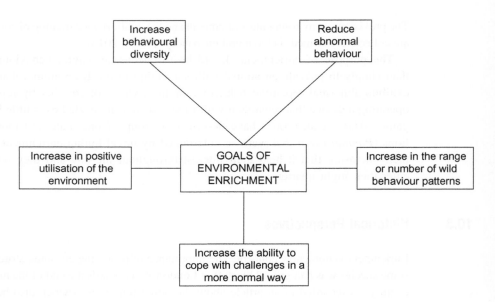

Fig. 10.1 The goals of environmental enrichment, based on information in Young (2003).

10.4   Is Enrichment a Scientific Concept?

Conventional environmental enrichment may take a number of forms, including nutritional, occupational, sensory, physical and social, and training may also be considered enriching. But is it scientific?

Early attempts at enrichment involved giving animals 'toys' or feeder devices that required them to work harder for their food, and there was little evaluation of their effectiveness. A model framework for assessing the efficacy of behavioural husbandry that may be used to assess the effectiveness of a training or enrichment programme has been described by Colahan and Breder (2003). This framework is known as SPIDER: goal-**S**etting, **P**lanning and approval process, **I**mplementation, **D**ocumentation/record-keeping, **E**valuation and subsequent programme **R**efinement.

Mellor et al. (2009) noted that early studies of enrichment in zoos suffered from poor experimental design partly because zoos typically keep small numbers of each species and keepers are motivated by the need to provide immediate benefits for the animals in their care rather than a desire to carry out controlled experiments. They pointed out that the term 'enrichment' refers to a concept that fills numerous roles for many different groups and more directly applies to prevailing social concerns about animal welfare than to the underlying empirical evidence.

Decisions about the type of enrichment that is suitable for a species are often based on subjective assessment and anthropomorphism. In the wild, Eastern fence lizards (*Sceloporus undulatus*) spend much of their time off the ground. Providing them with raised basking platforms would appear to be a suitable enrichment for these animals when kept in captivity. However, when provided with such platforms this did not affect their survival, baseline levels of plasma corticosterone, activity, time spent hiding, basking behaviour, growth or overall body condition (Rosier and Langkilde, 2011). This example highlights the importance of objectively testing the effectiveness of enrichment.

Providing enrichment in a zoo is time-consuming. Even something as straightforward as preparing a scatter feed can absorb a lot of time, especially if it means cutting up fruit for a herd of elephants! An international multi-institutional questionnaire survey of zoos holding mammals found that enrichment that was time-consuming to provide was made available less frequently than that which required less staff time and effort (Hoy et al., 2010). The time taken up by other husbandry tasks was the most important factor in limiting the implementation and evaluation of enrichment. Most animal care staff agreed that they would provide and evaluate more enrichment if time allowed.

Evaluating the effectiveness of enrichment is time-consuming and keepers have little time available for data collection during their normal daily routines. Canino and Powell (2010) studied a single polar bear (*Ursus maritimus*) at the Bronx Zoo in New York to determine if a novel method of collecting data could be used to assess the extent to which new enrichment items would reduce his pacing. They recorded his behaviour five times per day for 119 days as they passed his exhibit during their normal work routine. They found that the newer items were more effective than the previously used items at increasing play and decreasing pacing, and that their 'multi-point scan sampling' method was effective at collecting ample and reliable data.

10.5   Can Simple Objects Provide Enrichment?

The value of an object as enrichment does not depend upon its complexity. Pruetz and Bloomsmith (1992) studied the value of manipulable objects as enrichment for chimpanzees at the Science Park chimpanzee breeding facility of the University of Texas M. D. Anderson Cancer Center, and concluded that:

Texture, destructibility, portability, complexity and adaptability may be important in determining the object's value as effective enrichment. The destructible wrapping paper was a more worthwhile enrichment object than the indestructible Kong Toy™ for the captive chimpanzees in this study.

Commercially available enrichment devices are highly artificial in appearance and detract from the naturalistic appearance of modern zoo enclosures. They are also expensive; an unsinkable 85 kg ball containing flotation chambers suitable for entertaining a polar bear can cost around US$1,200.

Simple objects and actions have often been suggested as inexpensive enrichment 'devices', but the evidence that they alter behaviour is equivocal. Line et al. (1991) provided rhesus macaques (*Macaca mulatta*) with wooden sticks and four different dog toys. They found that high levels of use were recorded each time a new object was introduced, but use of all of the objects tested declined by the second day. The authors concluded that sticks and simple toys had limited effectiveness as environmental enrichment, but this could be increased by a regular schedule of object rotation.

Reinhardt et al. (1987) found that tree branches provided enrichment for individually caged rhesus macaques. This conclusion was based on a single five-minute observation of behaviour made two months after the introduction of the branch using 86 monkeys. Use of the branch – as indicated by wear – was recorded in 87 per cent of the animals. During the brief study period, 63 per cent of the subjects were seen holding, standing on, perching on, shaking or gnawing the branch (or engaging in a combination of these behaviours). The apparent high level of interaction with the branches was not surprising as the stainless steel cages were very small (0.62 m^3) and completely barren apart from food and a drinking nipple.

In another study, sticks cut from deciduous tree species were given to 28 individually caged adult rhesus macaques. Sixty-eight per cent of the macaques used their sticks, and of these 84 per cent used them continuously throughout the 12-week study period and the remaining 16 per cent stopped using them after 2–6 weeks. Ninety per cent of the stick usage involved gnawing and 10 per cent rolling. Enrichment with sticks reduced self-directed behaviour and increased activity in previously inactive animals (Champoux et al., 1987).

10.6   Enrichment and Welfare

In a meta-analytical review conducted by Shyne (2006) the author examined the effects of enrichment on stereotypic behaviour in mammals kept in zoos from 54 published studies. Most of the species examined were primates (mostly gorillas and chimpanzees), bears, felids (mostly big cats), seals and walruses, canids, giraffes and okapis. Shyne concluded that enrichment substantially reduced the frequency of stereotypic behaviour exhibited by mammals. However, most of the studies she examined collected data immediately after enrichment was introduced, but most did not report whether or not the enrichment was still effective after data collection stopped.

Feeding enrichment may increase foraging behaviour and reduce stereotypies, but often has no long-term effect and results in habituation (Schneider et al., 2014). In any event, most studies are of short duration so long-term effects cannot be studied.

Group-housed male squirrel monkeys (*Saimiri sciureus*) showed a decrease in the propensity to develop the foot and tail ulcers associated with floor contact when perches at several heights were added to their cages (Williams et al., 1988). In addition, overall morbidity decreased as a result of reduced aggression.

Hoehfurtner et al. (2021) conducted experiments to determine if an increase in environmental complexity was beneficial to the behaviour and welfare of corn snakes (*Pantherophis guttatus*). They found that the snakes used environmental enrichment when it was available and exhibited changes in behaviour that indicated improved welfare. While anxiety tests detected few effects of the provision of enrichment, the snakes showed a strong preference for the enriched enclosure over the standard enclosure when given a choice.

When lions and other large carnivores were kept in metal cages, food was pushed through the bars on a pole. Nowadays some zoos feed their big cats by placing food at the top of a tall pole (Fig. 10.2). The use and value of feeding poles as enrichment for tigers has been reviewed by Law and Kitchener (2020). They surveyed 19 zoos – mostly in the UK – and found that 79 per cent used or had used feeding poles, usually about once per week. The poles were 3–6 m high and they found no confirmed reports of serious injuries or deaths of cats resulting from their use. Evidence of a health benefit was reported as skeletons of tigers that did not use feeding poles had a mean arthrosis score four times that of those that used poles.

Experiments have shown that the welfare of many zoo-housed species (e.g. black-footed cats (*Felis nigripes*), gorillas (*Gorilla gorilla*), elephants) may be improved by auditory and olfactory stimulation (Wells and Egli, 2004; Wells et al., 2006).

Laws and Kitchener (2020) have suggested that zoos should manage predators alongside their prey to provide mutual visual and olfactory enrichments. However, if predators and prey are housed in adjoining enclosures this may have the undesirable effect of reducing the effective size of the prey species' enclosure as individuals distance themselves from potential predators, and attracting predators to a fence line where they pace stereotypically as their attempts to chase potential prey are frustrated. However, elevated platforms may provide some enrichment for predators if they allow them to view potential prey species in other enclosures from a distance.

The environment plays an important role in the development of learning and memory. Maintenance in enriched housing conditions results in adult animals learning faster and remembering tasks for a longer period than is the case for those kept in impoverished conditions (Mallory et al., 2016). Learning plays an important role in foraging behaviour and reproductive behaviour, both of which are important in the success of captive breeding and reintroduction programmes.

Environmental enrichment and exercise appear to have a number of elements in common in that they enhance neurogenesis, neurotransmitters, growth factors, learning and possibly synaptic plasticity in a similar way (Van Praag et al., 2000). Environmental enrichment provided to adult and aged animals may result in brain changes, opening up the possibility that movement from an impoverished to an enriched environment may be beneficial.

10.7   Social Interactions as Enrichment

For social animals to exhibit normal social behaviour they clearly need to be housed with conspecifics in a group with an appropriate sex ratio and age and social structure (Fig. 10.3). Some years ago I wrote a short article entitled 'Are elephant enrichment studies missing the point?' (Rees, 2000). The point I was making was that zoos around the world were providing elephants with a variety of toys to play with and novel

Fig. 10.2 (a) When big cats were kept in cages with vertical iron bars they were often fed meat by keepers using a pole (source: Whitsuntide holiday at the 'zoo': feeding the lions. *The Penny Illustrated Paper*, 30 May 1874, p. 348). (b) A lion (*Panthera leo*) at South Lakes Safari, England, feeding from a pole. Pole feeding confers improved skeletal health on captive big cats.

feeder devices, but appeared to have failed to notice that, in most cases, they were keeping elephants in small groups, thereby depriving them of the best enrichment of all: social contact with conspecifics. More than two decades later, and in spite of

Fig. 1.1 The Zoological Society of London has been at the forefront of the publishing of zoological research since 1830. (A black and white version of this figure will appear in some formats.)

Fig. 2.1 A visit to the zoo promotes family interactions, with parents coordinating the viewing of younger members. Western lowland gorillas *(Gorilla g. gorilla)*. (A black and white version of this figure will appear in some formats.)

Fig. 2.3 The giraffe enclosure at Detroit Zoo in Michigan (2011). Instead of mimicking a modern-day African savannah, this exhibit makes a link between giraffes and the culture of Ancient Egypt. (A black and white version of this figure will appear in some formats.)

Fig. 2.6 Blackpool Tower is a tourist attraction in Lancashire, England. It once housed a menagerie inside the building at its base. Some of the animals were used in the Blackpool Tower Circus, including Asian elephants. (A black and white version of this figure will appear in some formats.)

Fig. 2.12 The *Islands* exhibit at Chester Zoo in Cheshire, England, represents the South-East Asian islands of Sumatra, Papua, Sulawesi, Sumba, Bali and Panay. Visitors approach the exhibits through a simulated village representing the environs of the people of the region before reaching a boat ride and exhibits containing orangutans, tigers, sun bears and many other species representative of South-East Asia. (a) Traditional boats; (b) local shop. (A black and white version of this figure will appear in some formats.)

Fig. 2.14 Zoos are important places that remind adult visitors of their childhood. These photographs show structures of historical importance at Bristol Zoo, in England. (a) The last pole from the middle of the old bear pit. The pit itself was one of the original enclosures when the zoo was opened in 1836. (b) The Monkey Temple building was opened in 1928 and was home to around 100 rhesus macaques (*Macaca mulatta*) until the late 1980s. It is now a Grade II listed building. (c) These railings were once part of the bars of the gorilla house that was opened in 1956. (A black and white version of this figure will appear in some formats.)

Fig. 3.4 Interactive experiences in zoos have been shown to increase awareness and learning, and promote positive attitudes towards animals. (a) and (c) Burmese python (*Python bivittatus*); (b) curlyhair tarantula (*Tliltocatl albopilosus*). (A black and white version of this figure will appear in some formats.)

Fig. 3.5 Too much information. People visit zoos to see animals, not read signs. Elaborate signs may simply be ignored. (A black and white version of this figure will appear in some formats.)

Fig. 3.6 It has been shown that young people prefer vertical displays and older people prefer displays that are tilted. (A black and white version of this figure will appear in some formats.)

Fig. 5.5 Black and white colobus monkeys (*Colobus guereza*) are considered to be dangerous by the Dangerous Wild Animals Act 1976 in Great Britain and as such a licence is required by any member of the public who keeps them. (A black and white version of this figure will appear in some formats.)

Fig. 5.6 A female chimpanzee (*Pan troglodytes*) with her infant. Will chimpanzees and other great apes one day soon have the right not to be kept in captivity? In law the principle of *habeas corpus* requires that a person who imprisons another be brought before a court to prove that the detention is lawful. (A black and white version of this figure will appear in some formats.)

Fig. 6.2 Toys representing *Snowflake* in the gift shop at Barcelona Zoo, Spain, in 2018. He lived at the zoo for 37 years and was believed to be the only albino gorilla in captivity. Snowflake died as a result of skin cancer in 2003. (A black and white version of this figure will appear in some formats.)

Fig. 7.1 Baboon sculptures at the Tower of London. (A black and white version of this figure will appear in some formats.)

Fig. 8.7 (a) Chimpanzees (*Pan troglodytes*) and (b) bonobos (*P. paniscus*) often spend time near buildings or the entrance to their indoor accommodation and avoid being out in the open. (A black and white version of this figure will appear in some formats.)

Fig. 8.10 This polar bear (*Ursus maritimus*) at Detroit Zoo, Michigan, was performing stereotypic route-tracing whereby it repeatedly moved in 'circles', standing on the top of the tunnel, diving beneath the surface and swimming to an underwater window to a seal pool, and then returning to the tunnel and surfacing for air. (A black and white version of this figure will appear in some formats.)

Fig. 8.17 The viewing window to the dorcas gazelle (*Gazella dorcas*) enclosure at Barcelona Zoo in Spain is obscured by silhouettes of coloured leaves to reduce the effects of the presence of visitors. Inset: dorcas gazelle. (A black and white version of this figure will appear in some formats.)

Fig. 8.16 (a) A tiger (*Panthera tigris*) emerging from thick undergrowth at Edinburgh Zoo, Scotland. (b) Two tigers in an open area of their enclosure at Knowsley Safari, England. The extent to which animals are visible to visitors is in part determined by enclosure design. Thick vegetation may provide large predators with a naturalistic environment, but often at the expense of visibility. If enclosures are too open, their inhabitants are unable to escape the gaze of the public. (A black and white version of this figure will appear in some formats.)

Fig. 9.4 (a) Black rhinoceros (*Diceros bicornis*) and (b) banteng (*Bos javanicus*) are two of a number of species for which condition scoring systems are available. (A black and white version of this figure will appear in some formats.)

Fig. 9.6 Giraffids are prone to developing oral stereotypies in zoos. Okapi (*Okapia johnstoni*) (a) and giraffe (*Giraffa camelopardalis*) (b). (A black and white version of this figure will appear in some formats.)

Fig. 9.7 Greater one-horned rhinoceros (*Rhinoceros unicornis*) exhibiting stereotypic behaviour at the door to its indoor accommodation. Note the position of the head as the animal sways from side to side. (A black and white version of this figure will appear in some formats.)

Fig. 10.3 (a) A lone elderly female polar bear (*Ursus maritimus*) in an outdated enclosure at Berlin Zoo, Germany, in 2018. (b) Two of the eight polar bears at Yorkshire Wildlife Park, England, in 2022: four adult males, and in a separate enclosure, an adult female and her triplets. (A black and white version of this figure will appear in some formats.)

Fig. 10.8 Carcass feeding is important enrichment for many species. (a) King vulture (*Sarcoramphus papa*); (b) tiger (*Panthera tigris*); (c) Griffon vultures (*Gyps fulvus*); (d) African lions (*P. leo*). (A black and white version of this figure will appear in some formats.)

Fig. 10.9 An artificial termite mound can only act as an enrichment for chimpanzees (*Pan troglodytes*) if keepers have sufficient time to fill it with food. (a) An unrealistic and empty artificial termite mound; (b) a chimpanzee (*Pan troglodytes*) demonstrating tool use by removing food from a realistic artificial termite mound with a piece of grass. (A black and white version of this figure will appear in some formats.)

Fig. 10.13 (a) Protected contact training a bull Asian elephant (*Elephas maximus*). (b) Crate training an African elephant (*Loxodonta africana*). The elephant is being fed in a crate in preparation for transportation at a future date (source: Stephen Fritz). (A black and white version of this figure will appear in some formats.)

Fig. 10.3 (a) A lone elderly female polar bear (*Ursus maritimus*) in an outdated enclosure at Berlin Zoo, Germany, in 2018. (b) Two of the eight polar bears at Yorkshire Wildlife Park, England, in 2022: four adult males, and in a separate enclosure, an adult female and her triplets. (A black and white version of this figure will appear in some formats. For the colour version, please refer to the plate section.)

global concerns for the welfare of elephants in zoos, group sizes remain small (Rees, 2009b; 2021).

In a survey of 336 African elephants (*Loxodonta africana*) living in 104 zoos in 2006 the mean group size was 3.23 individuals, with one-fifth of elephants living alone or with one conspecific. Group size increased slightly to 3.85 by 2018 (based on 397 elephants in 103 zoos). In 2006 the mean group size of the 495 Asian elephants (*Elephas maximus*) living in 114 zoos was 4.34. By 2018 this total had increased to 750 animals in 162 zoos, but the mean group size of 4.63 indicated no statistically significant change (Rees, 2009b; 2021).

However, social interactions may have a negative effect on attempts at enrichment. In socially-housed animals enrichment options must take into account group dynamics as some techniques and devices may have negative consequences for some individuals. A large metal feeder ball that was available to a herd of eight (two males and six females) Asian elephants at Chester Zoo, England, was used almost exclusively by two of the adult females and the adult bull. The bull often monopolised the device to the extent that on one day he spent 56.3 per cent of his time interacting with it (Rees, 2009a). In addition to collecting food from the ball, he also spent time playing with the chain by which it was tethered.

In mixed-species exhibits, some individuals of one species may steal ('pirate') food intended for members of other species. The effect of varying food presentation on activity levels in eight ring-tailed lemurs (*Lemur catta*) and the extent to which they 'pirated' the food intended for the hyraxes (*Procavia capensis*) and porcupines (*Hystrix cristata*) with which they were housed was studied by Dishman et al. (2009) at the Santa Ana Zoo, California. They examined the effects of the presence or absence of browse in food boxes, and the clumping or spatial separation of food boxes. The presence of browse had some effect in increasing activity in the lemurs. Spatially separating food boxes had no effect on lemur activity, but reduced the 'pirating' of food intended for the hyraxes and porcupines by almost half.

Bildstein et al. (1993) showed that feeding behaviour is affected by aggression and dominance interactions in flamingos: the presence of American flamingos (*Phoenicopterus ruber ruber*) had an adverse effect on the feeding behaviour of Chilean flamingos (*P. chilensis*) at feeding trays.

10.8   Food Selection, Preference and Presentation

10.8.1   Time Spent Feeding

Herbivores spend a great deal of time feeding. Captive Sichuan takin (*Budorcas taxicolor tibetana*) spent 82.3 per cent of their time feeding, ruminating and resting (Powell et al., 2013). This is similar to other ruminants.

Predictable feeding times are helpful to visitors – because they can predict when they will see otherwise illusive animals – and to keepers because busy staff need a routine in order to get through the day's work. However, predictable feeding times can lead to changes in the behaviour of animals living in zoos as they anticipate feeding times and this may lead to an increase in stereotypic behaviour. This is a particular problem when animals receive their main feed at the end of the day in their indoor accommodation and are prevented from accessing this until then. The imposition of unnatural feeding regimes may also affect the physiology of some species and promote obesity, inactivity and stereotypic behaviour.

When African lions (*Panthera leo*) were adapted from a conventional feeding programme to a gorge-and-fast feeding schedule – similar to that found in the wild – digestibility increased and food intake and metabolisable energy intake both decreased (Altman et al., 2005). Repetitive active behaviours increased and no increase in agonistic behaviours was observed. Pacing occurred twice as frequently on feeding days compared with fast days. Switching to a more naturalistic feeding schedule thus increased activity and improved nutritional status in these lions.

10.8.2 The Effect of Natural Food on Behaviour

Wild slender lorises are almost exclusively insectivorous, but in captivity they are fed a primarily frugivorous diet. When live insects were fed to captive Northern Ceylon grey slender lorises (*Loris lydekkerianus nordicus*) significant changes in activity budgets were observed; inactivity was reduced and foraging levels increased to those seen in the wild. There was a significant increase in the adoption of foraging postures and a wider behavioural repertoire (Williams et al., 2015).

Many wild animals, especially herbivores, spend a great deal of time foraging. This is not generally the case in animals living in zoos unless they are grazers and provided with large enclosures. For some taxa, providing browse can provide an opportunity to increase the time spent feeding.

Feeding browse to gorillas increased the time spent feeding (from about 11 per cent of the day to 27 per cent) and decreased the incidence of regurgitation and reingestion behaviours (Gould and Bres, 1986). When orangutans (*Pongo pygmaeus*) were fed fresh browse at Chester Zoo, England, they exhibited a decrease in the time spent inactive in adults and infants and an increase in the time adults spent foraging in the woodchip floor beneath the branches of browse (Birke, 2002).

When browse was fed to African elephants (*Loxodonta africana*) a significant increase in time spent feeding was recorded along with significant decreases in drinking and inactivity (Stoinski et al., 2000). The authors concluded that browse increased species-typical behaviours in this species.

10.8.3 Food Presentation

For some species there has been little change in the method of feeding for well over a century. Penguins and sea lions are usually fed by throwing dead fish into their pool (Fig. 10.4).

Fig. 10.4 Californian sea lions at London Zoo in 1901 (source: 'The queer Christmas dinner of the sea lions at the zoo', *The Sphere*, 21 December 1901, p. 303).

Some zoos now make their animals work for their food by putting it out of reach. In some cases individuals will work to obtain food even when other food items are readily available. This behaviour is known as contrafreeloading (Fig. 10.5).

In the wild, food sources may be unreliable and their location may be unpredictable. Consequently, searching for food occupies a great deal of time and energy. In captivity food sources are often both reliable and easy to find. The way that food is presented to animals in zoos can significantly affect their behaviours and activity levels and consequently their welfare.

A study of the effect of three food enrichment items on the behaviour of black lemurs (*Eulemur macaco macaco*) and ring-tailed lemurs (*Lemur catta*) found that the item that required the most manipulation (a wire box containing whole grapes, apples or both, hidden in straw suspended from a tree branch) had the greatest effect on behaviour (Maloney et al., 2006). It caused a decrease in the incidence of resting and

Fig. 10.5 *Victor*, a male Asian elephant (*Elephas maximus*) at Berlin Zoo, Germany. Some animals will spend time working to obtain food when other food is freely available. This behaviour is known as contrafreeloading and is common in elephants.

increases in playing and grooming. There was, however, no significant effect on the incidence of feeding or foraging.

Self-operated food boxes were installed in the *Masoala Rainforest* exhibit at Zurich Zoo in Switzerland to encourage arboreal behaviour in three white-fronted lemurs (*Eulemur fulvus albifrons*) and a single Alaotran gentle lemur (*Hapalemur griseus alaotrensis*). When the boxes were present, overall activity and locomotion increased compared with when they were not, and visitors reported seeing the lemurs in the trees more often (Sommerfeld et al., 2006). The presence of the food boxes resulted in the behaviour of the lemurs approaching natural levels.

Giraffes (*Giraffa camelopardalis*) have a tendency to perform a number of oral stereotypies. A study of 214 giraffe and 29 okapi (*Okapia johnstoni*) living in 49 US zoos found that almost 80 per cent engaged in at least one type of stereotypy, the most common of which was the licking of non-food objects (72.4 per cent of individuals) (Bashaw et al., 2001). There is some experimental evidence that stereotypy can be reduced by encouraging giraffes to engage in more naturalistic foraging behaviour by using more complex feeders that require greater tongue manipulation to obtain food (Fernandez et al., 2008) (Fig. 9.6).

Some zoos allow visitors to participate in feeding browse to giraffes. A study of 30 giraffes at nine zoos – six with guest feeding programmes and three without – found that when time spent engaging with a guest feeding programme was added to the time spent eating routine diets, those individuals that spent more time engaged in

Fig. 10.6 Enrichment logs for brown bears (*Ursus arctos*) (inset) at the Welsh Mountain Zoo in North Wales.

total feeding behaviours performed less stereotypic behaviour (e.g. object-licking and tongue-rolling (Orban et al., 2016)). However, individuals that spent more time engaging in guest feeding programmes also spent more time idle.

Increasing the spatial unpredictability of food availability for Malayan sun bears (*Helarctos malayanus*) in Cologne Zoo, Germany, increased the time they spent foraging and resulted in a higher diversity of foraging behaviours (Schneider et al., 2014). No habituation was detected over a period of 12 consecutive days.

Studies of the enrichment value of honey-filled logs for bears – sloth bear (*Melursus ursinus*), American black bear (*Ursus americanus*) and brown bear (*U. arctos*) – showed that, although they habituated to the logs, this could be counteracted by refilling them with honey and providing multiple logs (Carlstead et al., 1991) (Fig. 10.6).

Skibiel et al. (2007) presented bones, frozen fish and spices to six felid species – lion (*Panthera leo*); tiger (*P. tigris*); jaguar (*P. onca*); cheetah (*Acinonyx jubatus*); ocelot (*Leopardus pardalis*); and cougar (*Puma concolor*) – at the Montgomery Zoo, Alabama, and found that all three enrichments increased active behaviours in these species. Spices and frozen fish, but not bones, significantly reduced the amount of time spent pacing. The effects of these enrichments on activity levels were not sustained seven days after their removal, but the effect of frozen fish on stereotypic behaviour was so sustained.

The effect of three feeding devices (food-filled baskets, polyvinyl chloride tubes and frozen ice pops) on the behaviour of moloch gibbons (*Hylobates moloch*) at

Belfast Zoo in Northern Ireland was studied by Wells and Irwin (2009). The authors concluded that such devices offered an effective enrichment for this species. Beneficial effects on behaviour have been reported from many other studies of feeding enrichment (Reinhardt and Roberts, 1997).

10.8.4 Effect of Chopping Food

It has long been the practice of zoos to chop fruit and vegetables into small pieces and scatter these in enclosures. Keepers spend a great deal of time chopping food for animals, but this may be counter-productive with regards to its effect on their behaviour.

The results of a study of ring-tailed coatis (*Nasau nasau*) at Beale Wildlife Park, England, suggested that whole-food diets may provide them with greater opportunities to express natural food manipulation behaviours than do chopped foods, and may also reduce the levels of aggression observed at meal times (Shora et al., 2018).

Blue and gold macaws (*Ara ararauna*) spent more time engaging with their food and less time inactive when whole food was provided compared with when the food was chopped (James et al., 2021).

Smith et al. (1989) showed that providing lion-tailed macaques (*Macaca silenus*) with whole fruit increased the amount of time they spent feeding and the total amount of food consumed. Although there is a common belief that chopping food into small pieces equalises access to the different food items among individuals in a group, this study found that providing whole foods increased mean dietary diversity.

10.8.5 Live Feeding and Carcass Feeding

Presenting predators living in zoos with live food or carcasses rather than pieces of meat seems intuitively a more enriching method of feeding than providing them with small pieces of meat. However, live feeding is problematic because it would be unacceptable to many visitors and in any event it is illegal in many jurisdictions.

One solution to this difficulty is to make prey animals appear to be alive. Gans and Mix (1974) described a sequential programmable insect dispenser that blows insects into a cage (or into the air) for feeding animals. These devices are useful for lizards and are a substitute for chasing live prey. Markowitz (1982) found that servals (*Leptailurus serval*) became more animated and more interesting to visitors when their food ('flying meatballs') was swung over their heads attached to a rope or rod.

Shepherdson et al. (1993) found that presenting a fishing cat (*Prionailurus viverrinus*) with live fish increased activity, behavioural diversity and enclosure utilisation. In a second study, multiple feedings of four leopard cats (*P. bengalensis*) increased exploratory behaviour from 5.5 per cent to 14 per cent (Fig. 10.7). In addition, it increased the diversity of behaviours and reduced the bout length and total duration of stereotypic pacing. These studies suggested that reducing the predictability of food

Fig. 10.7 Leopard cat (*Prionailurus bengalensis*).

availability and maximising the functional consequences of foraging behaviour in small cats can be effective enrichment.

Most enrichment studies are short term, and any of the long-term benefits from enrichment are not recorded. However, a long-term study of the behavioural impact of a switch from feeding pre-processed joints of meat to carcass feeding on a female Asiatic lion (*Panthera leo persica*) at Chester Zoo, England, found that pacing behaviour reduced significantly and resting behaviour increased 12 months after the change in diet (Finch et al., 2020). The average amount of time spent feeding increased both shortly after and 12 months after the diet change (Fig. 10.8).

The presentation of live fish to African lions (*Panthera leo*) and Sumatran tigers (*P. tigris sumatrae*) increased the variety and frequency of feeding behaviours (Bashaw et al., 2003). Fish reduced stereotypic behaviour in tigers from 60 per cent of scan samples to 30 per cent on the day of presentation. The presentation of horse leg bones increase the frequency of feeding behaviours, reduced stereotypic behaviour and increased non-stereotypic activity in both species. Both of these feeding enrichments had a beneficial effect on behaviour of these felids that lasted for at least two days after presentation.

10.9   Cognitive Enrichment

Enrichment may be classified as cognitive if it provides opportunities to solve problems and control some aspect of the environment, and is correlated with validated measures of wellbeing (Clark, 2011a).

Fig. 10.8 Carcass feeding is important enrichment for many species. (a) King vulture (*Sarcoramphus papa*); (b) tiger (*Panthera tigris*); (c) Griffon vultures (*Gyps fulvus*); (d) African lions (*P. leo*). (A black and white version of this figure will appear in some formats. For the colour version, please refer to the plate section.)

Six Western lowland gorillas (*Gorilla gorilla gorilla*) (one male and five females) at Bristol Zoo, England, were provided with cognitive enrichment in the form of a modular cuboid puzzle maze containing food rewards (Clark et al., 2019). Individuals could have removed food (nuts) from the interconnected modules with their fingers or tools, but only the three tool-using gorillas were successful in doing this. The only male in the group did not use the puzzle maze. Use of the device did not diminish over time, but it had no significant effect on overall activity budgets.

The puzzle maze used by gorillas at Bristol Zoo was designed to be used by one gorilla at a time. For species that naturally exhibit high levels of cooperation, enrichment can be used to encourage this behaviour. Researchers at Ocean Park, Hong Kong, have reported the use of the first multi-partner cognitive enrichment devices by a group of male bottlenose dolphins (*Tursiops*) (Matrai et al., 2022). Alliance formation is an important aspect of the lives of male dolphins. The novel devices used facilitated simultaneous actions for two, three or four dolphins. They were constructed from PVC tubes, fittings and caps and fitted with rope handles. The tubes were filled with fish and ice and could only be opened when the handles were pulled simultaneously. The dolphins received no training on the use of the devices and they successfully cooperated to open the three-way devices in 10/12 trials and the four-way devices in 10/12 trials.

The use of puzzle feeders is most often associated with primates and elephants because they require manipulation with hands or trunks. However, they have also been used for other species, including octopuses. A study of the effect of enrichment on giant Pacific octopuses (*Enteroctopus dofleini*) at Cleveland Metroparks Zoo, Ohio, found that enrichment – including prey puzzles – reduced the amount of time they spent on resting and locomotion and increased activity and behavioural diversity (Brady et al., 2010).

Although there has been an increase in the use of cognitive enrichment in zoos, it has only recently been quantified. An international survey of 177 respondents showed that cognitive enrichment is universally perceived as very important for animal welfare, even though there was little agreement on what constitutes cognitive enrichment (Hall et al., 2021). Carnivores were reported as receiving most cognitive enrichment (76.3 per cent), while fishes and amphibians received the least (16.9 per cent). The authors of the study concluded that keepers' lack of time and resources may be impeding the development and use of cognitive enrichment and that specific job roles should be created in this area.

10.10   Television, Computers and Motion Illusions

A survey of the use of technology to enhance the welfare of primates in zoos found that the most prevalent types of technological enrichment used were computers, television, radio and sprinklers (Clay et al., 2010).

Ten chimpanzees (*Pan troglodytes*) were shown videotapes of chimpanzees, other animals and humans by Bloomsmith et al. (2000). The animals watched the monitor for 38.4 per cent of the time available and individually housed chimpanzees watched it more than those kept in a social group. Although the chimpanzees' behaviour was not altered extensively by watching the videos, it occupied a significant proportion of their activity budget, suggesting that it may be a useful enrichment.

The rotating snake (RS) illusion is an optical illusion consisting of concentric circles of bands of colour that produce the illusion of movement in a group of coiled snakes. Several species have been shown to perceive this illusion (rhesus macaques (*Macaca mulatta*): Agrillo et al., 2015; fishes: Gori et al., 2014; domestic cats (*Felis catus*): Bååth et al., 2014) and consequently Regaiolli et al. (2019) investigated the use of this illusion as an enrichment for lions (*Panthera leo*) at Parco Natura Viva, a zoo in Italy. Three lionesses were exposed to the RS stimulus and two control stimuli. They found that two of the animals interacted with the RS stimulus more than with the two control stimuli, suggesting that they were attracted by the illusory motion. One of the lionesses exhibited a reduction in self-directed behaviours and an increase in attentive behaviours compared with the period when none of the stimuli was present. The authors suggest that this could indicate a welfare improvement.

10.11 Music and Auditory Stimulation

The welfare benefits of music therapy in livestock production are well established. The effect of music on cattle, pigs and poultry has been reviewed by Ciborowska et al. (2021). A number of studies have examined the potential value of auditory stimuli to zoo-housed animals, but it is difficult to draw general conclusions from the work that has been published.

The effect of auditory stimuli intended for Western lowland gorillas (*Gorilla g. gorilla*) on the behaviour of birds in a shared exhibit at Buffalo Zoo, New York, was investigated by Robbins and Margulis (2016). Three species of African birds (superb starlings (*Lamprotornis superbus*), speckled mousebirds (*Colius striatus*) and Lady Ross's turacos (*Mussophaga rossae*)) were observed during exposure to natural sounds, classical music and rock music. The average frequency of flying in all three species increased with natural sounds and decreased with rock music. Vocalisations increased in response to all three auditory stimuli in the starlings and mousebirds, but the turacos only exhibited an increased frequency of duetting in response to rock music.

An earlier study that used the same methodology to study the effect of auditory stimuli on the gorillas in the exhibit at Buffalo Zoo found that exposure to natural sounds resulted in a decrease in stereotypic behaviour. Exposure to classical and rock music increased stereotypic behaviours (Robbins and Margulis, 2014). However, when instrumental classical music was played for three weeks to moloch gibbons (*Hylobates moloch*) at Howletts Wild Animal Park in Kent, England, no significant difference in behaviour was seen compared with that recorded in the three control weeks when music was not played (Wallace et al., 2013). Three Sumatran orangutans (*Pongo abelii*) who were able to use a touchscreen to choose to either listen to music or experience silence either preferred silence or were indifferent, and none of the three animals could reliably discriminate between 'music' and 'scrambled music' (Ritvo and MacDonald, 2016).

A study of four female Asian elephants (*Elephas maximus*) at Dublin Zoo, Ireland, suggested that listening to classical music may help to reduce the time spent engaged in stereotypic behaviours (Wells and Irwin, 2008). However, Wallace et al. (2017) demonstrated that music played to chimpanzees (*Pan troglodytes*) had neither an enriching effect or a negative effect on their welfare.

Piitulainen and Hirskyj-Douglas (2020) developed an animal-centred interactive system to allow white-faced saki monkeys (*Pithecia pithecia*) at Korkeasaari Zoo, Finland, to play audio. In addition to allowing the animals to select different sounds, their interactions with the device were recorded. The monkeys triggered traffic audio more than silence, rain, zen and electronic music, and the authors concluded that audio is a promising way to provide enrichment for small primates. However, they did not discuss any behavioural or physiological benefits to the monkeys as the purpose of their study was to evaluate the device itself and determine which sounds the animals preferred.

10.12   Enrichment and Personality

Although the purpose of environmental enrichment is primarily to enhance the welfare of animals, it may also be used as a tool to assess personality. Four giant pandas (*Ailuropoda melanoleuca*) at Zoo Atlanta, Georgia, and the Smithsonian's National Zoological Park (or National Zoo, Washington, DC) were exposed to 10 novel enrichment items and their behaviours were recorded (Powell and Svoke, 2008). In addition, keepers rated each individual on 23 behavioural characteristics. The study found some consistency between the results from the two assessment methods, and the authors concluded that an analysis of responses to novel environmental enrichment may provide a tool for the rapid assessment of personality.

10.13   Enrichment and Invertebrates

Until relatively recently environmental enrichment in zoos was almost exclusively associated with mammals. Increasing numbers of studies are being published concerned with the environmental enrichment of other vertebrate taxa and invertebrates, although the number of studies of the latter is still very small.

There is some evidence that very simple invertebrates can benefit from enrichment in ways that could have implications for conservation. Male fruit flies (*Drosophila melanogaster*) reared in an enriched environment were found to be twice as successful at acquiring mates as those reared in standard conditions (Dukas and Mooers, 2003). The most important factor in increasing reproductive success was the larger space available to those flies in the enriched conditions.

Environmentally enriched house crickets (*Acheta domesticus*) were found to have an increased number of newborn cells in their mushroom bodies – the main integrative structures of the insect brain – compared with conspecifics housed in cages with an impoverished environment where they were deprived of most visual, auditory and olfactory stimuli. The effect of enrichment on neurogenesis seemed to be confined to the beginning of imaginal (adult) life (Scotto-Lomassese et al., 2000). Further studies of this species have shown that crickets that experienced an enriched rearing condition as young adults – regardless of nymphal rearing conditions – performed better on a memory task than individuals that experienced an impoverished condition (Mallory et al., 2016).

Yasumuro et al. (2016) showed that enrichment (exposure to novel substrates) accelerated the development of cryptic behaviour – the ability to mimic the patterns in their environment – in the pharaoh cuttlefish (*Sepia pharaonis*). Further experiments with the same species showed that 3D objects promoted the maturation of learning, memory and depth perception, and isolation inhibited the maturation of memory and depth perception. Hunting success in the cuttlefish was always highest in individuals from the enriched environment and hunting success and number of prey captured were always lowest in the isolated cuttlefish (Yasumoro and Ikeda, 2018).

The effects of rearing environment on the development of behavioural tendencies in tarantulas have been studied by Bengston et al. (2014). They found that captive basal tarantulas (*Brachypelma smithi*) kept in enriched conditions developed a behavioural syndrome – a correlated suite of behavioural traits – that included exploratory and bold/shy behaviours, but those kept in restricted (minimal) conditions did not develop a behavioural syndrome. The authors concluded that early environment can induce behavioural syndromes in some populations of spiders and that continual stress may break down normal behavioural development. They suggested that their study 'provides a cautionary tale for those studying behavioural syndromes in captivity' but it may also have implications for captive breeding programmes.

Enrichment for giant Pacific octopuses (*Enteroctopus dofleini*) has been discussed by Anderson and Wood (2001). They are often exhibited in aquariums, and enrichment may include the use of live prey such as crabs, fish, lobsters, clams or freshwater crayfish. Food may be hidden or placed in a puzzle box, and invertebrates and some fish species that the octopuses would normally encounter in the wild could be added to the tank.

The destructive behaviour of an octopus kept in an aquarium has been recounted by Anderson (2005). She apparently occupied her time by attacking her tank: destroying her water filter, moving rocks around the tank, scratching the glass and blowing gravel from the bottom. Mather and Anderson (2007) concluded that this animal was in need of enrichment to keep her occupied.

Anderson and Wood (2001) suggest that the environments of smaller octopuses may be enriched by introducing complexity into their environment, providing a proper den or lair space, and the use of toys. They can also be provided with 'complex' food or feeding strategies. Although they provide a detailed discussion of *potential* enrichment for octopuses, Anderson and Wood concluded that:

As yet, we have no evidence that enrichment is beneficial for captive GPOs [giant Pacific octopuses]. We have no indication that captive octopuses need enrichment.

However, a study of the tropical octopus *Callistoctopus aspilosomatis* suggested that the quality of their environment affected their behaviour (Yasumoro and Ikeda, 2011). Octopuses were more exploratory in an enriched environment (coral skeleton and artificial sea grass on a sandy bottom) and a standard environment (sanded bottom) than in poor environments (containing neither objects nor sand). The animals only showed bold body patterns in the poor environment, and when a pipette was introduced into the water the octopuses only attacked it in the standard and enriched environments. Enrichment has been shown to be beneficial in increasing the learning ability of common cuttlefish (*Sepia officinalis*) (Dickel et al., 2000).

10.14   Sometimes Enrichment Does Not Work or Goes Wrong

It cannot be assumed that elaborate feeding devices necessarily provide behavioural enrichment.

When the behaviour of insectivorous tree runner lizards (*Plica plica*) fed using a delivery device that slowly released crickets into their enclosure was compared with that observed when crickets were provided by standard scatter feeding, the device was found to cause a decrease in activity (Januszczak et al., 2016). In comparison, scatter feeding promoted activity and increased hunting difficulty partly due to the complex physical environment of the enclosure, which made it difficult to catch prey.

Even hiding food may not be enriching for some animals. Hiding food (peanuts) above a stone border in their outdoor enclosure did not enhance searching behaviour in Asian elephants (*Elephas maximus*), so this method of food hiding did not constitute an environmental enrichment (Wiedenmayer, 1998).

Enriched environments, by their very nature, are likely to present more potential hazards for animals than sterile environments. These potential hazards have been discussed by Hare et al. (2007) based on anecdotal evidence. They include animals trapping body parts (e.g. heads and legs) in tubes and tyres, entanglement in ropes and bungee cord, entrapment in sprung devices, items lodged in the mouth, ingestion of non-food items and using enrichment devices as tools to cause damage. Particular attention was drawn to the risk of allowing animals access to enrichment devices at night and during other unsupervised times.

Hare et al. (2007) describe a male orangutan (*Pongo* sp.) using a length of PVC pipe to reach and break all the light bulbs in the fittings located above the mesh roof of his enclosure, thereby showering the enclosure with shards of glass. They also warn against the possible transfer of disease-causing materials among animals (e.g. carnivore faeces used as enrichment for potential prey species) and the possibility that some scents may cause irritation of the mucous membranes of the gums and eyes.

Hare et al. also list a number of examples of animals throwing enrichment items from their enclosures. For example, a fur seal (*Arctocephalus*) threw an ice block containing frozen fish from its enclosure and hit a child's head; a male brown bear (*Ursus arctos gyas*) threw a truck tyre over a crowd of visitors; and a gorilla (*Gorilla g. gorilla*) threw ice blocks at visitors when she felt threatened. Animals have also used enrichment devices as tools to escape their enclosures. A bull African elephant (*Loxodonta africana*) used an untethered feeder ball as a stepping stone and walked out of his enclosure. Care must be taken not to allow trees to grow large enough to produce branches that overhang fences and potentially facilitate the escape of arboreal species.

10.15   Why Are Enrichment Practices Difficult to Implement?

Anyone visiting a zoo is likely to see enclosures containing putative enrichment devices that do not appear to be being used while keepers clean windows and prepare food for their animals (Fig. 10.9). When there is clear evidence that at least some

Fig. 10.9 An artificial termite mound can only act as an enrichment for chimpanzees (*Pan troglodytes*) if keepers have sufficient time to fill it with food. (a) An unrealistic and empty artificial termite mound; (b) a chimpanzee (*Pan troglodytes*) demonstrating tool use by removing food from a realistic artificial termite mound with a piece of grass. (A black and white version of this figure will appear in some formats. For the colour version, please refer to the plate section.)

enrichment practices provide benefits to animals living in zoos, why do zoo staff find these practices so difficult to implement?

Big cat keepers were surveyed in an attempt to understand the impediments to providing enrichment for this group of animals (Tuite et al., 2022). Interviews were used to collect quantitative data from 23 keepers working in 12 zoos distributed across Europe, the United States, Australia, New Zealand, South Africa and South-East Asia. The impediments identified included uncertainty about which enrichment practices are effective, conflicting priorities (e.g. sick animals) and concerns about visitor perceptions (because they do not like to see unnatural objects in enclosures). Consequently, stimulation of the animals and the provision of enrichment were often not seen as important. Keepers reported that they were 'purely surviving', 'struggling to understand the goal' and 'judge the effectiveness' of enrichment. The authors of the study suggested that there needs to be greater clarity regarding the objectives of enrichment and a means of assessing its effectiveness.

10.16 Training

10.16.1 Introduction

Zoos train their animals for a variety of reasons. Training can have many benefits for the animals themselves and for their keepers. Training animals to participate in health checks, cooperate with routine veterinary procedures and enter transportation crates can reduce the amount of stress they experience. Training them to return to their indoor accommodation when called can reduce the number of staff required to round them up from a large enclosure, and protected contact training of large dangerous

Fig. 10.10 Capybara (*Hydrochoerus hydrochaeris*) training. The keeper is using operant conditioning to train this animal by pairing the sound of a whistle (a bridge) with the required behaviour and then providing a food reward.

mammals can improve keeper safety. The provision of animal shows in zoos allows trainers to show off the natural abilities of their charges and provide information to visitors. This may involve exhibiting natural behaviour (such as balancing or feeding) or performing tricks. This would not be possible without a considerable investment of time in training. Some animals are trained to participate in research and in some cases training may act as enrichment.

Regardless of its purpose, the basic method of training animals involves operant conditioning whereby the desired behaviour is associated with positive reinforcement, often a small food reward (Fig. 10.10).

10.16.2   Training Medical Behaviours

The medical behaviours in which animals may be trained have been divided into three categories by Mattison (2012), based on the ease with which they may be trained and the time required. They are: foundation behaviours, which are easy to train and take little time (e.g. crate and scale training); intermediate behaviours, which are a little more difficult to train and take longer (e.g. nail trimming and restraint procedures); and advanced behaviours, which take longer and require the animal to endure some pain or discomfort (e.g. injection or masking for anaesthesia without restraint).

Several primates at Disney's Animal Kingdom, Florida, have been trained to perform a wide range of behaviours, including gorillas (*Gorilla gorilla*), mandrills (*Mandrillus sphinx*) and gibbons (Colahan and Breder, 2003). These behaviours relate to husbandry (e.g. presenting the face, foot, arm, hand, knee, open mouth, tongue); veterinary treatment (e.g. accepting an ear thermometer, injection, oral medication, wound-cleaning, stethoscope and ultrasound examination); and research (semen and urine collection). Sea lions and seals are routinely trained to allow staff to examine their teeth. An orangutan (*Pongo* sp.) at Blackpool Zoo, England, was trained to show her teeth to zoo staff to facilitate dental examinations and continued to do this when visitors approached her enclosure (Fig. 10.11)

Wienker (1986) has described the training of giraffes (*Giraffa camelopardalis*) at Woodland Park Zoo in Seattle, Washington, to enter a restraint device (squeeze cage) voluntarily. The cage was used for hoof trimming and routine veterinary examinations, including X-rays, blood collection and injections. After conditioning to the cage it was used 63 times in four years without serious injury to animals or staff. The habituation of antelope and bison to cooperate with veterinary procedures has been described by Grandin (2000).

The training of birds and small mammals for medical behaviours to facilitate veterinary treatment and examination at Point Defiance Zoo and Aquarium,

Fig. 10.11 This orangutan (*Pongo* sp.) has been trained to show her teeth to zoo staff and practices this behaviour regularly for bemused visitors.

Tacoma, Washington, has been discussed by Mattison (2012). She describes a tawny frogmouth (*Podargus strigoides*) that was caught and placed in a cardboard box for weighing for more than 15 years before being trained to stand on a scale, and a green-winged macaw (*Ara chloropterus*) trained to allow a towel to be wrapped around its body to facilitate restraint. The same macaw was trained to place its head in a mask for anaesthesia.

We tend to associate the capacity to be trained with mammals – especially those that are generally considered to be intelligent, such as monkeys, apes, elephants, sea lions and cetaceans – and birds such as parrots and macaws. However, lower vertebrates can also be trained.

The use of operant conditioning and desensitisation to facilitate the veterinary care of reptiles has been discussed by Hellmuth et al. (2012). The training of a Nile crocodile (*Crocodylus niloticus*) at the Bronx Zoo in New York to facilitate examination and veterinary treatment has been described by Hellmuth et al. (2012). The crocodile was conditioned to a whistle bridge, then target trained, habituated to a number of people and desensitised to touch on his tail. This training eventually resulted in the animal being able to be stationed on a scale while blood was collected. Training began in May 2008 and blood was successfully collected in March 2010.

The use of classical and operant conditioning in training Aldabra tortoises (*Geochelone gigantea*) for venipuncture and other husbandry procedures has been reported by Weiss and Wilson (2003) (Fig. 10.12). They trained the tortoises to associate the sound of a clicker with food (classical conditioning) and went on to

Fig. 10.12 Classical and operant conditioning have been used to train Aldabra tortoises (*Geochelone gigantea*).

use operant conditioning to train them to move towards and hold at a target consisting of a red ball fixed to the end of a dowel.

In a separate study, zoo-housed Galapagos and Seychelle tortoises (Testudinidae) successfully underwent operant conditioning and learned faster when trained in the presence of a group than when trained individually. Tortoises also learned to distinguish colours in a two-choice discrimination task. When tested nine years after the initial training the animals still retained the operant conditioning, but needed to relearn the discrimination task (Gutnick et al., 2020).

Corwin (2012) has noted that the learning ability of fishes has been studied for over 200 years and described the use of operant conditioning (especially positive reinforcement) to train behaviours to aid in the management of their diets, capture techniques and medical procedures.

The use of positive reinforcement to train four adult zebra sharks (*Stegostoma fasciatum*) at the Downtown Aquarium in Denver, Colorado has been described by Marranzino (2013). The sharks were desensitised to staying within a closed holding tank off the main exhibit, tactile stimulation and the presence of multiple trainers in the holding tank. One of the sharks was also desensitised to the presence of a stretcher in the holding tank.

10.16.3 Training and Safety

Dangerous animals need to be trained to reduce, or avoid, physical contact between the animals and their keepers. Elephant keepers have traditionally handled their charges by entering their enclosures and freely associating with them in much the same way as did elephant handlers in circuses. Some training occurred so that the elephants could be, for example, physically examined, chained at night and easily moved between enclosures. They have also been trained to enter transportation crates (Fig. 10.13b).

The US Department of Labor Statistics has reported that the occupation of 'elephant handler' has the highest fatality rate of any documented occupation. Lehnhardt (1991) recorded 15 elephant-related deaths between 1976 and 1991, approximately one death per year. In some years two elephant keepers have been killed. This equates to a fatality rate of 333 per 100,000 workers, but this figure is misleading because the actual number of keeper deaths is low and zero in some years (Toscano, 1997). In England, three elephant keepers were killed in three different zoos over a period of just 20 months between February 2000 and October 2001. Most attacks are deliberate and occur when a keeper is in direct contact with an elephant. Many recorded attacks are serious, with 23.5 per cent resulting in the death of a keeper (Gore et al., 2006).

As a result of safety concerns and in an attempt to reduce the extent to which elephant behaviour was being affected by human contact, a system of protected contact has been widely adopted whereby elephants and keepers are separated by a protected contact wall (Fig. 10.13a). The animals are target trained to stand in specific positions and present different parts of their bodies to the keeper (or veterinarian) for

Fig. 10.13 (a) Protected contact training a bull Asian elephant (*Elephas maximus*). (b) Crate training an African elephant (*Loxodonta africana*). The elephant is being fed in a crate in preparation for transportation at a future date (source: Stephen Fritz). (A black and white version of this figure will appear in some formats. For the colour version, please refer to the plate section.)

inspection or treatment. Increased use of protected contact management techniques should greatly reduce keeper injuries and deaths (Roocroft, 2007).

10.16.4 Is Training Enrichment?

Many zoo professionals have suggested that training is an enriching process for animals (Fig. 10.14). However, lack of clarity over the meaning of 'enriching' has hampered research on this subject. Melfi (2013) has suggested five hypotheses whereby animals would be considered to be enriched by training, and tested these using the available published data. These hypotheses were:

1. If it affords learning opportunities, as learning is considered to be enriching.
2. If it can achieve the same results as conventional environmental enrichment (CEE).
3. If it increases human–animal interactions.
4. If it provides a dynamic change in the animals' day.
5. If it facilitates the provision of CEE.

From the data, Melfi concluded that training could be considered enriching according to hypothesis 1 while the animal is still learning. It could also be enriching according to hypothesis 2 if the ultimate consequence of training was considered itself enriching. There were insufficient data to evaluate hypothesis 3 and the data did not support that training was enriching in and of itself (according to hypotheses 4 and 5).

Melfi concluded that training was not considered to be an appropriate alternative to the provision of CEE and recommended that both training and CEE should be provided to ensure an integrated holistic captive animal management strategy which will meet an animal's needs.

Fig. 10.14 Is training enrichment?

Westlund (2014) took a different approach in her analysis of the value of training as enrichment. She suggested four criteria by which an intervention may be considered enrichment. The putative enrichment should

1. give the animal more control over its environment;
2. add behavioural choices;
3. promote species-appropriate repertoires; and
4. empower the animal to deal adequately with challenges.

After an analysis of the available literature, Westlund concluded that formal training using operant conditioning fulfils all of her proposed criteria and that some training may thus be regarded as environmental enrichment. She concluded that the application of a comprehensive training programme in zoos will help animals achieve better welfare than is possible using CEE alone by the addition of other training techniques such as counter-conditioning and systematic desensitisation. These techniques would, respectively, alter an animal's response to an unpleasant stimulus or make it more tolerant of an unpleasant stimulus by initially exposing it to the stimulus at a very low intensity.

A study of Magellanic (*Spheniscus magellanicus*) and Southern rockhopper penguins (*Eudyptes chrysocome*) showed that it is possible to train animals to interact with enrichment devices rather than relying solely on interactions resulting from trial-and-error encounters (Fernandez, 2019) (Fig. 10.15). Fish were placed in two enrichment devices during training sessions. This resulted in several hits to the devices – compared with zero hits during baseline conditions – and an increase in time spent swimming.

Fig. 10.15 Magellanic penguins (*Spheniscus magellanicus*) have been trained to interact with enrichment devices rather than rely on chance encounters.

When baseline conditions were reintroduced (without fish), contact with the devices rapidly declined and swimming time in the rockhoppers decreased. When the enrichment devices were reintroduced with fish but without training, the rockhoppers made the greatest number of device contacts and exhibited the highest percentage of time swimming.

In a study of 47 dolphins (*Tursiops truncatus* and *T. aduncus*) at 25 facilities around the world, individuals that were trained on a predictable schedule exhibited higher behavioural diversity than those on a semi-predictable schedule (Miller et al., 2021). In addition, behavioural diversity increased with the number of accessible habitats, but was negatively correlated with the maximum habitat depth.

10.17 Conclusion

Some enrichment practices can improve the welfare of animals in zoos. However, keepers need to be convinced of their value and be given enough time to implement taxon-specific appropriate enrichment. It is also important to undertake scientific studies of the value of particular enrichment techniques, and this also requires time and staff resources to be allocated to this activity. Many studies of enrichment have been conducted over very short periods of time and it is clear that long-term studies are needed to measure the real value of any new enrichment device or technique. Training of many species is essential to their welfare because it facilitates veterinary procedures and reduces the need for keepers to employ handing techniques that cause stress. In very dangerous species training is essential to keep keepers and other zoo staff safe. In some cases there is evidence that training can act as a form of enrichment.

11 Conservation Breeding, Reproduction and Genetics

11.1 Introduction

Attitudes to wildlife have changed irreversibly since the mid-twentieth century:

we were also wanting to collect snakes and chimpanzees, antelopes and sunbirds, in fact we hoped to take back to London a representative collection of the whole of the animal life of this part of Africa.

These are not the words of a professional animal trapper, but those of the naturalist and broadcaster Sir David Attenborough in a television documentary he made in collaboration with London Zoo for the BBC Natural History Unit, referring to a collecting trip to Sierra Leone that was first transmitted in 1955 (BBC, 1955).

In the middle of the twentieth century, zoos were still taking animals from the wild. Few zoo professionals would have given that a second thought. They mounted animal collecting expeditions to remote places or purchased animals from animal dealers or other zoos and circuses.

Gerald Durrell, the founder of Jersey Zoo, described his six-month animal collecting trip to British Cameroon (now part of modern-day Nigeria and Cameroon) in 1957 in his book *A Zoo in My Luggage* (Durrell, 1960). His zoo, now known as *Durrell*, has an international reputation for work in the conservation of endangered species and the training of conservationists working across the world, especially in developing countries.

Conservation breeding is a relatively new activity for zoos. Just a few decades ago, if an animal died in a zoo it was easily replaced with another taken from the wild or purchased from a dealer. George Chapman was a famous London animal dealer with an office in Tottenham Court Road. *Chapman's Monthly Notes* from December 1926 advertised a long list of available species from aardvarks to zebra, Shetland ponies to 'tame red squirrels' (Fig. 11.1). His business closed in 1935.

The role of wild populations of animals in supplying zoos declined significantly in the 1970s with the coming into force of the Convention on International Trade in Endangered Species of Wild Fauna and Flora 1973 (CITES).

Although some animals, such as African lions (*Panthera leo*), have always bred readily in captivity, other species have not, either because there have been too few individuals kept in captivity (and few movements between zoos) or because

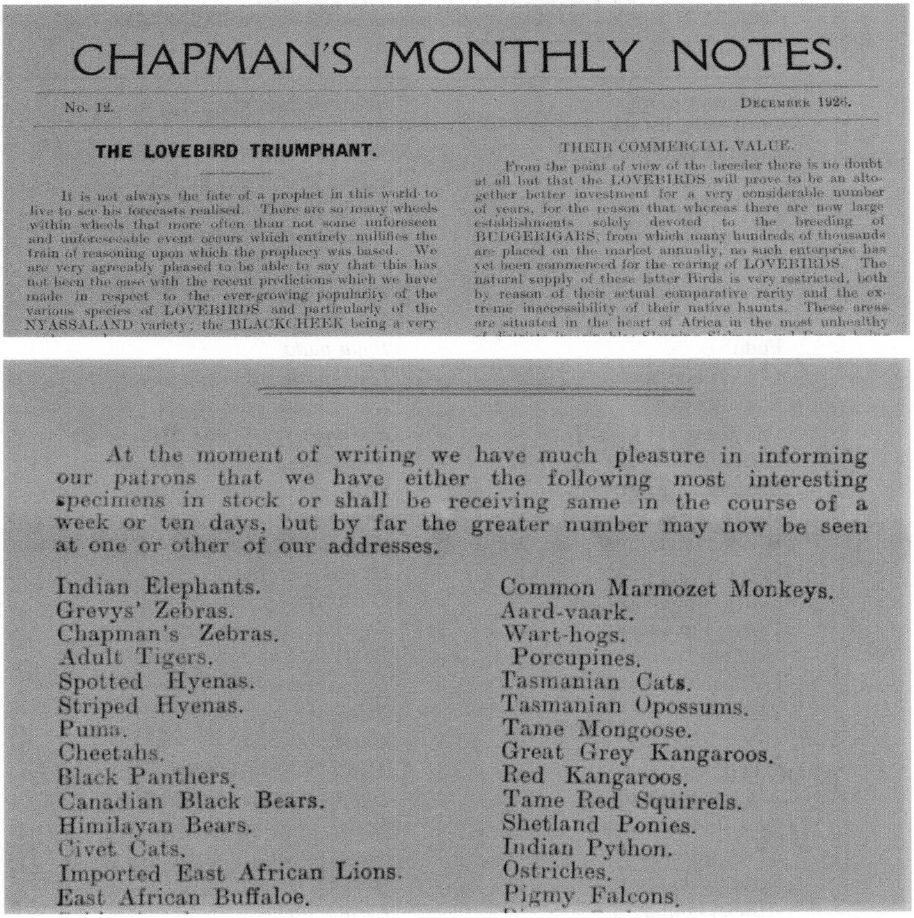

CHAPMAN'S MONTHLY NOTES.

No. 12. DECEMBER 1926.

THE LOVEBIRD TRIUMPHANT.

It is not always the fate of a prophet in this world to live to see his forecasts realised. There are so many wheels within wheels that more often than not some unforeseen and unforeseeable event occurs which entirely nullifies the train of reasoning upon which the prophecy was based. We are very agreeably pleased to be able to say that this has not been the case with the recent predictions which we have made in respect to the ever-growing popularity of the various species of LOVEBIRDS and particularly of the NYASSALAND variety; the BLACKCHEEK being a very

THEIR COMMERCIAL VALUE.

From the point of view of the breeder there is no doubt at all but that the LOVEBIRDS will prove to be an altogether better investment for a very considerable number of years, for the reason that whereas there are now large establishments solely devoted to the breeding of BUDGERIGARS, from which many hundreds of thousands are placed on the market annually, no such enterprise has yet been commenced for the rearing of LOVEBIRDS. The natural supply of these latter Birds is very restricted, both by reason of their actual comparative rarity and the extreme inaccessibility of their native haunts. These areas are situated in the heart of Africa in the most unhealthy

At the moment of writing we have much pleasure in informing our patrons that we have either the following most interesting specimens in stock or shall be receiving same in the course of a week or ten days, but by far the greater number may now be seen at one or other of our addresses.

Indian Elephants.
Grevys' Zebras.
Chapman's Zebras.
Adult Tigers.
Spotted Hyenas.
Striped Hyenas.
Puma.
Cheetahs.
Black Panthers.
Canadian Black Bears.
Himilayan Bears.
Civet Cats.
Imported East African Lions.
East African Buffaloe.

Common Marmozet Monkeys.
Aard-vaark.
Wart-hogs.
Porcupines.
Tasmanian Cats.
Tasmanian Opossums.
Tame Mongoose.
Great Grey Kangaroos.
Red Kangaroos.
Tame Red Squirrels.
Shetland Ponies.
Indian Python.
Ostriches.
Pigmy Falcons.

Fig. 11.1 *Chapman's Monthly Notes* No. 12, December 1926, offered a wide variety of species for sale (source: Chetham's Library, Manchester, England).

appropriate species-specific husbandry had not been developed. This has necessitated the development of cooperative breeding programmes for threatened species.

11.2 Cooperative Breeding Programmes

Cooperative breeding programmes involving exchanges of animals between zoos are of relatively recent origin. The first international studbook (ISB) for a wild species was established for the European bison (*Bison bonasus*) in 1932. The second was begun 25 years later, in 1957, for Père David's deer (*Elaphurus davidianus*). There was no ISB for the giant panda (*Ailuropoda melanoleuca*) until 1976 or for the Komodo dragon (*Varanus komodoensis*) until 1995 (Table 11.1).

Table 11.1 Selected examples of international studbooks maintained by the World Association of Zoos and Aquariums (WAZA) as at 22 May 2022.

Common name	Scientific name	Established
European bison	*Bison bonasus*	1932
Przewalski's horse	*Equus ferus przewalskii*	1959
Persian onager	*Equus hemionus onager*	1961
Pygmy hippopotamus	*Choeropsis liberiensis*	1966
Okapi	*Okapia johnstoni*	1966
Indian rhinoceros	*Rhinoceros unicornis*	1966
Bonobo	*Pan paniscus*	1967
Pudu	*Pudu puda*	1969
Golden lion tamarin	*Leontopithecus rosalia*	1970
Snow leopard	*Uncia uncia*	1971
Bush dog	*Speothos venaticus*	1972
Giant panda	*Ailuropoda melanoleuca*	1976
Chinese alligator	*Alligator sinensis*	1982
White-naped crane	*Grus vipio*	1982
Hooded crane	*Grus monacha*	1985
Muskox	*Ovibos moschatus*	1985
Sumatran rhinoceros	*Dicerorhinus sumatrensis*	1986
Cotton-top tamarin	*Saguinus oedipus*	1986
Babirusa	*Babyrousa babyrussa*	1987
Hartmann's mountain zebra	*Equus zebra hartmannae*	1987
Black lemur	*Eulemur macaco*	1987
Cuvier's gazelle	*Gazella cuvieri*	1987
Addax	*Addax nasomaculatus*	1989
Black lion tamarin	*Leontopithecus chrysopygus*	1989
Great Indian hornbill	*Buceros bicornis*	1990
Mhorr gazelle	*Nanger dama*	1990
Red-billed curassow	*Crax blumenbachii*	1991
African hunting dog	*Lycaon pictus*	1991
Rodrigues fruit bat	*Pteropus rodricensis*	1992
Caracal	*Caracal caracal*	1994
Komodo dragon	*Varanus komodoensis*	1995
Kori bustard	*Ardeotis kori*	1996
Partulid snails	*Partulidae*	1996
Blue-throated macaw	*Ara glaucogularis*	1997
Blue-billed curassow	*Crax alberti*	2007
Javan leopard	*Panthera pardus melas*	2014
Lear's macaw	*Anodorhynchus leari*	2019
Raggiana bird-of-paradise	*Paradisaea raggiana*	2020

Source: www.waza.org/priorities/conservation/waza-international-studbooks (accessed 22 May 2022).

In addition to ISBs, regional zoo associations also operate breeding programmes. European Endangered Species Programmes (EEPs) began in 1985 and were established by the European Association of Zoos and Aquaria (EAZA). They are now known as EAZA Ex-situ Programmes, but have retained the acronym EEP. In North

America, Species Survival Plan® (SSP) programs were established in 1981. At the beginning of 2021 there were almost 500 SSPs (AZA, 2021). The contributions that North American zoos have made to endangered species recovery in the United States have been discussed by Che-Castaldo (2018) (see Section 12.3.4).

Two decades ago, Tribe and Booth (2003) noted that the challenge for zoos was to transform themselves from traditional animal displays to interactive, entertaining conservation centres that bridge the gap between their captive collections and wild populations. They acknowledged the obstacle created by the conflict between the need to be commercially successful and achieving professional conservation credibility in the light of the great expense associated with species recovery programmes.

The conservation of insects has received considerable recent interest, especially those taxa which are of economic interest and have lately suffered significant population declines, such as pollinators (Samways, 2019). However, zoos have played a relatively minor role in the conservation of invertebrate taxa to date. Pearce-Kelly et al. (1991) have described the display, culture and conservation of invertebrates at London Zoo since it began displaying marine species in the Fish House (opened in 1853). The Zoo's Insect House opened in 1881 and began exhibiting both terrestrial and freshwater invertebrates. The conservation value of *ex-situ* insect breeding programmes has been discussed by Pearce-Kelly et al. (2007).

11.3 Population Biology of Zoo Animals

11.3.1 Population Characteristics

Animal populations possess distinct characteristics such as density, a pattern of dispersion, a growth rate, a death rate, an immigration rate and an emigration rate. This is as true for captive populations as it is for wild populations, although clearly animal populations kept in zoos are artificially managed by people.

Zoo populations increase due to births and immigrants (from other collections or the wild) and decrease due to deaths and emigrants (those sent to other collections or reintroduced to the wild). Births may be natural or due to assisted reproduction. As well as natural deaths, animals may be euthanised when sick or injured, or for population control.

It is possible to examine the social structure of groups of a particular species held in zoos on a global scale using data from studbooks or the Zoological Information Management System (ZIMS) (formerly the International Species Information System (ISIS)).

Mortality rates and survival rates for animal populations may be calculated using a life table, a mortality schedule for a population which indicates the probability that an individual will die between one age class and the next – for example, between the age of three and four years. From such a table it is possible to derive a survivorship curve which shows the decline in survival with increasing age. These and other demographic tools are widely used to study the dynamics of animal populations in the wild and have also been used to study zoo populations. Zoo studies have included analyses of populations of chimpanzees (*Pan troglodytes*) (Dyke et al., 1995), squirrel monkeys (*Saimiri sciureus*) (Zimbler-DeLorenzo and Dobson, 2011) and captive fishes (e.g. Comfort, 1962). In some cases studies survivorship in wild populations has been compared with that in zoo populations (e.g. in chimpanzees (Courtenay and Santow, 1989), elephants (Clubb et al., 2008) and orangutans (Wich et al., 2009)).

It must be remembered that survivorship curves tell us what has happened in the past and reflect the environmental conditions of the past; they cannot tell us about the future. Survivorship curves cannot be used to determine the effect of the current zoo environment on the welfare and survival of animals. If survival in a long-lived species was poor in neonates 40 years ago, this will have had an effect on the age structure and survival that is still apparent today. For many captive species, survivorship has improved as a result of improved husbandry, but this will not show up in survivorship curves for many years.

Kohler et al. (2006) used the ISIS database to compare mortality rates in 42 species of captive animals. The species selected were mostly mammals (apes, small primates, carnivores, hoofstock and kangaroos) but also included crocodilians, ratites and raptors. In total, the study included 35,229 individuals. Mortality of wild-born individuals did not differ significantly from that of captive-born animals, and in most of the groups studied, above age five years female survivorship significantly exceeded that of males. In most groups there was a typical pattern of reduced life expectancy with age. However, in raptors life expectancy at 15 years of age was almost the same as it was at 1, 5 and 10 years. A similar pattern was observed in crocodilians.

11.3.2   Group Size and Sex Ratio

It is still possible to find zoos that hold animals individually or in inappropriate group sizes and sex ratios. This may result in poor welfare – due to the lack of opportunities for social contact and social learning – and reduce opportunities for reproduction.

Hussain et al. (2015) reported that Lahore Zoological Garden in Pakistan kept a single specimen of many species, including the following: African elephant (*Loxodonta africana*), white rhinoceros (*Ceratotherium simum*), agile wallaby (*Macropus agilis*), jungle cat (*Felis chaus*), red fox (*Vulpes vulpes*), marsh mongoose (*Atilax paludinosus*) and one each of nine bird species. In addition, the zoo held a number of other highly social species in small groups, including two giraffes (*Giraffa camelopardalis*), two chimpanzees (*Pan troglodytes*), two hippopotamuses (*Hippopotamus amphibius*) and two vervet monkeys (*Chlorocebus pygerythrus*).

Considerable concern has been expressed about the small size of elephant groups held in zoos and the zoo community's inability to produce self-sustaining captive populations (Olsen and Wiese, 2000; Wiese, 2000; Hutchins and Keele, 2006; Rees, 2009b; 2021; Prado-Oviedo et al., 2016).

In 2006 I examined the global distribution of Asian elephants (*Elephas maximus*) and African elephants (*Loxodonta africana*) living in zoos using the ISIS database (now ZIMS) (Rees, 2009b). Most were kept by zoos in Europe or North America, and

Table 11.2 The global distribution of ISIS zoos (2006) and ZIMS zoos (2018) holding elephants.

	Loxodonta africana		*Elephas maximus*	
	2006	**2018**	**2006**	**2018**
Region	**Zoos (%)**	**Zoos (%)**	**Zoos (%)**	**Zoos (%)**
Europe	42 (40.4)	45 (43.7)	54 (47.4)	76 (46.9)
North America	53 (51.0)	40 (38.8)	44 (38.6)	30 (18.5)
Africa	2 (1.9)	4 (3.9)	0 (0.0)	1 (0.6)
Central America	2 (1.9)	0 (0)	2 (1.8)	0 (0)
South America	2 (1.9)	6 (5.8)	1 (0.9)	3 (1.9)
South-East Asia	2 (1.9)	8 (7.8)	8 (7.0)	47 (29.0)
Australia	1 (1.0)	0 (0)	5 (4.4)	5 (3.1)
Total	104 (100)	103 (100)	114 (100)	162 (100)

Source: based on Rees, 2021.

many zoos were still keeping small numbers of elephants that did not reflect their social structure in the wild (Table 11.2). Cows outnumbered bulls four to one (*Loxodonta*) and three to one (*Elephas*) and groups contained seven or fewer individuals: mean, 4.28 ($\sigma = 5.73$). Some 20 per cent of elephants lived alone or with one conspecific; 46 elephants (5.5 per cent) had no conspecific. The minimum group sizes recommended by regional zoo association guidelines were ignored by many zoos. At that time, AZA in the United States recommended that breeding facilities should keep herds of 6–12 elephants, and the British and Irish Association of Zoos and Aquariums (BIAZA) recommended keeping together at least four cows over two years old. In 2006, over 69 per cent of Asian and 80 per cent of African cow groups – including those under two years old – consisted of fewer than four individuals.

Twelve years later I compared the ISIS data from 2006 with the data held in ZIMS in 2018 to examine subsequent changes in elephant mean group sizes and sex ratios (Rees, 2021). There appeared to have been a small increase in mean group size in both genera between 2006 and 2018, on a global basis. However, only the increase in group size of *Loxodonta* was statistically significant. Regional inconsistencies were apparent. In Europe, the mean group size increased very slightly in *Loxodonta*, but in *Elephas* it fell by some 9.5 per cent. In North America the mean group size increased in both genera – 60.6 per cent in *Loxodonta* and 18.1 per cent in *Elephas* – but mean group sizes were still only 4.45 and 3.33, respectively, in 2018.

These data suggest there has been little progress towards increasing the sizes of elephant group in zoos, with the exception of *Loxodonta* in North America, but this population started from a low base.

Elephants taken from the wild in the past produced a skewed sex ratio in favour of females because these were preferred by zoos and circuses as they were considered to be less dangerous than bulls. Between 2006 and 2018 there was improvement in the sex ratios in zoos. In 2006 the global ratio of males to females was 1:3.93 in *Loxodonta* and 1:3.14 in *Elephas*. These changed to 1:2.98 and 1:2.32, respectively,

Table 11.3 The global distribution of elephants in zoos recorded by ISIS in 2006 and ZIMS in 2018.

	Loxodonta africana				*Elephas maximus*			
	2006		**2018**		**2006**		**2018**	
Region	$\male{:}\female$	*Sex ratio*	$\male{:}\female$	*Sex ratio*	$\male{:}\female$	*Sex ratio*	$\male{:}\female$	*Sex ratio*
Europe	40.126	1:3.15	47.135	1:2.87	51.188	1:3.69	89.216	1:2.43
North America	22.125	1:5.68	40.138	1:3.45	23.101	1:4.39	17.83	1:4.88
Africa	2.4	1:2.00	2.5	1:2.50	0.0	–	0.1	0:1.00
Central America	1.3	1:3.00	0.0	–	0.4	0:4.00	0.0	–
South America	1.1	1:1.00	4.8	1:2	1.0	1:0.00	0.5	0:5.00
South-east Asia	2.5	1:2.50	6.9	1:1.50	40.67	1:1.68	112.203	1:1.81
Australasia	0.3	0:3.00	0.0	–	4.14	1:3.50	8.16	1:2.00
Totals	68.267	1:3.93	99.295	1:2.98	119.374	1:3.14	226.524	1:2.32
	335*		394+		493**		750	
	Totals for both species: **2006** = 828; **2018** = 1,144							

One *Loxodonta** and two *Elephas*** specimens whose sex was not recorded have been omitted from the 2006 data. Three *Loxodonta*+ whose sex was not recorded have been omitted from the 2018 data.
Source: ISIS, 2006a; 2006b; ZIMS, 2018a; 2018b.

by 2018, probably reflecting the increased success of captive breeding programmes in zoos (Table 11.3). However, while sex ratios improved in both genera in Europe, there was a slight deterioration in *Elephas* in North America, from 1:4.39 in 2006 to 1:4.88 in 2018.

11.3.3   Bachelor Groups

The propensity of most animal species to produce equal numbers of male and female offspring results in an oversupply of males in those species that naturally live in polygamous harem groups and are kept in naturalistic groups in zoos. One solution to this problem is to establish bachelor (all-male) groups capable of supplying males to other social groups when necessary. However, this solution poses challenges in relation to their accommodation and management.

Gorillas (*Gorilla gorilla*) naturally live in polygamous harem groups usually consisting of one male living with several females and their young. When they are housed as breeding units this inevitably leads to an excess of males, and the challenges of housing excess males in bachelor groups in zoos have been extensively studied. This work has included studies of the feasibility of housing all-male groups and the factors influencing their formation and maintenance (Stoinski et al., 2001; 2004); facility design for bachelor groups (Coe et al., 2009); the effects of age and group type on social behaviour in bachelor and mixed-sex groups (Stoinski et al., 2013); behavioural monitoring of interactions between bachelor and breeding groups (Grand et al., 2013); and comparisons of male–male interactions within bachelor and breeding groups (Pullen, 2005). Other studies have examined aggression and group cohesion (Leeds et al., 2015); the

Fig. 11.2 Przewalski's horses (*Equus ferus przewalskii*) have been the subject of a number of studies of behaviour in bachelor groups.

effects of overnight separation on the incidence of wounding (Gartland et al., 2021); determinants of dominance behaviours (Medina, 2010); the effects of altering group composition (Gartland et al., 2018); the response to the death of silverbacks in multi-male groups (Less et al., 2010); personality and wellbeing in bachelor groups (Schaefer and Steklis, 2014); and the reaction to large crowds of visitors (Kuhar, 2008).

Research conducted on bachelor herds of Asian elephants and the management of bulls has included studies of newly integrated individuals and groups (Hambrecht and Reichler, 2013; Schreier et al., 2021); the facilitation of the social behaviour of bulls (Hartley et al., 2019); and an analysis of the future challenges of managing this species in zoos (Sach et al., 2019; Schmidt and Kappelhof, 2019). Keeping bachelor groups of elephants is not without risk. In 2021 an adult bull African elephant (*Loxodonta africana*) killed a 12-year-old African bull at Noah's Ark Zoo Farm, Somerset, England (BBC, 2021). The zoo kept four bulls in the same enclosure at the time of the attack.

Studies of bachelor groups of Przewalski's horse (*Equus ferus przewalskii*) have included research on the social structure of a bachelor group at Minnesota Zoological Gardens in the United States (Tilson, 1988); the behaviour of stallions during bachelor group formation (Draganova and Gurnell, 2004); a comparison of stallion groups (*E. ferus prezwalskii* and *E. callabus*) kept under natural and domestic conditions (Christensen et al., 2002); and the behaviour of stallions in an enclosure at the Askania-Nova Biosphere Reserve in Ukraine (Zharkikh and Andersen, 2009) (Fig. 11.2).

Hamadryas baboons (*Papio hamadryas*) in zoos have traditionally been kept in single male groups, but more recently they have been housed in multi-male groups, with surplus males held in bachelor groups. The social compatibility of an all-male group of Hamadryas baboons was studied by Koot et al. (2016). Ten social groups of these baboons were monitored by Wiley et al. (2018) across seven North American zoos for one year. They found that there was no difference in wounding rates between group types or by sex, suggesting that the welfare of males is not compromised by keeping them in either multi-male or bachelor groups.

In a study of social spacing in a bachelor group of related captive brown woolly monkeys (*Lagothrix lagotricha*), similar behavioural spacing mechanisms appeared to be operating in captivity and in the wild, where males remain with their natal group and females disperse (White et al., 2003). This philopatry allows the males to experience long periods of familiarity, have a common developmental experience and the possibility of shared kinship. In this study a group of four monkeys was merged with a pair to form a single bachelor group. Contact between the individuals was rare and this appeared to accommodate the introduction of the two new monkeys into the group.

In the wild, male cheetahs live in coalitions, usually littermates that have remained together. Chadwick et al. (2013) described the successful creation of a naturalistic coalition of three male cheetahs (*Acinonyx jubatus soemmeringii*) at Chester Zoo, England (Fig. 11.3). Two coalitions each containing two males were combined and sociograms indicated that each individual initially continued to associate most often with the animal with which it was originally paired. When one male was removed the remaining individuals formed a coalition of three similar to those observed in the wild (Caro, 1994).

There has been little research to date on the social impact of managing surplus male pinnipeds in bachelor groups. Emmett et al. (2021) studied four related male South American fur seals (*Arctocephalus australis*) housed at Bristol Zoo, England, and found that positive social interactions were more frequent than negative interactions – there were no excessive negative interactions – and negative interactions did not increase with the onset of the breeding season.

The chemical control of aggression in bachelor groups has been discussed for antelopes (Penfold et al., 2007); rock hyrax (*Procavia capensis*) (Raines and Fried, 2016) (Fig. 11.4); fringe-eared oryx (*Oryx gazella callotis*) (Patton et al. 2001); capybara (*Hydrochoerus hydrochaeris*) (Yu et al., 2021); Amur leopards (*Panthera pardus orientalis*) (Harley, 2019) and collared lemurs (*Eulemur collaris*) (Ferrie et al., 2011). A study of the social housing of surplus male Javan langurs (*Trachypithecus auratus*) established that castration can be used as a management tool to allow surplus males to remain as subordinate follower males in bachelor pairs and in mixed-sex groups as companions (Dröscher and Waitt, 2012).

11.3.4   National and Global Studies of Zoo Populations

The existence of global databases of animal holdings in zoos makes it possible to undertake studies of the global distribution and demography of particular species held

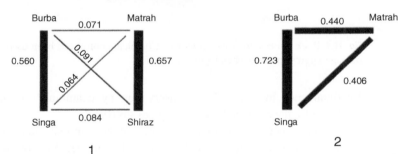

Fig. 11.3 (a) Cheetahs (*Acinonyx jubatus soemmeringii*) at Chester Zoo, England. (b) Sociograms showing the relationships within a group of four male cheetahs. The thickness of each line indicates the strength of association. These animals were previously kept as pairs; Burba with Singha and Matra with Shiraz. (1) When Shiraz was removed from the group a coalition of three was formed (2) (source: Sociograms based on Chadwick et al., 2013).

within the zoo community. Such studies make it possible to examine, for example, whether or not zoos are keeping particular taxa in appropriate social groups compared with those observed in natural conditions or with reference to zoo husbandry guidelines (e.g. Rees, 2009b).

Species inventories are also an important resource for researchers studying the role of zoos in conservation. For example, Azevedo et al. (2011) investigated the role of

Fig. 11.4 Rock hyrax (*Procavia capensis*). Chemical control has been used in this species to manage aggression in bachelor groups.

Brazilian zoos in *ex-situ* bird conservation by examining the annual inventories produced by the Sociedade de Zoológicos do Brasil.

Breeding success and mortality rates of zoo animals may be assessed using studbook records, some of which are available online. Wielebnowski (1996) examined the relationship between juvenile mortality and genetic monomorphism in captive cheetahs (*Acinonyx jubatus*), and Ryan et al. (2002), studied the effect of hand-rearing on the reproductive success of Western lowland gorillas (*Gorilla gorilla gorilla*). Both studies used the relevant studbook data for the North American zoo populations.

Although zoo databases may be useful for population studies, Hosey et al. (2012) have cast doubt on the usefulness of zoo records in answering behavioural questions.

11.3.5 The Effect of Social Factors on Reproduction

Some species breed more successfully when kept in large numbers, while others breed less well when their density is high or when incompatible individuals are kept together.

Fig. 11.5 Chilean flamingos (*Phoenicopterus chilensis*).

Studies have shown that reproductive success in flamingos is more likely in large flocks than in small flocks. A study of flamingo flocks in zoos in Britain and Ireland found that breeding flocks were significantly larger than nonbreeding flocks and that larger flocks bred more frequently than smaller flocks (Pickering et al., 1992).

A six-year study of Caribbean flamingos (*Phoenicopterus ruber*) at the Smithsonian's National Zoological Park (or National Zoo) in Washington, DC showed that the addition of birds to the flock caused an increase in group display activity (Stevens and Pickett, 1994). This led the authors to suggest that an alternative strategy for promoting reproduction might be to separate birds in a flock and reunite them prior to the breeding season. The results of a study of a flock of Chilean flamingos (*P. chilensis*) at Dublin Zoo in Ireland also suggested that new additions to a colony may trigger breeding in the subsequent year (Farrell et al., 2000) (Fig. 11.5).

DeMatteo et al. (2006) studied the reproductive biology of bush dogs (*Speothos venaticus*) and found that the presence of a male shortened the interoestrous intervals in females and increased the number of oestrous cycles. Males appeared to exhibit year-round, non-seasonal, sperm production (Fig. 11.6). Female sloth bears (*Ursus ursinus*) were most social when exhibited with a familiar male, and social behaviour and non-social activity decreased in the presence of an unfamiliar male (Forthman and Bakeman, 1992)

A study of reproductive behaviour and success in giant pandas (*Ailuropoda melanoleuca*) held at two breeding centres with different population densities during gestation, birth and the postpartum period found that high density increased cub rejection and decreased maternal care (Ciminelli et al., 2021). This may have been the result of stress caused by the high density of neighbouring conspecifics or the

Fig. 11.6 Bush dog (*Speothos venaticus*).

increased keeper activity in the high-density facility. The relationship between behaviour and reproductive success in giant pandas has also been discussed by Kleiman (1983) and Zhang et al. (2004).

The reproductive success of black and gold howler monkeys (*Alouatta caraya*) in the European captive population has been poor (Fig. 11.7). In a study of the effect of social factors on reproduction in this species, Farmer et al. (2011) found that females living in family groups produced more offspring than those housed in pairs, and more of these offspring survived. Males living in family groups were also more successful at having offspring than those living in a pair. Males who performed a higher howl rate had increased reproductive success, as did females who regularly heard the howls of familiar conspecifics.

11.3.6   Effect of Climate on Reproduction and Survival

Many *ex-situ* captive breeding programmes for endangered species are operated by zoos located in parts of the world where the local climate is quite different from that experienced by these species in the wild. Although some zoos create artificial climatic conditions for some of the species they hold – for example, warm and moist conditions for tropical forest species and cold conditions for polar species – many are exposed to the local conditions. For some species this does not have an effect on their reproduction, but for others it may affect their breeding success.

Princée and Glatston (2016) used data from the global studbook for the red panda (*Ailurus fulgens*) to demonstrate that this species – which is adapted to a cold, damp

Fig. 11.7 Black and gold howler monkeys (*Alouatta caraya*) at *Durrell*. Females of this species living in family groups in zoos produced more offspring than those housed in pairs.

climate – does not reproduce well in zoos located in warmer climates. They were able to show that the climate at the location of birth had an impact on neonate survival.

The effect of climate on the age-specific survival of Asian elephants (*Elephas maximus*) in Myanmar was studied by examining historical records for 1,024 semi-captive elephants from four different sites (Mumby et al., 2013). Maximal survival for elephants between 1 month and 17 years of age was observed at approximately 24°C and any departures from this increased mortality. Neonates and mature individuals had maximal survival at lower temperatures. Mumby et al. suggested that low temperatures could contribute to the low survival of elephants in zoos where the lowest temperatures are, on average, equal to or lower than the coldest months in Myanmar.

The effect of climate change on the reproduction and survival of animals in the wild and in captivity may become critically important as climatic zones shift as a consequence of climate change.

11.4   Assisted Reproductive Technologies

11.4.1   Introduction

Assisted reproductive technologies (ARTs) include artificial insemination (AI), assays of steroid metabolites, semen collection and processing, *in vitro* fertilisation and embryo transfer, and cloning.

Almost four decades ago Marshall (1984) warned of the difficulties inherent in developing a semen storage programme at a zoo because the factors involved in freezing semen and insemination procedures would likely vary between species. In spite of this, great strides continue to be made in AI technology for rare species.

The use of model domesticated species to assist in the development of techniques for improving reproduction in rare species has been helpful in some taxa. The use of the domestic cat (*Felis catus*) as a model for assisted reproduction procedures in wild felids has been discussed by Sowińska (2021), and the relevance of the horse (*Equus callabus*) as a model for optimising cryopreservation and embryo production in wild equids has been considered by Smits et al. (2012).

Although the use of AI techniques for rare mammals is frequently reported in the literature, the techniques have also been used in other taxa – for example, AI technology for ratites has been reviewed by Malecki et al. (2008).

11.4.2 Detecting Oestrus and Hormone Monitoring

Hormone monitoring is widely used in zoos to monitor reproductive status. Measuring steroid hormone (oestrogen and progesterone) metabolites in the faeces of animals allows zoos to study their reproductive cycles and to determine if and when individuals become pregnant (Schwarzenberger et al., 1996). Such changes may be correlated with changes in the frequency of mounting and other behaviours.

DeMatteo et al. (2006) validated a non-invasive faecal hormone monitoring technique for bush dogs (*Speothos venaticus*) that may provide population managers with tools for optimising the breeding potential of this species. The application of endocrine monitoring to the management of giant pandas has been discussed by Czekala et al. (2003).

Odour-detection dogs have been widely used to identify drugs, cancer, explosives and for tracking, and to detect oestrus in dairy cows (Fischer-Tenhagen et al., 2011). The use of a trained two-year-old beagle to detect pregnancy in polar bears (*Ursus maritimus*) was investigated by Curry et al. (2021). The dog was trained to discriminate between faecal samples from pregnant and non-pregnant bears using 300 samples from zoo-housed bears. During training evaluation the dog's true positive rate (sensitivity) and true negative rate (specificity) were both 1.00. However, in real-time tests carried out during two consecutive cubbing seasons performed on 16 female polar bears in season 1 and 17 in season 2, the beagle's sensitivity was zero in both years and specificity was 0.97 and 1.00 in seasons 1 and 2, respectively. The authors suggested that, among other things, the reduced sensitivity of the dog to real-time situations may have been due to a failure to generalise the target odour to novel pregnancies and that there may have been a degradation of some of the samples. Exposure to many more unique conditions was probably required, but this is problematic when sample sizes are limited.

A novel method of detecting oestrus in a slow loris (*Nycticebus* sp.) using bioacoustics has been described by Schneiderová and Vodička (2021). They monitored a single female for 21 months while housed singly or with one of two males to determine whether or not her oestrous cycle could be detected from her vocal activity. The authors discovered that a regular cycle of increased whistle production corresponded to a previously described oestrous cycle of slow lorises, whether or not a male was present. Vaginal smears collected when the female was close to her peak period of vocal activity showed signs of proestrus and oestrus. This study demonstrated that bioacoustics is a promising non-invasive method of anticipating the oestrous cycle in the slow loris.

11.4.3 Radiography and Ultrasonography

Developments in radiography and ultrasonography have been important in devising non-invasive methods for the study of reproductive anatomy, pathology and physiology in recent years.

Ultrasonography has been used to assess the urogenital tracts of elephants. Hildebrandt et al. (2000a) examined 280 captive and wild African and captive Asian female elephants and concluded that uterine tumours, endometrial cysts and ovarian cysts that resulted in acyclicity were the primary pathological lesions that influenced reproductive rates. Male elephants of both species were found to have low (14 per cent) observable reproductive tract pathology, even in older bulls, but apparent infertility of non-organic cause in these otherwise healthy bulls was high (32 per cent) (Hildebrandt et al., 2000b). Oocyte recovery has been achieved in black and white rhinoceroses using transrectal ultrasonography (Hermes et al., 2009b).

The use of these methods to study reproductive cycles in captive Asian yellow pond turtles (*Mauremys mutica*) has been demonstrated by Cheng et al. (2010). They used radiography to monitor clutch size and ultrasonography to measure ovarian follicle growth. Ultrasound has also been used to evaluate reproductive cycles in neotropical boid snakes (Garcia and de Almeida-Santos, 2022).

11.4.4 Artificial Insemination

Artificial insemination is an important tool for captive breeding programmes because, for example, it removes the need for animals to be moved between collections to mate, facilitates the use of stored sperm after the death of male animals and allows genetic diversity to be introduced into captive populations by the collection of sperm from wild individuals.

The first successful elephant pregnancy resulting from AI was reported by Dickerson Park Zoo, Springfield, Missouri (Schmitt, 1998), resulting in the birth of an Asian elephant calf in November 1999. AI in elephants has been made possible by the development of electroejaculation and the study of semen characteristics (Howard et al., 1989; Mar et al., 1995), along with the use of ultrasonography to assess reproductive function (Hildebrandt et al., 2000a).

A female beluga (*Delphinapterus leucas*) was artificially inseminated with semen at *SeaWorld San Diego*, California, and subsequently gave birth to a calf in 2008 (O'Brien et al., 2008). This was the first time that AI had been successfully used in this species.

The semen was collected by manual stimulation using an artificial vagina consisting of a circular polystyrene rim of 10 cm diameter covered by a latex liner.

Semen Collection

Melville et al. (2008) described the collection of semen by manual stimulation in a black flying fox (*Pteropus alecto*), in which motile spermatozoa were obtained in 17 out of 34 attempts. Semen quality (e.g. concentration and motility) was similar to that previously obtained by electroejaculation, with the exception of volume. The collection of semen in a lar gibbon (*Hylobates lar*) by manual stimulation and its subsequent successful cryopreservation have been described by Takasu et al. (2016).

Electroejaculation has been used for a wide range of species, including chimpanzees, bonobos and gorillas (Bader, 1983); lesser Malay chevrotain (*Tragulus javanicus*) (Haron et al., 2000); La Plata three-banded armadillo (*Tolypeutes matacus*) (Herrick et al., 2002); Siberian tigers (*Panthera tigris altaica*) (Fukui et al., 2013); agouti (*Dasyprocta leporina*) (Mollineau et al., 2010); gaur (*Bos gaurus*) (Napier, 2011); cheetah (*Acinonyx jubatus*) (Marrow et al., 2015); coati (*Nasau nasau*) (Lima et al., 2009) and the viviparous lizard *Sceloporus torquatus* (Martínez-Tores et al., 2019). Electroejaculation has also been used to obtain semen from giant pandas (*Ailuropoda melanoleuca*) at Ueno Zoo in Japan (Masui et al., 1989) and Anubis baboons (*Papio anubis*) at the National Zoo in Cuba (Luis et al., 2009).

Penile vibrostimulation was used to obtain semen in the common marmoset (*Callithrix jacchus*) by Schneiders et al. (2004). The ejaculates obtained contained on average 3–4 times as many total and motile spermatozoa as found in samples obtained by rectal probe electroejaculation.

The consistent production of ejaculates from rhinoceroses (*Rhinoceros unicornis*, *Diceros bicornis* and *Ceratotherium simum*) using a specially designed rectal probe that accurately fitted their anatomy has been reported by Roth et al. (2005). However, high-quality samples were only obtained in 50 per cent of attempts.

Electroejaculation traditionally involves high voltages (3–15 volts at 900 mA maximum) and considerable discomfort for the subject. The use of a novel low-voltage (2–3 volts at 500 mA maximum) procedure that produced good-quality semen samples and less stress has been described for Lanyu miniature pigs (*Sus barbatus sumatrensis*) living in the Livestock Research Institute in Taiwan (Chen et al., 2020).

Artificial vaginas have been used to collect semen in a range of taxa, including a Sumatran orangutan (*Pongo pygmaeus abelii*) (Vandevoort et al., 1993); a greater one-horned rhinoceros (*Rhinoceros unicornis*) (Schaffer et al., 1990); a cheetah (*Acinonyx jubatus*) (Durrant et al., 2001); a bonobo (*Pan paniscus*) (Yoshida, 1997); a Javan banteng (*Bos javanicus*) (Yoelinda et al., 2020) and a Bactrian camel (*Camelus bactrianus*) (Mosaferi et al., 2005) (Fig. 11.8).

The use of pharmacological methods of enhancing penile erection for manual semen collection in the white rhinoceros (*Ceratotherium simum simum*) has been reported by Silinski et al. (2002). In a double-blind trial using alpha-adrenergic agents they found improvements in the strength and duration of the erection and more

Fig. 11.8 Bactrian camel (*Camelus bactrianus*). Semen from this species has been collected by using an artificial vagina.

frequent pulsatile contractions of the penis muscles, but no significant improvement in seminal fluid collection.

Collecting semen from large mammals can be hazardous for zoo staff and veterinarians. A specially designed restraint chute for the manual collection of semen in white rhinoceroses has been described, which incorporates easy access to the genital area to facilitate manual stimulation (Walzer et al., 2000).

Semen Quality, Quantity and Sex Sorting

A study of lifetime semen production in a cheetah (*Acinonyx jubatus*) found that sperm quality improved from age three years to age eight years, then declined with age (Durrant et al., 2001). The authors were surprised to find that the animal did not reach peak sperm production until age eight years as average life expectancy is approximately seven years.

Inbreeding has been documented for many years in the captive dusky gopher frog (*Lithobates sevosus*), and this has resulted in severely reduced sperm quality: concentration, viability, total mobility and forward progressive motility (Hinkson and Poo, 2020).

The sex sorting of semen results in semen which contains a single type of sex chromosome. This is a chromosome-sorting technique used to control the sex of offspring produced by AI by separating X chromosomes from Y chromosomes.

Sorting of fresh and frozen–thawed spermatozoa from Western lowland gorillas (*Gorilla g. gorilla*) at the Henry Doorly Zoo in Omaha, Nebraska has been described by O'Brien et al. (2005). Other studies of sex sorting have been conducted using the sperm of Asian elephants (*Elephas maximus*) (Hermes et al., 2009a); black and white rhinoceroses (*Diceros bicornis* and *Ceratotherium simum*) (Behr et al., 2009); bottlenose dolphin (*Tursiops truncatus*) (O'Brien et al., 2006); wood bison (*Bison bison athabascae*) (Zwiefelhofer et al., 2021) and non-human primates (O'Brien et al., 2004).

Ovarian Superstimulation

The ovaries may be induced to increase their release of eggs by chemical stimulation. Ovarian superstimulation has been achieved in anoestrous rhinoceroses (*D. bicornis* and *C. simum*) using deslorelin acetate, a gonadotropin-releasing hormone analogue (Hermes et al., 2009b). Ovarian superstimulation has been used in a number of other species, including pig-tailed macaques (*Macaca nemestrina*) (Cranfield et al., 1989) and wood bison (Palomino et al., 2020).

Cryopreservation of Gametes

The cryopreservation of gametes is important for both the short-term transportation of gametes and their long-term storage in gene banks. Thiangtum et al. (2006) cryopreserved semen from the fishing cat (*Prionailurus viverrinus*) and established that good-quality ejaculates could be obtained containing spermatozoa that exhibited adequate function after cryopreservation for *in vitro* fertilisation procedures.

The collection and cryopreservation of sperm from polar bears (*Ursus maritimus*) living in zoos has been described by Wojtusik et al. (2021, 2022). They established that sperm can be collected efficiently by urethral catheterisation and that sperm can be rescued post mortem from the epididymides and vasa deferentia.

Experiments on AI of Galliformes were performed using fresh and frozen–thawed semen collected from birds kept by Clères Zoological Park, France, by Saint Jalme et al. (2003). The results of this work demonstrated the feasibility of using a simple, inexpensive method to achieve AI and cryopreservation of semen in endangered pheasant species.

The ability to freeze sperm in the field is important if wild populations are to be used to support the maintenance of genetic diversity in zoo populations. The importance of developing links between *in-situ* and *ex-situ* populations of felid species has been considered by Swanson et al. (2007). They noted that the *in-situ* collection and cryopreservation of semen from free-living populations offers a viable option for introducing founder genes into captive populations without taking individuals from the wild.

The cryopreservation of mammalian ovaries and oocytes has been described by Jewgenow et al. (2011), along with the establishment of the Felid Gametes Rescue Project within EAZA zoos.

Recent studies have examined: the cryopreservation of giraffe epididymal spermatozoa using different extenders and cryoprotectants (Hermes et al., 2022); the optimisation of cryopreservation in black-footed (*Spheniscus demersus*) and gentoo (*Pygoscelis papua*) penguins using dimethylacetamide and dimethylsulphoxide (Marti-Colombas et al., 2022); and semen collection, evaluation and cryopreservation in the bonobo (*Pan paniscus*) (Gerits et al., 2022).

Studying Surrogate Species to Assist in AI Development

For some rare species it has been necessary to study the reproductive biology of surrogate species to inform the development of AI techniques. For example, when the last remaining black-footed ferrets (*Mustela nigripes*) were taken into captivity in the United States, the deficit of scientific data on their reproductive biology resulted in scientists studying the European polecat (*M. putorius*) and the Siberian polecat (*M. eversmanii*) as model species (Santymire et al., 2014).

Possible Negative Consequence of AI

Although AI may be a valuable tool in the conservation of some species, there may be negative behavioural consequences in other species. There is some evidence that the development of normal sexual behaviour in elephants has a social facilitation element that requires juvenile males to be reared in a group containing a sexually active adult male (Rees, 2004b). This sexual behaviour might not develop normally if a male calf is born by AI and not subsequently exposed to adults mating.

A study of giant pandas (*Ailuropoda melanoleuca*) found that females inseminated by AI were more likely to reject cubs than those who were naturally mated (Li et al., 2022). The authors suggested that AI females' lack of information on males may reduce parental investment and recommended that panda conservation programmes should prioritise natural mating.

11.4.5 Cloning

Somatic cell nuclear transfer (SCNT) is a method of cloning whereby the genetic material from a body cell from one individual is transferred to an egg cell from a different individual, from which the genetic material has previously been removed. This cell is then stimulated with an electric current to make it start dividing like a normal embryo. It is then implanted into the uterus of a surrogate mother who later gives birth normally. The first mammal to be cloned from an adult somatic cell (from a sheep's udder) was *Dolly* the sheep in 1996 at the Roslin Institute in Scotland (Campbell et al., 1996).

The cloning of a gaur (*Bos gaurus*) using interspecies nuclear transfer has been reported by Lanza et al. (2000) and the birth of African wildcat kittens (*Felis silvestris lybica*) from domestic cats was reported by Gómez et al. (2004).

Zoos have been at the forefront of the development of ARTs for rare species. However, for most endangered species efficient protocols for ARTs simply do not exist (Herrick, 2019). Clulow et al. (2022) have drawn attention to the fact that, although many species of reptiles and amphibians are threatened with extinction, the development of ARTs for these taxa has been slow, particularly for reptiles.

11.5   Population Management

Zoo populations must be managed to prevent over-population, genetic decline as a result of inbreeding and the inheritance of undesirable characteristics.

11.5.1   Population Control and Contraception

Zoo populations need to be managed to maintain genetic diversity and control the numbers of species that produce large numbers of offspring that cannot be accommodated. In some cases, animals can be kept in separate enclosures or single-sex groups to prevent breeding. In other cases they are either culled or provided with contraception.

The use of deslorelin implants as a contraceptive in North American zoos and aquariums has been reviewed by Agnew et al. (2021). It can interrupt both ovulation and sperm production, but has been used mostly in females. Deslorelin has been widely used in mammals but, more recently, also in some birds, reptiles and fishes. Efficacy has been high in mammals, with failures resulting in offspring in 1.3 per cent of individuals. Higher failure rates have been recorded in birds (14.7 per cent of individuals producing eggs). The numbers of reptiles and fishes treated were too low to warrant analysis. Deslorelin was considered by the authors of the review to be effective and safe for female mammals, with the possible exception of carnivores where it may result in ovulation and cause endometrial pathology. It may not be effective in males of some mammalian species.

11.5.2   Managing Genetic Diversity

Genetic diversity must be maintained in the animals used for captive breeding to ensure the long-term genetic health of the population. It is important to manage populations as metapopulations for genetic and other reasons. Ballou (1993) proposed that zoos should adopt metapopulation strategies that establish geographically separate populations to reduce the possibility of catastrophic loss caused by an infectious disease epidemic.

The management of genetic diversity in captive breeding and reintroduction programmes has been discussed by Ralls and Ballou (1992). They concluded the following:

1. A captive breeding programme should be part of a comprehensive conservation strategy for a taxon that also addresses the problems it faces in the wild.

2. Such a captive breeding programme should be founded well before the wild population has been severely reduced in size.
3. Captive breeding programmes make their most effective conservation contribution if they are demographically and genetically managed.
4. Genetic management should focus on maintaining genetic diversity in order to:
   i. minimise undesirable genetic changes;
   ii. avoid the effects of inbreeding depression;
   iii. maintain future options for genetic management.
5. The numbers of captive-bred individuals available for reintroduction will be limited by genetic and demographic factors.
6. Genetic and demographic management techniques are well developed and not taxon-specific. However, husbandry and reintroduction techniques tend to be taxon-specific and require research and funding.

Genetic fingerprinting is an important tool in the conservation work of zoos because it allows the identification of close relatives – so that inbreeding can be avoided – and it can help to identify evolutionarily significant units (ESUs). Zoo populations, especially those of extremely rare species, are often descended from a small number of founders and therefore susceptible to the deleterious effects of inbreeding. To alleviate this situation, animals – or sometimes just their sperm – may be taken from wild populations to increase the genetic diversity of the gene pool. For example, a large-scale frozen repository of sperm from wild-caught cheetahs (*Acinonyx jubatus*) is available to *ex-situ* populations as a result of developments in the cryopreservation of spermatozoa (Crosier et al., 2006; Comizzoli et al., 2009).

Sometimes this situation is reversed and there is potential for a zoo population to assist in increasing the genetic diversity of a wild population. The wild grey wolf population (*Canis lupus*) in Sweden is highly inbred and descended from just five individuals. Inbreeding depression has reduced litter size and there is a high frequency of spinal disorders. Jansson et al. (2015) compared pedigree data from zoo and wild populations and found that, although the populations had two founders in common, there was potential to almost double genetic variation from 11.2 to 21.1 founder alleles by introducing genes from the zoo population.

Kumar et al. (2016) assessed the genetic diversity of the red panda population (*Ailurus fulgens*) in Padmaja Naidu Himalayan Zoological Park, Darjeeling, India by generating genotypes of 15 individuals. Population viability analysis showed that the population had a very low survival probability (<2 per cent) and would rapidly experience a drop in genetic diversity from 68 per cent observed heterozygosity (H_O) to 37 per cent largely due to its small population size and a male-biased sex ratio. They suggested that supplementation of the zoo's population with a pair of adult individuals every five years would increase survival probability to 99 per cent and genetic diversity to 61 per cent. They predicted that this would allow the future harvesting of pandas for reintroduction and inter-zoo exchanges.

Asexual reproduction reduces the potential to maintain genetic diversity in a population. Although Komodo dragons (*Varanus komodoensis*) normally reproduce sexually, parthenogenesis has been recorded in captive animals. Watts et al. (2006) used genetic fingerprinting to determine that parthenogenetic offspring had been produced by two female dragons that had been kept at separate institutions and isolated from males. Zoos commonly keep only females, moving males between collections for breeding purposes. Watts et al. suggested that the sexes should be kept together to avoid triggering parthenogenesis, which will decrease genetic diversity within the captive population.

Managers of zoo captive breeding programmes expect to see demographic and genetic declines in the species they manage as a result of the limited number of founders, breeding constraints and other characteristics of small populations. Che-Castaldo et al. (2021) examined a data set spanning almost 20 years relating to more than 400 *ex-situ* vertebrate populations managed as SSP programs by AZA and found no change in the demographic and genetic characteristics of the majority of these populations. About 87 per cent of the populations analysed consisted of fewer than 200 individuals and 51 per cent were descended from fewer than 20 founders. Although some zoo and aquarium populations that were unsustainable at the time of the analysis were identified, cooperative management was helping to slow down or prevent declines in the health of many species.

11.5.3   Which Species Should We Save?

Some would argue that saving a species that is the only extant member of a particular genus is more important that saving one from a genus containing many extant species. All other things being equal, this viewpoint would value a cheetah (the only extant species of *Acinonyx*) more than a leopard (one of several species of the genus *Panthera*).

Some scientists argue that the limited funds available for conservation should be used to protect the habitats that we have left (and the species they support) rather than focusing our efforts on a relatively small number of species. This argument is especially convincing when the choice is between protecting an area of tropical forest or breeding a rare species in zoos with no immediate prospect of reintroduction to the wild.

11.5.4   Subspecies, ESUs and Conservation

In the past, animals of related species were routinely housed together, resulting in hybrids between, for example, lions (*Panthera leo*) and tigers (*P. tigris*) (ligers and tigons). Hybrids between polar bears (*Ursus maritimus*) and brown bears (*U. arctos*) at Łódź Zoo in Poland have been discussed by Kowalska (1969). Needless to say, such hybrids were subsequently of no use to captive breeding programmes.

Modern zoos no longer routinely house related species together so interspecies hybrids are rare. However, the recognition of new subspecies identified by DNA analysis has resulted in the identification of hybrids between subspecies in zoo populations. In some cases subspecies have been reclassified as good species and separate breeding programmes have been established for them.

With unlimited resources and access to appropriate breeding stock the conservation community might find the conservation of all – or at least most – intraspecies genetic variation a desirable objective. One of the objectives of the UN Convention on Biological Diversity 1992 is the conservation of genetic diversity between and within species. Article 2 of the Convention provides a broad definition of biodiversity:

'Biological diversity' means the variability among living organisms from all sources including, inter alia, terrestrial, marine and other aquatic ecosystems and the ecological complexes of which they are part; this includes diversity within species, between species and of ecosystems.

Article 9 states the obligations of Contracting Parties in relation to *ex-situ* conservation:

Each Contracting Party shall, as far as possible and as appropriate, and predominantly for the purpose of complementing in-situ measures:

(a) Adopt measures for the ex-situ conservation of components of biological diversity, preferably in the country of origin of such components;

(b) Establish and maintain facilities for ex-situ conservation of and research on plants, animals and micro-organisms, preferably in the country of origin of genetic resources;

(c) Adopt measures for the recovery and rehabilitation of threatened species and for their reintroduction into their natural habitats under appropriate conditions;

(d) Regulate and manage collection of biological resources from natural habitats for ex-situ conservation purposes so as not to threaten ecosystems and in-situ populations of species, except where special temporary ex-situ measures are required under subparagraph (c) above; and

(e) Cooperate in providing financial and other support for ex-situ conservation outlined in subparagraphs (a) to (d) above and in the establishment and maintenance of ex-situ conservation facilities in developing countries.

The reality, of course, is that interspecies variation is difficult to conserve in the relatively small populations held by the zoo community, and in any event, for many species, experts often disagree on the number of subspecies present. In fact, attempts to conserve subspecies *ex-situ* may hinder attempts to produce self-sustaining captive populations by excluding hybrids and preventing gene flow between putative subspecies.

The conservation breeding of some species may be assisted if fewer subspecies are recognised. If zoos hold 100 individuals of a species consisting of 10 individuals of each of 10 subspecies, each subspecies would be harder to conserve than if just two subspecies were recognised and the zoos held 50 individuals of each. Breeding programmes for Asian elephants (*Elephas maximus*) do not recognise the putative subspecies (Shoshani and Eisenberg, 1982) because these programmes have access to too few individuals to make breeding subspecies a viable proposition. However, there has been recent concern about the number of hybrids in the Asian elephant EEP and it has been suggested that their presence could have adverse effects on any future reintroduction potential of the EEP population (Schmidt and Kappelhof, 2019).

Wilting et al. (2015) analysed variation among the nine putative subspecies of tiger using a multiple-trait test that considered molecular, morphological (skull and pelage)

Table 11.4 Putative subspecies of tiger (*Panthera tigris*).

Vernacular name	Scientific name
Bengal tiger	*Panthera tigris tigris*
Caspian tiger*	*P. t. virgata*
Amur tiger	*P. t. altaica*
Javan tiger*	*P. t. sondaica*
South Chinese tiger	*P. t. amoyensis*
Balinese tiger*	*P. t. balica*
Sumatran tiger	*P. t. sumatrae*
Indochinese tiger	*P. t. corbetti*
Malayan tiger	*P. t. jacksoni*

* Recognised as extinct by the IUCN/SSC Red List of Threatened Species.
Source: based on Wilting et al., 2015.

and ecological data (Table 11.4). Their results supported the recognition of only two subspecies: *Panthera tigris sondaica* (the Sunda tiger) and *P. t. tigris* (the continental tiger), which consists of northern and southern management units (ecotypes). They suggest that their results have profound implications for the management of *in-situ* and *ex-situ* tiger populations and facilitate a pragmatic approach to tiger management by allowing zoo-bred subspecies hybrids to be used in captive breeding programmes.

11.5.5   Collection Planning

Many zoos use valuable space to house animal species of little conservation interest – because they are relatively common in the wild – thereby reducing their capacity to accommodate more conservation-sensitive species and more individuals of these species. Furthermore, there is considerable bias towards the keeping of large and attractive species, regardless of their conservation status.

Frynta et al. (2013) studied mammals held by zoos worldwide and found that zoos focused on keeping large and attractive species. The size of zoo populations of boid snakes (boas and pythons) were correlated with attractiveness to humans, body size and, to a lesser extent, taxonomic uniqueness, but not rarity. As perceived attractiveness is important to the public and affects some components of conservation effort, it has been argued that it should be included in conservation reasoning (Marešová and Frynta, 2008).

A study of the threatened terrestrial vertebrate species on the IUCN Red List held in the ISIS network of zoos found that 695 (23 per cent) of the 3,955 species kept were classified as threatened (Conde et al., 2013). In the ISIS zoo network the representation of species that may require conservation breeding programmes was low and their geographical distributions among zoos would make management as metapopulations difficult.

An analysis of AZA cooperative breeding programmes suggests that many more founders and tens of thousands more spaces for animals need to be created in order to make these populations viable if all of the species currently bred are to be maintained (Powell, 2019). The limited resources available to zoos and aquariums and the increasing public concern regarding their value suggests that the principles, practices and philosophies relating to collection planning at institutional, regional and global levels needs to be re-examined. Powell argues that strategic and more scientifically defensible decisions need to be made by regional zoological associations and breeding programmes about which species should be safeguarded so that society could be given a 'Promise List' of species whose extinction zoos would commit to preventing, even if suitable habitats for them disappear. He fears that, as situations in the wild worsen, the zoo community may be asked to work with more species in the future as their conservation status deteriorates. Planning for these future demands on resources must be addressed as part of the process of developing the Promise List with reference to the IUCN/SSC 'Guidelines on the Use of *Ex Situ* Management for Species Conservation' (IUCN/SSC, 2014).

Although it could be argued that common species kept in zoos have no conservation role, it has been suggested that a non-threatened species may be used as a flagship for a related, morphologically similar threatened species that is absent from zoos: a 'proxy species'. This could enhance the value of existing collections and support conservation objectives (Kerr, 2021).

The manner in which AZA Taxon Advisory Groups (TAGs) have been constructed has resulted in some being concerned with a small number of species (e.g. the Elephant TAG, Bear TAG and Ape TAG) while others cover a large number of species (e.g. the Amphibian TAG, Aquatic Invertebrate TAG and Marine Fishes TAG) (Table 11.5). Two decades ago the problems relating to the creation of regional collection plans (RCPs) for the more speciose taxonomic groups managed by TAGs were identified by Smith et al. (2002). They suggested that speciose TAGs should develop actions plans to direct the RCP process so that they can limit the number of taxa they need to consider while ensuring that the plans are directly relevant to conservation.

11.5.6 Population Management and Target Setting

Setting targets for population sizes of rare species is a complex undertaking. Sanderson (2006) has suggested a four-tier system for setting incrementally higher population target levels: demographic sustainability, ecological integrity, sustainable use and ultimately the restoration of historical numbers. He noted that densities may need to be specified in many cases in addition to overall population size, and stressed the importance of using extant or historical reference ecosystems for setting target levels.

At around the same time it had become clear that the zoos affiliated to the World Association of Zoos and Aquariums (WAZA) had failed to manage their populations sustainably in spite of considerable scientific input and organisational effort (Lees and Wilcken, 2009). A decade later Powell et al. (2019) – three researchers from Lincoln Park Zoo, St Louis Zoo and AZA) – recognised that 'most of our collaboratively managed animal populations are not viable for the long term'.

Table 11.5 AZA Taxon Advisory Groups 2021.

Amphibian	Lizard
Anseriformes	Marine Fishes
Antelope, Cattle, Giraffid and Camelid	Marine Mammal
Ape	Marsupial and Monotreme
Aquatic Invertebrate	New World Primate
Bat	Old World Primate
Bear	PACCT (Passerines)
Canid and Hyaenid	Pangolin, Aardvark and Xenartha
Caprinae	Parrot
Charadriiformes	Penguin
Chelonian	Piciformes
Ciconiiformes/Phoenicopteriformes/Pelecaniformes	Prosimian
Columbiformes	Raptor
Coraciiformes	Rhinoceros
Crocodilian	Rodent, Insectivore and Lagomorph
Deer (Cervid/Tragulid)	Small Carnivore
Elephant	Snake
Equid	Struthioniformes
Felid	Terrestrial Invertebrate
Freshwater Fishes	Turaco/Cuckoo
Galliformes	Hippo, Peccary, Pig and Tapir
Gruiformes	

Source: www.aza.org/list-of-taxon-advisory-groups (accessed 25 March 2022).

The concept of minimum viable population (MVP) is widely used by zoos and other conservation organisations. It is defined as the minimum population size required to provide some specified probability that the population will survive for a given period of time. Some studies have suggested MVPs of more than 5,000 are necessary for long-term persistence, regardless of the species and the environmental conditions. However, recent studies have cast doubt on the general applicability of this figure and suggest it may not be useful for conservation planning. Population viability analysis (PVA) is an assessment of a population made using a computer program that provides a quantifiable means of predicting the probability that a population will become extinct, which can be used for prioritising conservation needs. It allows the calculation of an MVP for a species, taking into account both deterministic factors (e.g. habitat loss or overexploitation) and stochastic (random) factors (e.g. demographic, environmental and genetic factors).

Bustamante (1996) used the computer program VORTEX to perform a PVA of bearded vultures (*Gypaetus barbatus*) held in European zoos and those released to the wild. The model showed that the most effective way to increase the release rate of birds without increasing the size of the captive population was to improve hatching success. The population viability of captive Asian elephants (*Elephas maximus*) in Lao PDR has been discussed by Suter et al. (2014) and the inability of zoos to establish a sustainable population of this species has been noted by Rees (2003).

Yuan et al. (2021) examined inbreeding depression in, and conducted a PVA of, South China tigers (*Panthera tigris amoyensis*) in captivity. They estimated that the captive population had a probability of extinction of 1 per cent within 100 years and recommended a breeding plan that would reduce the effects of inbreeding and increase the percentage of females breeding.

The inability of most zoo populations to achieve sustainability has prompted zoo professionals to re-examine the value of MVPs and the setting of viability targets. Putnam et al. (2022) suggested that the inability of *ex-situ* populations to meet specific viability targets obscures the benefits such populations receive from rigorous, science-based management. They suggest a decoupling of viability measures and the way they predict population persistence from the benefits of science-based management because viability measures are not good indicators of which populations may benefit most from such management. Putnam et al. point out that successful management strategies produce reproductively robust populations that limit inbreeding and retain genetic diversity better than unmanaged populations.

11.5.7   Predicting Population Growth

Demographic data available on zoo populations make it possible to model future changes. Glatston and Roberts (1988) studied the population status of red pandas (*Ailurus fulgens*) in zoos using data from the international studbook for the species. They made both demographic and genetic analyses of the captive-born population at the time and simulated future trends in the population structure.

The population dynamics of zoo populations of spectacled bears (*Tremarctos ornatus*) have been modelled by Faust et al. (2004) using data from animals maintained in AZA institutions. In particular, they examined the possible effects of birth sex ratio (BSR) bias. At the time of the study the population contained 65 per cent males. Such a bias could make it difficult to form a sufficient number of monogamous pairs to grow the population in the future. To examine the effects of BSR bias they simulated the effects of a nine-year run of bias populations (either 70 per cent male or 70 per cent female) and compared these with no bias populations (50 per cent of each sex) over a time scale of 30 years. As would be expected, runs of the simulation with a female bias produced slightly faster population growth than those with a male bias or no bias. However, the predicted trajectories of the population in the presence of biased BSR did not differ greatly from the baseline result and under all three scenarios the population reached its target of 100 bears after 30 years.

11.5.8   Breeding Recommendations

Breeding and transfer plans (BTPs) are essential for the proper operation of captive breeding programmes. Gray et al. (2021) have examined the reasons why breeding and transfer recommendations are sometimes not fulfilled by zoos and aquariums. These recommendations consisted of instructions to 'Breed With', 'Do Not Breed', 'Hold' and 'Send To'. They distributed 2,335 surveys to collect information on breeding recommendations, and received 1,307 responses from 167 zoos and aquariums from 2007 to 2019. This represented a response rate of 56 per cent. Common reasons for 'Breed With' recommendations remaining unfulfilled were related to the status of an individual and a pair's breeding behaviour. For other types of recommendation, reasons often related to management or logistical factors. More than 55 per cent of 'Breed With' recommendations were attempted but did not result in any offspring, eggs or detectable pregnancy. This was the result of pair incompatibility or the lack of time allowed for breeding. Recommendations to 'Hold' or 'Do Not Breed' were often unfulfilled because BTP recommendations had been updated.

In an earlier study of a database of over 110,000 breeding and transfer recommendations issued from more than 200 SSPs from 1999 to 2013, Faust et al. (2019) established that 'Breed With' recommendations were fulfilled at a rate of just 20 per cent before the next BTP recommendation. 'Send To' recommendations were filled in 56.8 per cent of cases, 'Do Not Breed' in 95.7 per cent of cases and 'Hold' at a rate of 92.9 per cent.

11.6   Do We Always Need an *Ex-Situ* Breeding Programme?

Zoos promote captive breeding as being critical to the survival of many species, often emphasising the importance of 'flagship' and 'umbrella' species. Captive breeding may be essential for some species, but for many more there are other solutions.

In 2016 the giant panda (*Auloripoda melanoleuca*) was downlisted from Endangered to Vulnerable by the IUCN. Staggering amounts of money have been spent on *ex-situ* breeding programmes in China and on new facilities in Western zoos to house pandas loaned from China to attract more visitors and in the hope of producing cubs. Edinburgh Zoo in Scotland received two pandas in December 2011 and housed them in a new exhibit. In 2019 they were moved to a larger exhibit that was funded by the Scottish government and cost £2.5 million, in addition to the cost of leasing the animals, which is around £1 million per year (McKenna, 2019). At the time of writing (December 2022), and after many attempts at producing cubs by AI, Edinburgh Zoo still has not bred giant pandas. Meanwhile, in China, the wild population has been downlisted as a result of the government's conservation policies and the restoration and protection of the bamboo forests (Zhiyun et al., 2002; Swaisgood et al., 2018; Wei et al., 2018).

Conservationists often justify spending extraordinary amounts of money on conserving particular well-known species because they say that this indirectly helps to protect other species living in the same ecosystem. Zoos often refer to particular species that they hold as important for this reason. Logically, conserving so-called 'umbrella species' should help protect other species, but a recent study has cast doubt on the validity of this concept as a general conservation principle. Although the *in-situ* conservation of giant pandas has been a great success in recent years, this success has not benefited the large predators with which they are associated.

Li and Pimm (2016) reported that 'investing in almost any panda habitat will benefit many other endemics'. However, others have shown that, although the giant panda is considered to be both a flagship and an umbrella species, and is one of the species whose conservation has been supported by huge financial resources and scientific effort, this work has not translated into improvements in the conservation status of other species (Li et al., 2020). Specifically, the authors reported the

wide distribution range retreat of the leopard (*Panthera pardus*, 81% loss), snow leopard (*P. uncia*, 38%), wolf (*Canis lupus*, 77%) and dhole (*Cuon alpinus*, 95%) from protected areas in the giant panda distribution range since the 1960s.

There has never been a captive breeding programme for mountain gorillas (*Gorilla beringei beringei*). Nevertheless, their numbers have recovered in the wild. Robbins et al. (2011) reported an increase from around 250 individuals in 1981 to approximately 400. Using demographic data for 1967–2008 they showed that an annual decline of 0.7 ± 0.059 per cent occurred in unhabituated gorillas that received intensive levels of conventional conservation, but habituated gorillas that also received extreme conservation measures showed a population increase of 4.1 ± 0.088 per cent. These extreme measures included daily monitoring, increased protection and the facilitation of veterinary treatment. However, the availability of suitable habitat is important in this success. Sudden increases in mountain gorilla density may cause population decline due to increased male deaths resulting from violent encounters between social units and an increase in infanticide (Caillaud et al., 2020).

11.7 De-extinction: The Resurrection of Lost Species

A number of attempts have been made to breed back animals that have been lost to extinction using animals from zoos and other captive populations.

Lehocká et al. (2018) analysed the genetic diversity of 545 Barbary lions (*Panthera leo leo*) recorded in studbooks between 2011 and 2017. They found the population to be highly inbred – as a result of the long-term use of specific lines and families for mating – and estimated the effective population size to be just 26.66. Prior to this analysis, Black et al. (2010) analysed the genetic health of Moroccan royal lions and Burger et al. (2006) published an urgent call for further breeding of the relic zoo population. Barnett et al. (2006) have argued that the loss of the Barbary line from the zoo population of lions would erode lion genetic diversity and for this reason zoo lions should be managed to prevent this.

The last quagga (*Equus quagga quagga*) – a subspecies of the plains zebra – died in Artis Zoo in Amsterdam, the Netherlands, in 1883 (Fig. 11.9). Some 100 years later, in the 1980s, the Quagga Project was conceived in South Africa with the purpose of breeding back the subspecies by selectively breeding from plains zebras with the appropriate physical features such as a chestnut body colour with decreased body stripes and unstriped legs. The work of the project has been described by Parsons et al. (2007) and Harley et al. (2009).

Fig. 11.9 A quagga (*Equus q. quagga*) at London Zoo in 1870. The last individual died in Artis Zoo in Amsterdam, the Netherlands, in 1883, but attempts are now being made to breed the subspecies back in South Africa (source: Biodiversity Heritage Library – http://biodiversitylibrary.org/page/28201475).

Three pathways for de-extinction have been discussed by Shapiro (2017): breeding back, cloning via SCNT and genetic engineering. She argues that, because the phenotype of an organism is the result of interactions between its genome and the environment, even animals with cloned nuclear genomes will not be exact copies of the extinct species on which they are modelled. The technical challenges faced by scientist attempting to bring back extinct species have been considered by Richmond et al. (2016). Shapiro concludes that de-extinction can only ever be a means of creating ecological proxies for extinct species.

11.8   Zoos and Plant Conservation

Although seed banks are now well established as a method of conserving plant diversity *ex-situ*, this method is not an option for many threatened plant species (Wyse et al., 2018). For these 'exceptional species' – those that either do not produce seeds or are recalcitrant – maintaining a living plant collection is the most common *ex-situ* conservation method utilised.

Fant et al. (2016) outlined seven steps describing how the botanical community can build on the approaches used by the zoological community to maintain the *ex-situ* diversity of threatened exceptional plant species currently managed in living collections.

The viability of the use of a coordinated metacollection approach to the conservation of plants in multiple botanical collections – based on the pedigree-focused approach used by zoos to support the long-term viability of captive animal populations – has been assessed by Wood et al. (2020). They found that current practices are limited in their inability to compile, share and analyse plant collection data at the individual level and original provenance data are difficult to obtain. They suggested the development of a central database to aggregate and track unique individual plants using zoo-style studbooks could transform the *ex-situ* conservation of threatened plant species. The conservation value of metacollections of plants has also been discussed by Griffith et al. (2020).

11.9 Conclusion

The days when zoos could replace their animals with individuals caught in the wild have largely been confined to history. Most of the animals now living in zoos have been born in captivity and many are part of cooperative breeding programmes whereby zoos exchange individuals, thereby creating metapopulations split across a large number of sites. For many species this approach has been very successful, but for others it has not yet been possible to create sustainable zoo populations. A number of ARTs have been developed by zoos and others for those species that are either very rare or do not breed well in captivity. Some of these hold great promise, but they are expensive and success is rarely guaranteed. Zoos have attempted to achieve minimum sustainable populations but have been unsuccessful for many species. However, there is now some recognition that even if this is not achieved, population management has many important benefits that improve the genetic health and viability of captive populations. Although captive breeding may be essential for the survival of some species, it is important to acknowledge that for others concerted *in-situ* conservation efforts can sometime produce spectacular results without the intervention of zoos.

12 Restoration, Rehabilitation and *In-Situ* Conservation

12.1 Introduction

Zoos are increasingly involved with reintroduction projects and *in-situ* conservation. Once a sufficient number of animals have been bred in cooperative breeding programmes, appropriate individuals must be selected for release, and trained to avoid predators and to find food and shelter to increase their post-release survival rates. After release they should be monitored for disease and so that survival rates may be determined. *In-situ* conservation projects may involve zoos providing partners with expertise, training, community education, equipment, funding and other resources to support indigenous species and ecosystems. This chapter is about these activities and begins with a consideration of the purpose of reintroductions.

12.2 Reintroduction

12.2.1 What Is the Purpose of Reintroduction Programmes?

In 2001 I published a paper concerned with the legal obligation of states to reintroduce species back into the wild (Rees, 2001). In this paper I discussed the legal commitments made by states by virtue of having ratified certain international treaties (e.g. the Berne Convention 1979 and the UN Convention on Biological Diversity 1992) and the obligations of the Member States of the European Union under the Wild Birds Directive and the Habitats Directive (Box 12.1).

I also discussed the constraints on reintroductions and particularly the political, legal, social and ecological difficulties of reintroducing large predators. In this paper I argued that the purpose of reintroductions was often far from clear. It is disingenuous for conservationists to suggest that they are restoring lost ecosystems because often they are adding some species that have been lost, but not others, on a piecemeal basis, and usually to a habitat that has been substantially modified or degraded. The resultant ecosystems will not be like any that existed previously. For example, beavers (*Castor fiber*) were reintroduced into Scotland under licence in 2009, having been extirpated by the sixteenth century (Kitchener and Conroy, 1997). But unless the lynx (*Lynx lynx*), brown bears (*Ursus arctos*) and other species that were part of the ecological community present at that time are also restored, the addition of beavers is only a

Box 12.1 International Legal Obligation to Reintroduce Native Species
The Convention on the Conservation of European Wildlife and Natural Habitats 1979 (the Berne Convention) was the first wildlife treaty to encourage its Parties to reintroduce native species as a method of conservation. Under Article 11(2) of the Convention the Contracting Parties undertake:

(a) to encourage the reintroduction of native species of wild flora and fauna when this would contribute to the conservation of an endangered species, provided that a study is first made in the light of the experiences of other Contracting Parties to establish that such reintroductions would be effective and acceptable;
(b) to strictly control the introduction of non-native species.

More recently, the UN Convention on Biological Diversity 1992 has reaffirmed an international commitment to the recovery of species. The preamble to the treaty states that:

the fundamental requirement for the conservation of biological diversity is the in-situ conservation of ecosystems and natural habitats and the maintenance and recovery of viable populations of species in their natural surroundings.

Article 8(d) creates an obligation upon Contracting Parties to:

Promote the protection of ecosystems, natural habitats and the maintenance of viable populations of species in natural surroundings;

and Article 8(f) creates a further obligation to:

promote the recovery of threatened species.

Article 9(c) creates an obligation to reintroduce threatened species, requiring that:

Each Contracting Party shall, as far as possible and as appropriate, and predominantly for the purpose of complementing in-situ measures:

[...]

(c) Adopt measures for the recovery and rehabilitation of threatened species and for their reintroduction into their natural habitats under appropriate conditions.

small step towards producing an ancient lost ecosystem. In July 2022 three female European bison (*Bison bonasus*) were released into a woodland in Kent where one of them subsequently produced a calf (Wildwood Trust, 2022). These animals are ecological analogues of the steppe bison (*B. priscus*) that once inhabited the British Isles.

In contrast to this piecemeal approach, some scientists have advocated rewilding schemes on a grand scale. Perhaps the most fanciful was that of Donlan et al. (2006), who suggested that a Pleistocene rewilding could be achieved in the Great Plains of the United States by introducing analogues of species that are now extinct, thereby reinstituting lost ecological processes. They proposed, for example, that elephants (*Elephas maximus* and *Loxodonta africana*) could be used to replace the five species of North American proboscideans lost to extinction. Needless to say, this idea provoked considerable criticism and debate (Toledo et al., 2011).

Sanderson (2006) has argued that the ultimate purpose of setting target populations for threatened species should be the restoration of historical numbers in the wild. But for most species we simply do not have reliable estimates of these historical numbers and, in any event, habitats across the world have been depleted to such an extent that this would be impossible. Imagine for a moment how the great herds of bison could be restored to their historical levels across North America.

12.2.2   Early Reintroduction Projects

The use of reintroductions to restore lost species to their former habitats is not a new approach to wildlife conservation. Many early reintroductions clearly had a commercial motive as they involved fish and game animals (e.g. beaver (*Castor fiber*): Radford, 1906; elk (*Cervus elephus*): Gerstell, 1936; Atlantic salmon (*Salmo salar*): McCrimmon, 1950). The following is an extract from a document produced by the Pennsylvania Game Commission (2013) relating to a project to reintroduce elk:

History of Pennsylvania Elk

Through the early 1800s, elk (*Cervus elaphus*) inhabited much of Pennsylvania. In the mid-1800s, as human settlements increased, the elk population declined. By the late 1800s, elk had been totally eliminated from their last stronghold in areas around Elk County. In 1913, the Pennsylvania Game Commission (PGC) began reintroducing elk. Elk from Yellowstone National Park, South Dakota, and a private preserve in Pennsylvania were released until 1926.

This project did not appear to involve zoos. In fact, the captive breeding and release of animals into the wild by zoos for conservation purposes is a relatively new phenomenon. The reintroduction of the Arabian oryx (*Oryx leucoryx*) into several countries in the Middle East was the first attempt to re-establish a population of a species that was extinct in the wild using zoo-bred animals (Stanley-Price, 1989). The last wild individual is believed to have been shot in 1972. The species was first reintroduced into the desert in central Oman in January 1982 (Spalton, 1993; Spalton et al., 1999).

Many of the early attempts at restoring wild animal populations were unsuccessful. In the 1980s it was estimated that about half of 1,000 cases of bird reintroductions, translocations and introductions had failed (Long, 1981 as cited in Kleiman, 1989), and it was suggested that perhaps 4 or 5 of the fewer than 20 mammal reintroduction using captive-bred individuals had established viable populations (Wemmer and Derrickson, 1987).

However, there have also been some great successes. By 1889 the American bison (*Bison bison*) had declined to just 835 individuals in the United States, from a population of over 60 million. In 1907 a herd of 15 captive-bred bison from the

Fig. 12.1 European bison or Wisent (*Bison bonasus*).

Bronx Zoo in New York was introduced into a reserve in Oklahoma and subsequently others were released in South Dakota, Nebraska and Montana (Kleiman, 1989). In 2021 the total population of bison in North America was estimated to be over 360,000 (National Bison Association, 2021). Success has also been achieved with the European bison or Wisent (*B. bonasus*) following releases of animals produced by captive breeding in zoos and breeding stations (Perzanowski and Olech, 2013) (Fig. 12.1).

12.2.3 Reintroduction Terminology

When is a reintroduction not a reintroduction? The term has been used by some authors to mean any form of putting animals back into the wild. Beck et al. (1994) defined reintroduction – for the purpose of their survey of the reintroduction of captive-born animals involving zoos – as:

the intentional movement of captive-born animals into or near the species' historical range to re-establish or augment a wild population.

Notwithstanding the appropriateness of this definition for the study to which it applied, it is important to be clear about the precise nature of the process that is taking place when animals are put back into the wild (Rees, 2001).

Nechay (1996) has drawn attention to the importance of distinguishing between the reintroduction, restocking, reinforcement, translocation and introduction of species. He notes that the term 'reintroduction' is often used in practice to describe the release of animals in order to enhance their existing populations: a process that he prefers to call 'restocking' or 'reinforcement'. Nechay also emphasises that the term 'reintroduction' should not be used where a species is placed in an ecosystem where it has not previously existed, as in these circumstances the term 'introduction' is more precise. He concludes that whether a species is the subject of introduction, reintroduction or reinforcement may have serious implications for its future for a number of reasons, including the uncontrolled transfer of genes from one area to another and the possible spread of disease.

The International Union for Conservation of Nature (IUCN) has defined reintroduction as:

an attempt to establish a species in an area which was once part of its historical range, but from which it has been extirpated or become extinct. (IUCN, 1995)

More recently it has been defined as:

the intentional movement and release of an organism inside its indigenous range from which it has disappeared. (IUCN/SSC, 2013)

It is interesting to note that the IUCN Reintroduction Specialist Group has recently changed its name to the Conservation Translocation Specialist Group.

12.2.4   The Role of Zoos in Reintroductions

The role of zoos in the reintroduction of species to the wild is either insignificant – because so few reintroductions have been successful – or essential – because zoos have saved (or helped to save) a number of species that would otherwise be extinct – depending upon whether you believe zoo professionals or the anti-zoo community. The truth is probably somewhere in between.

Captive breeding takes time. Early success is unlikely but, for some species, decades of persistence have paid dividends and animal have been successfully reintroduced. However, Reading et al. (2013) have noted that reintroduction success has been particularly low for threatened and endangered species reintroduced from captivity to the wild and that success rates are especially low for those species that are intelligent, require greater training during development and live in complex societies.

Zoo-bred animals are generally not the major source of animals used in reintroduction projects; however, notable exceptions include the European bison (*Bison bonasus*), golden lion tamarin (*Leontopithecus rosalia*), Hawaiian goose (*Branta sandvicensis*), Père David's deer (*Elaphurus davidianus*) and Mauritius kestrel (*Falco punctatus*) (IUDZC and the Captive Breeding Specialist Group of IUCN/ SSC, 1993; Frankham et al., 2002). For many species little more than a protective zoo environment was required in order to encourage reproduction, but in others (e.g. the California condor *Gymnogyps californianus*) novel rearing techniques have been developed and there can be no doubt that a conservation benefit accrued as a result.

12.2.5 The Challenges of Putting Animals Back into the Wild

A total of 293 animal translocation case studies published in the IUCN's Global Re-introduction Perspective Series were examined and the difficulties identified were categorised for analysis by Berger-Tal et al. (2020). Over 1,200 difficulties were reported, the most frequent of which were, in frequency order, animal behaviour issues (106 reports; 27.6 per cent), monitoring difficulties (96 reports; 32.8 per cent), lack of funding (95 reports; 32.4 per cent), quality of release habitat (77 reports; 26.3 per cent), lack of baseline knowledge (64 cases; 21.8 per cent) and lack of public support (61 cases; 20.8 per cent).

Behaviour difficulties included animal movements and dispersal, foraging and social interactions; in several cases more than one behavioural difficulty was reported for a single case study. So, although behaviour difficulties were the most reported (in terms of total reports), monitoring difficulties were reported in the highest percentage of case studies (32.8 per cent). Of the behavioural difficulties reported, 45.3 per cent were related to movement and dispersal, 16 per cent concerned learning, foraging accounted for 10.4 per cent, territoriality 8.5 per cent and social behaviour 8.5 per cent. When animals are bred in captivity it is essential that the potential for behavioural problems to threaten post-release survival are recognised in advance and addressed.

12.2.6 Defining Zoos' Involvement in Reintroductions

How should we define whether or not a zoo is involved in a reintroduction? Beck et al. (1994) counted a zoo as being involved in a reintroduction project:

if at least one of the reintroduced captive-born animals, or at least one of its documented ancestors, lived in a zoo.

This is a very broad definition and would allow a zoo to claim involvement in reintroduction if it contributed just a single individual animal that fulfilled this criterion. Beck and his co-workers acknowledged that:

Some zoos and some zoo critics exaggerate the importance of reintroduction as a zoo conservation function ... state and federal wildlife agencies are the major proponents and managers of reintroduction.

Based on the definition above, Beck et al. estimated that only about one in five of the 350 zoos and aquariums in the developed world at the time (1991–1992) had been involved in reintroductions. It is important to note that, at the time, the lead author was employed by the National Zoological Park, Washington, DC (the National Zoo), and two of the others were members of the IUCN/SSC Reintroduction Specialist Group.

Zoos often refer to their role in saving species such as the black-footed ferret (*Mustela nigripes*), the Arabian oryx (*Oryx leucoryx*) and Partula snails (*Partula* spp.). Critics claim that the ferret was saved by a captive breeding programme led by the US

Fish and Wildlife Service, the project to reintroduce the oryx has not had long-term success and that large zoos filled with large animals are not required to breed snails. All of these claims are true, at least in part. However, this does not mean that zoos have not made some significant contributions to the reintroduction of rare species back into the wild.

The IUCN Amphibian Conservation Action Plan (ACAP) was launched in 2007. Data from the Amphibian Ark database and the IUCN Red List and information gathered from the academic literature indicated that the number of amphibian species involved in captive breeding and reintroduction projects increased by 57 per cent in the seven years after the launch of the ACAP compared with 1966–2006 (Harding et al., 2016). However, there were relatively few reintroductions in the period after the ACAP launch, with most programmes focusing on securing captive assurance populations (i.e. species taken into captivity as a precaution against extinction in the wild) and conservation-related research. About half of the programmes involved zoos and aquariums and the remainder were run by specialist government facilities or NGOs (Fig. 12.2). Harding and Griffiths suggested that the irreversibility of many current threats to amphibians may make reintroduction an impractical goal for many captive breeding projects and proposed that 'imaginative solutions' may be needed to enable amphibians to survive alongside current, emerging and future threats.

In recent years, zoos around the world have been slowly recognising that they do not have the space and appropriate facilities for housing elephants in conditions that are compatible with their complex behavioural and social needs (Rees, 2021). Some have

Fig. 12.2 The Gaherty Reptile and Amphibian Conservation Centre, *Durrell*, Jersey.

been moved to elephant sanctuaries from zoos (and circuses) to live the remainder of their lives in naturalist settings and large, open spaces in the United States, Europe and South America. Now at least one zoo is being even more ambitious.

In July 2021 the operators of Howletts Wild Animal Park in Kent announced their plan to translocate 13 African elephants, all but one of which were born in the park, to Kenya. The project is being undertaken by the Aspinall Foundation, the Kenya Wildlife Trust and the Sheldrick Wildlife Trust, with the objective of releasing the animals to the wild after a period of acclimatisation (*The Guardian*, 2021). If the project succeeds it will be the first time captive-bred African elephants raised in a zoo have been reintroduced into the wild, and it will set an important precedent for other zoos around the world that still hold elephants. At the time of writing, December 2022, the elephants were still living at Howletts.

12.2.7 Success of Reintroductions of Zoo-Bred Animals

There is no universally accepted definition of the success of reintroduction projects. Judgements are often subjective; however, Robert et al. (2015) have suggested that the IUCN Red List criteria for remnant populations could be used.

In an analysis of intentional introductions or reintroductions of native birds and mammals to the wild in Australia, Canada, Hawaii, New Zealand and the United States between 1973 and 1986 using wild-caught animals, Griffith et al. (1989) found that 75 per cent of 163 were successful. When projects using captive-reared animals were considered, only 38 per cent of 34 were successful. However, whether or not a project had been successful was determined by the project managers' judgements of their own projects.

Only 16 (11 per cent) of the 145 reintroduction projects examined by Beck et al. (1994) were successful. Of the 129 projects for which the relevant data were available, 76 (59 per cent) used zoo-bred animals or their descendants, although some projects released both zoo-bred and captive-born individuals that had not been born in a zoo. The authors were unable to determine how many zoo-born animals had been introduced, but concluded that zoos did not appear to be the primary proponents, animal providers, funders or managers of reintroduction programmes.

A comparison of the success of carnivore reintroduction projects using wild-caught and captive-bred individuals found that the former were much more likely to survive than the latter (Jule et al., 2008). Humans were the direct cause of death in over half of all fatalities, and captive-born carnivores were particularly susceptible to starvation, unsuccessful predator/competitor avoidance and disease after release.

Scientists – including two from the Monterey Bay Aquarium in California – have used sentiment analysis to study the success of species reintroductions (Van Houtan et al., 2020). This method analyses text to quantify affective states and subjective information by searching for keywords such as 'success'. The authors used machine learning to automate the analysis of 4,313 scientific abstracts published over four decades and concluded that reintroduction projects have become less variable and increasingly successful over time.

However, a study undertaken by the Zoological Society of London's (ZSL) Institute of Biology and others drew rather less favourable conclusions. They examined 361 peer-reviewed papers concerned with reintroductions published over two decades (1995–2016) and concluded that they remained largely focused on short-term population establishment (Taylor et al., 2017). Over 60 per cent of the papers addressed questions at the population establishment level and about one-third at the population persistence level, with just 4 per cent and 3 per cent considering metapopulation and ecosystem questions, respectively. The authors found no appreciable increase in the proportion of studies that provide direct support for management decisions by explicitly comparing alternative management actions. Importantly, they found little evidence that conservation science had been embedded in practice by developing clear *a priori* management-relevant questions explicitly comparing different management options. The authors identified a need for more decisions to be made based on, for example, predictive modelling and the analysis of experimental and modelling data.

Although conservation biologists have learned a great deal about how to improve the success of animal reintroductions and translocations, mistakes are still being made. As part of the conservation effort to protect Tasmanian devils (*Sarcophilus harrisii*) from a transmissible cancer known as devil facial tumour disease (DFTD) that was rapidly depleting the species in its natural range, devils were introduced to Maria Island in Tasmania in 2012 in an attempt to establish an insurance population (Scoleri et al., 2020). Four years later they had completely extirpated the indigenous short-tailed shearwater (*Puffinus tenuirostris*) population and a population of around 3,000 little penguins (*Eudyptula minor*) completely disappeared.

12.2.8 Zoos and Native Species Population Supplementation and Reintroduction

Some zoos have worked with conservation agencies to supplement and protect local biodiversity, including initiating and cooperating with population supplementation and reintroduction programmes.

The usefulness of captive rearing in promoting the recovery of butterfly populations has been discussed by Crone et al. (2007). In 1999 Oregon Zoo (Portland, Oregon) and Woodland Park Zoo (Seattle, Washington) began working with the US Fish and Wildlife Service and The Nature Conservancy to captive breed the Oregon silverspot butterfly (*Speyeria zerene hippolyta*) and to return their pupae to sites on the Oregon Coast (Hays and Stinson, 2019). Many thousands of pupae have been released to supplement the wild populations and the two zoos have cooperated to produce a husbandry manual for the species (Andersen et al., 2010).

The Royal Zoological Society of Scotland (RZSS) – operators of Edinburgh Zoo and the Highland Wildlife Park – has been involved in the reintroduction of Eurasian beavers (*Castor fiber*) from Norway to Knapdale in Argyll and Bute in the Scottish Highlands (Goodman et al., 2012). Beavers were once common in Scotland but are thought to have become extinct between the twelfth and sixteenth centuries (Kitchener and Conroy, 1997). In 2007 the Scottish Wildlife Trust and the RZSS formed the

Fig. 12.3 The Scottish wildcat (*Felis silvestris grampia*) interbreeds in the wild with domestic cats. This hybridisation threatens the future survival of the species in Scotland.

'Scottish Beaver Trial'. The release of the beavers was authorised by Scottish Natural Heritage (now .known as NatureScot), the statutory nature conservation agency for Scotland. The RZSS has been involved with a number of aspects of the reintroduction, including captive care and welfare (Campbell-Palmer and Rosell, 2015) and health surveillance (Goodman et al., 2017), and the Wildlife Conservation Research Unit (WildCRU) of the Zoology Department of the University of Oxford undertook supporting ecological work (Macdonald et al., 2000; South et al., 2000). The RZSS is also involved in a project to prevent the disappearance of the Scottish wildcat (*Felis silvestris grampia*) as a result of hybridisation with domesticated cats (Senn et al., 2019) (Fig. 12.3)

Chester Zoo in England has successfully reintroduced a number of native species to areas in the north-west of England from zoo-bred populations: barn owls (*Tyco alba*), sand lizards (*Lacerta agilis*), water voles (*Arvicola amphibius*) and harvest mice (*Micromys minutus*) (Reid, 2007).

12.2.9 Preparing Animals for Reintroduction

Many early reintroductions were unsuccessful because the individuals released had been bred in captivity and were unprepared for life in the wild. Only 35 per cent of the reintroduction projects examined by Beck et al. (1994) used pre-release training to prepare zoo-reared animals for reintroduction (Table 12.1).

There is much more to a reintroduction programme than simply captive breeding – or otherwise acquiring – animals, transporting them to a release site and then letting

Table 12.1 Examples of training given to captive-bred animals prior to release.

Taxon	Training received	Reference
Golden lion tamarin (*Leontopithecus rosalia*)	Searching for hidden food	Beck et al. (1991)
Black-footed ferret (*Mustela nigripes*)	Hunting and killing prey in large enclosures	Oakleaf et al. (1992)
Thick-billed parrots (*Rhynchopsitta*)	Encouragement to fly	Wiley et al. (1992)
Nile crocodile (*Crocodylus niloticus*)	Catching live fish in holding pool	Morgan-Davies (1980)
Elf owl (*Micrathene whitneyi*)	Catching insects	Beck et al. (1994)
Masked bobwhite quail (*Colinus virginianus ridgwayi*)	Cross-fostering to a wild-caught related species prior to release as family groups	Carpenter et al. (1991)

them go. Animals that have been bred and reared in captivity may adapt to their protected environment and suffer fitness losses, including a loss of anti-predator behaviour, food-finding behaviour, locomotor and navigation skills, normal social behaviour (including breeding and nesting and/or refuge use) and appropriate habitat selection (Shier, 2016).

Captive-bred animals may become habituated to the presence of humans. They may lose any fear of people and even be attracted to them because they may have learned to associate humans with the provision of food. Arboreal animals reared in cages with fixed climbing structures may find it difficult to balance on moving branches, and individuals reared in flat enclosures may be unsteady walking over uneven surfaces. Prey species may have no knowledge of their natural predators and animals that have been provided with food that is not part of their diet in the wild may not be able to find sufficient suitable food to sustain themselves after release.

If an acceptable post-release survival rate is to be assured, zoos must make detailed and time-consuming preparations before animals can be reintroduced, and arrange for their post-release monitoring.

Enrichment and Habitat Selection

Enrichment is important to the general physical and psychological welfare of many species kept in zoos. It may also be important to the success of reintroduction programmes. Reading et al. (2013) have argued that behavioural factors are the most important factors affecting post-release survival, and suggest that strategic enrichment programmes targeted at developing specific survival skills promise improvements in success (Fig. 12.4). They also noted that physical enrichment can improve the physical condition of animals and offers the prospect of improved welfare pre- and post-release.

Head-started hawksbill turtles (*Eretmochelys imbricata*) did not appear to choose resting places that protected them from hazardous conditions or locations that were

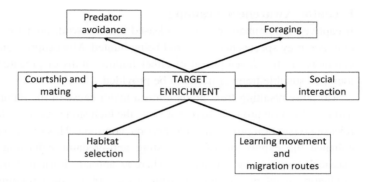

Fig. 12.4 Target enrichments for reintroduction success (source: based on information in Reading et al., 2013).

adequate for efficient resting. Okuyama et al. (2010) suggested that they should be exposed to ledges and other structures in the rearing tank during pre-release training so that they learn to use similar structures in the wild.

Wild black-footed ferrets (*Mustela nigripes*) use prairie dogs (*Cynomys* spp.) for food and shelter in their burrows. Post-release survival in these ferrets may be greatly influenced by rearing methods, especially pre-release conditioning. Biggins et al. (1998) studied 282 black-footed ferret kits that were released into prairie dog colonies in Wyoming, Montana and South Dakota in the United States. They found that ferret survival rates were highest among those reared from an early age in outdoor pens with simulated prairie dog habitat and lowest for cage-reared ferrets with no pen experience.

Environmental enrichment is not always the most important factor determining post-release survival. A study of the effects of head-starting success in captive-bred Eastern box turtles (*Terrapene carolina*) found that the duration of captive rearing was more important than environmental enrichment for enhancing survival. Enrichment had little effect on either post-release behaviour or survival, but turtles that had been reared for longer (21 months) – and were therefore larger – generally had a higher survival rate than those reared for a shorter period (9 months) (Tetzlaff et al., 2019b). The authors of the study concluded that the attainment of greater body size before release allowed greater movement and reduced susceptibility to predation.

When animals are released to the wild it is important to select a habitat with appropriate characteristics. Rowlands et al. (2021) studied burrowing in captive Desertas wolf spiders (*Hogna ingens*), a species that is endemic to Vale da Castanheira, Desertas Grane Islands, Madeira. They performed choice experiments at Bristol and Whipsnade Zoos to investigate preferences for soil types and found that the type and depth of the substrate had an impact on the construction of burrows. Spiders favoured light, loosely packed substrate and the optimum depth was over 50 mm. This knowledge is important for both *ex-situ* conservation and for informing habitat selection for future reintroductions.

Predator-Awareness Training

If captive-bred animals are to be released into the wild, predators need to be able to hunt and prey species need to avoid being hunted. Most captive environments are not conducive to the development of either hunting skills or predator-avoidance behaviours, so suitable training needs to be provided.

Mortality resulting from predation is a major cause of the failure of reintroduction and translocation projects. In recent years the inclusion of anti-predator training in pre-release preparation procedures has become common. This training has the potential to enhance the expression of pre-existing anti-predator behaviour and may include classical conditioning procedures whereby animals learn that model predators are predictors of adverse events (Griffin et al., 2000). Such training usually involves pairing a predator cue with an unpleasant stimulus. Predator-awareness training may aid predator recognition and improve post-release survival rates.

A review of predator-avoidance training found that 42 per cent of published studies concerned fish, 29 per cent involved mammals, 20 per cent were conducted on birds, 7 per cent on amphibians and 2 per cent on reptiles (Edwards et al., 2021). The methods of training used were highly varied and, although most studies reported success, only one-third of studies involved animals that were released after training to establish how it had affected survival.

An assessment of predator-awareness training in terrestrial vertebrates by Rowell et al. (2020) located just 34 publications (describing 40 studies of 29 species). They found it difficult to evaluate the outcomes of training due to the diversity of methods, the different measures of success used and the small sample sizes.

The training of numbats (*Myrmecobius fasciatus*) at Perth Zoo in Western Australia involved exposure to bird warning calls, a hand-tethered live bird of prey and an overhead bird of prey silhouette on a wire-and-pulley system (Jose et al., 2011).

When animals are captive-bred for several generations they may lose their natural responses to predators. When the responses of captive-bred and wild-caught populations of oldfield mice (*Peromyscus polionotus subgriseus*) were compared, individuals from populations that had been in captivity for multiple generations sought refuge less often than did those from wild populations (McPhee, 2004). Variance in predator-response behaviours increased with the number of generations in captivity. This phenomenon would require an increase in the number of individuals released to reach the target population size to compensate for mortality increases.

Meerkats (*Suricata suricatta*) communicate the presence of predators to each other using alarm calls (Fig. 12.5). Captive-born meerkats use the same repertoire of alarm calls as wild individuals and can recognise potential predators through olfactory cues. Hollén and Manser (2007) found that individuals could distinguish between the faeces of potential predators (carnivores) and non-predators (herbivores) and concluded that this ability may have been retained in captive animals because of the recency of relaxed selection on these populations. However, it is clear that some captive-born animals have lost – at least temporarily – some of the behaviours necessary for survival in the wild.

Fig. 12.5 Meerkats (*Suricata suricatta*) communicate the presence of predators to each other using alarm calls.

A study of 17 greater rheas (*Rhea americana*) at Belo Horizonte Zoo, Brazil, that were exposed to three predator and three non-predator models found that, although they increased alert and wary behaviours, they were unable to discriminate between predators and non-predators (Azevedo et al., 2012). Anti-predator training of captive-born greater rheas resulted in released female birds avoiding tall vegetation, which makes predator detection difficult, but the movement of males was influenced more by the reproductive season and the species' complex mating system (Cortez et al., 2018). The anti-predator training of 12 captive-bred greater rheas by exposing them to a stuffed puma (*Felis concolor*) – a natural predator – paired with an aversive stimulus and a chair (an innocuous stimulus) that had not been so paired, did not protect them from predation (Cortez et al., 2015). Eight months after release no birds survived and only one had been killed by a puma. The remainder were killed by dog attacks and as a result of poaching. These birds clearly did not recognise humans or dogs as potential predators.

Captive-born collared peccaries (*Pecari tajacu*) failed to discriminate between predator models (canids and felids) and non-predator models (non-predatory animals), indicating a need for anti-predator training if these animals were to be released into the wild (De Faria et al., 2018).

Head-starting programmes for hellbenders (*Cryptobrancus alleganeinsis*) have successfully used classical conditioning to train the larvae of these large aquatic salamanders to recognise and show a fright response to the scent of introduced trout, a potential predator (Crane and Mathis, 2011).

Tetzlaff et al. (2019a) conducted a meta-analysis of 41 studies of the effects of anti-predator training, environmental enrichment and soft release on wildlife translocations and found that pre-release behavioural conditioning improved the survival of translocated animals by 50 per cent (compared with unconditioned individuals) and reduced movements, and individuals were three times more likely to show site fidelity. Survival was also improved by anti-predator training, environmental enrichment and soft release. Juveniles released from captivity derived the greatest benefit from conditioning. When the survival rates of different taxa were compared, fishes exhibited the greatest benefit from conditioning. The survival of mammals and reptiles also benefited from conditioning, but reports of the results for bird translocations suggested they required improvement.

Hunting Training

There are few published accounts of the successful training and release of captive-bred or orphaned predators compared with those concerned with predator-avoidance training.

Three confiscated cheetahs (*Acinonyx jubatus*) and one leopard (*Panthera pardus*) – which had been taken from the wild in Botswana – were released after limited pre-release training that resulted in successful hunting in all three individuals (Houser et al., 2011). The training occurred at the Jwaneng base camp of Cheetah Conservation Botswana and consisted of exposure to various live and dead prey, including poultry, rabbits and impala (*Aepyceros melampus*). The behaviour of the animals was monitored before and after their release.

Walker et al. (2022) reported success in the release of wild-born, captive-raised orphaned cheetahs in Namibia following strategic pre- and post-release management. This included the selection of appropriate individuals for release, forming artificial coalitions (see also Chadwick et al., 2013), balancing habituation levels in captivity, selecting appropriate release sites and providing strategic support during post-release monitoring. The authors noted that annual survival estimates for the rehabilitated individuals were comparable to those of wild conspecifics and that some released individuals reproduced with wild cheetahs.

Fifteen out of 18 orphaned African wild dogs (*Lycaon pictus*) (three litters) confiscated by the Namibian Ministry of Environment, Forestry and Tourism, were successfully captive reared by the Cheetah Conservation Fund, Namibia, and then translocated for soft release (Marker et al., 2021) (Fig. 12.6).

Food-Finding Behaviour

Captive-bred animals should be exposed during rearing to food items that their wild counterparts prefer, if they are to be adequately prepared for future release. Feeding captive-bred individuals on inappropriate food may make it difficult for reintroduced animals to identify and locate appropriate foods from those available in the wild. However, prey analysis of a head-started hawksbill turtle (*Eretmochelys imbricata*) suggested that this species can make feeding adaptations to the natural environment (Okuyama et al., 2010).

Fig. 12.6 Captive-reared African hunting dogs (*Lycaon pictus*) have been successfully released to the wild.

The reintroduction of captive-bred black-footed ferrets (*Mustela nigripes*) in the United States has been very successful. This species depends on prairie dogs for food in the wild. It has been shown experimentally that early diet in black-footed ferrets affects later food preference. Vargas and Anderson (1996) exposed three groups of kits during the assumed sensitive period for olfactory imprinting (60–90 postnatal days) to different diets: no prairie dog, prairie dog three times per week and prairie dog daily. The kits were individually tested at age five months using a cafeteria food choice trial and a higher preference for prairie dogs was found in those animals exposed to higher quantities of this food in their early diet.

Locomotor Activity and Mobility

In the first year after release reintroduced captive-born golden lion tamarins (*Leontopithecus rosalia*) exhibited an improvement in locomotor skills: less frequent falling, increased time spent on natural substrates and travelling at greater heights (Stoinski and Beck, 2004) (Fig. 12.7). These changes appeared to affect survival. For example, animals that did not survive for six months spent more time on artificial substrates than did surviving individuals. The authors of this study suggested that this and other species should be exposed to complex environments early in their development.

When captive-raised voles (*Microtus*) were released to the wild in Finland they moved very little from the release site in the first 3–6 days (Banks et al., 2002). Predation was highest during the first three days, with no mortality after they began

Fig. 12.7 When captive-born golden lion tamarins (*Leontopithecus rosalia*) are released into the wild, individuals that have been exposed to complex environments show increased survival rates.

moving substantially beyond their release site. This study suggests that the innate fear response which limits the movements of naïve individuals in a novel environment makes them more susceptible to chemo-sensing predators due to odours from accumulated waste. The authors of the study suggested that pre-release conditioning is required to encourage individuals to overcome initial release site fidelity.

Social Behaviour and Cultural Transmission

It is important that animals learn appropriate social behaviours from conspecifics and that conditions that facilitate this learning are maintained in captivity.

Wallace (1994) noted that three main techniques have been used in attempts to achieve self-sustaining wild populations of threatened species: parent-rearing, cross-fostering and isolation rearing (i.e. rearing in isolation from humans). He emphasised the importance of considering the behavioural aspects of preparing birds for release, together with a consideration of life history strategies and an understanding of the degree of interspecific and intraspecific sociality. These factors are important in the development of effective behavioural preparation of individuals for release.

There is some evidence that social facilitation is important in the development of normal sexual behaviour in Asian elephants (*Elephas maximus*) (Rees, 2004b) (Fig. 12.8). If this is true, it is important to keep juvenile elephants of both sexes within a social group containing sexually active adult males and females so that the youngsters may imitate the courtship and mating behaviours of the adults. Keeping elephants in all-female groups with their calves and using artificial insemination to increase the population size may deprive young bulls of the opportunity to learn some sexual behaviour.

Fig. 12.8 A young bull Asian elephant (*Elephas maximus*) 'mounting' a female calf shortly after observing an adult bull mounting an adult female.

Social facilitation affected suckling behaviour in plains zebra (*Equus quagga*), mountain zebra (*E. zebra*) and Grevy's zebra (*E. grevyi*) at Dvůr Králové Zoo, Czech Republic (Levá and Pluháček, 2020). Foals suckled for longer when seeing other foals suckling and synchronised suckling bouts were less often terminated by the mother than unsynchronised bouts.

Social facilitation also affects egg-laying in African village weaverbirds (*Ploceus cucullatus cucullatus*) (Victoria and Collias, 1973), male sexual behaviour in the guppy *Poecilia reticulata* (Farr, 1976) and may be a factor in the reproductive success of captive-born female Southern white rhinoceroses (*Ceratotherium simum simum*) (Swaisgood et al., 2006).

Cultural differences in vocalisations and other behaviours may develop in populations of the same species that are geographically separated. The loss of culture in some species may be a precursor to population decline and eventual extinction. In a study of regent honeyeaters (*Anthochaera phrygia*) the songs of captive-bred birds were found to differ from those of all wild birds. In wild males, 27 per cent sang songs that differed from the regional cultural norm and 12 per cent of males living in areas of low population density failed to sing any species-specific songs, and sang the songs of other species instead. Males singing atypical songs were less likely to pair or nest than those that sang the regional cultural norm and suffered reduced fitness as a result (Crates et al., 2021).

Reintroduced animals may benefit from being released in social groups – for example, by the social transmission of foraging skills in a new environment – but the extent to which animals maintain familiar groups and their social bonds following translocation is not well understood. Franks et al. (2019) have suggested that sociality prior to translocation may not be important and that those individuals most able to adapt and form new associations at a new site are most likely to be the surviving founders of a reintroduced population.

Fitness costs may result from the disruption of social relationships when solitary mammals are translocated. Stephens' kangaroo rats (*Dipodomys stephensi*) translocated with their neighbours travelled shorter distances before establishing territories and had higher survival rates and reproductive success than those translocated without neighbours (Shier and Swaisgood, 2012). Immediately after release, kangaroo rats translocated with neighbours spent more time foraging and creating burrows and less time fighting than those moved without neighbours.

A review of the effects of changes in social groups and their influence on post-release survival in translocated animals has been provided by Franks et al. (2019).

Reintroduction and Personality

The personality of individual captive-bred animals of the same species affects their chances of survival after release. The value of ratings of animal personality as a tool for improving the breeding, management and welfare of mammals living in zoos has been discussed by Tetley and O'Hara (2012). More recently, Azevedo and Young (2021) have explored the importance of considering animal personality in conservation.

Captive-bred swift foxes (*Vulpes velox*) that died in the six months following their release had previously been judged as 'bold' (Bremner-Harrison et al., 2004). In captivity, in the presence of novel stimuli, they left their dens more quickly, approached novel stimuli more closely and showed activities indicating low fear compared with those individuals that survived. The authors of the study concluded that candidates for release should be selected on the basis of behavioural variation to improve introduction success.

A study of the effect of personality traits on the behaviour and survival of Blanding's turtles (*Emydoidea blandingii*) bred at Detroit Zoo, Michigan, and subsequently released into the Shiawassee National Wildlife Refuge was conducted by Allard et al. (2019). They found that more exploratory individuals had higher survival rates and travelled longer distances after release, but neither boldness nor aggression was related to survival. The use of muskrat (*Ondatra zibethicus*) dens was associated with higher survival and both bolder and more exploratory turtles made use of these. Exploratory and aggressive turtles were more often found basking outside of water, while bold turtles were more often found at the water's surface. Both of these behaviours increased the risk of predation and were considered to be a trade-off between risk and physiological health.

Preparing the Ground: Public Consultation

Public consultations concerned with wildlife reintroductions may attract considerable interest and comment from those individuals and organisations 'for' reintroduction and those 'against'. It is essential that local people support a reintroduction, especially if there is potential for those that oppose it to jeopardise its success. In many, if not most, countries reintroductions must be licensed by the government, and soliciting the views of the public via a formal public consultation is an important part of the licensing process.

In the 1990s a proposal to reintroduce grizzly bears (*Ursus arctos*) in central Idaho and western Montana in the United States resulted in seven public hearings and a five-month public comment period that drew more than 24,000 comments (Idaho Fish and Game Commission, 1997). In a 2007 public consultation concerning a proposal to reintroduce black-footed ferrets (*Mustela nigripes*) in Logan County, Kansas, the US Fish and Wildlife Service received over 16,000 comments – largely from supporters of the proposal – over a period of 30 days (USFWS, 2007).

The European beaver (*Castor fiber*) was once widespread in Eurasia, but was almost hunted to extinction by the beginning of the twentieth century (Fig. 12.9). The species was highly prized for its fur and a secretion (castoreum) used to scent mark that is utilised in the manufacture of some perfumes. The beaver has subsequently been reintroduced to much of its former range.

An extensive public consultation took place in Scotland before a trial reintroduction of beavers; in May 2008, after a long period of procrastination, the Scottish government gave permission to the RZSS and the Scottish Wildlife Trust for a scientifically monitored trial reintroduction of beavers to Knapdale Forest in Mid Argyll in 2009. As part of this process an extensive consultation process took place (Table 12.2).

Although the process of public consultation has been part of the reintroduction process for some considerable time, achieving an appropriate level of participation is still sometimes problematic. Hawkins et al. (2020) have discussed the factors that constrained public participation in consultation, information sharing and the

Fig. 12.9 European beavers (*Castor fiber*) have been released into Scotland.

Table 12.2 A summary of the responses obtained from organisations consulted prior to the reintroduction of European beavers (*Castor fiber*) to Knapdale, Mid Argyll, Scotland.

Organisation	View
Argyll & Bute Council	Supportive
Argyll Bird Club	Supportive
Argyll District Salmon Fishery Board	Has concerns and would require safeguards to be put in place, supports AFT position
Argyll Fisheries Trust (AFT)	Has concerns and would require safeguards to be put in place
Association of Salmon Fishery Boards	Against and require clarification on issues raised
Association of Scottish Visitor Attractions	Supportive
British Waterways	Against and require clarification on issues raised
Confederation of Forest Industries	Against
National Union of Farmers Scotland (NFUS)	Has major concerns, wants clear exit strategy and safeguards in place
Ramblers Scotland	Supportive
Royal Society for the Protection of Birds (RSPB)	Supportive
Scottish Environmental Protection Agency	Has concerns and would require safeguards to be put in place
Scottish Rural Properties and Businesses Association	Against
Scottish Water	Has concerns (particularly post-trial) and would require safeguards to be put in place
Wild Scotland	Supportive
Woodland Trust Scotland	Supportive

Source: Scottish Beaver Trial, 2007.

transparency of communications regarding the possibility of reintroducing the Eurasian lynx (*Lynx lynx*) into England.

Post-Release Monitoring and the Prevention of Disease Transmission

The success of many early reintroductions was hampered by the fact that little or no scientific information about animals was collected post-release. The importance of gathering information on growth, survival rate, disease and other aspects of the biology of reintroduced animals is now widely acknowledged and post-release monitoring is now regarded as essential to the process.

Following the successful captive breeding of oriental pied hornbills (*Anthracoceros albirostris*) in Khao Kheow Open Zoo, Thailand, GPS receivers were attached to the backs of some of the birds prior to reintroduction (Chaiyarat et al., 2012). The activity patterns of these birds compared with those of other individuals released without GPS units were not significantly different. The GPS units allowed the home range size to be determined as, on average, 0.13 km^2.

A research team, two of whom were affiliated to the Memphis Zoological Society, Tennessee, monitored juvenile captive-bred giant salamanders (*Andrias davidianus*) obtained from two farms in China following their release into two rivers in the Qinling Mountains of central China (Zhang et al., 2016). A VHF radio transmitter was surgically implanted in each animal and, in addition, they were marked with passive integrated transponder (PIT) tags to allow identification beyond the life of the transmitter battery. Body mass had no effect on survival, but survival was higher in individuals held for a longer period following surgery. The study concluded that the animals could survive in the wild for one year following release, provided adequate surgery recovery time was allowed.

Great care must be taken to reduce the possibility of disease transmission between captive and wild populations. The release site selection protocol for the release of captive-bred black-footed ferrets (*Mustela nigripes*) in the United States excluded areas where sylvatic plague (*Yersinia pestis*) was endemic in local wildlife (Ballou, 1993).

When golden lion tamarins (*Leontopithecus rosalia*) (Fig. 12.7) were reintroduced to the Atlantic coastal rainforest of Brazil they were released in areas uninhabited by conspecifics. This was an attempt to prevent disease transmission from captive to wild populations, although there was some subsequent contact and interbreeding (Ballou, 1993; Montali et al., 1995).

The establishment of a health surveillance programme for Eurasian beavers (*Castor fiber*) reintroduced into Scotland has been discussed by Goodman et al. (2012).

12.2.10 Zoos' Contributions to Reintroductions and the 'One Plan' Approach

Who Are the Main Players?

Zoos have made, and continue to make, useful contributions to the restoration of species to the wild and the supplementation of existing populations. However, they are not the main players in the restoration process – government agencies and other NGOs are far more important.

The contributions of zoos and aquariums to reintroductions in the context of changing conservation perspectives have been discussed by Gilbert et al. (2017). They found that the majority of species in member institutions of the European Association of Zoos and Aquaria (EAZA) were not globally threatened. However, more than half of the 156 reintroduced species and 260 projects supported by EAZA members involved species that were classified by the IUCN Red List as Threatened – that is, Critically Endangered, Endangered or Vulnerable – Near Threatened or Extinct in the Wild. Although EAZA members provided some animals for release in reintroduction projects, their greatest contributions to such projects were the provision of funds, staff, expertise, equipment and project coordination. Gilbert et al. concluded that zoos and aquariums have an important role to play in reintroductions as emphasis shifts towards integrated conservation management of species, combining *in-situ* and

Table 12.3 Taxonomic breakdown of contributions made by zoos to North American releases of captive-bred individuals.

Taxon	%
Amphibians	42
Terrestrial invertebrates	29
Mammals	19
Birds	17
Reptiles	15
Fishes	2
Marine invertebrates	0

ex-situ efforts in the so-called One Plan Approach. It should be noted that this is not an independent assessment of the role of zoos in conservation: three of the four authors indicated their affiliations at the time of publication as Marwell Wildlife, a zoo in England.

An analysis of the 1,863 articles in the North American Conservation Translocations database published prior to 2014 on translocations in the United States, Canada, Mexico, the Caribbean and Central America found that conservation translocation involved captive breeding for 162 (58 per cent) of the 279 animal species translocated and just 54 zoos contributed animals for release (Brichieri-Colombi et al., 2019). The 40 species bred by zoos for release represented just 14 per cent of all the animal species for which conservation translocations were published (Table 12.3). The proportional involvement of zoos in captive breeding programmes for release increased from 1974 to 2014, as did the proportion of scientific papers focused on translocation that were co-authored by zoo professionals. The authors of the study recommended that there should be increased dialogue between the zoo community, academics and governments to optimise the role zoos play in translocations.

One Plan Approach

The One Plan Approach is a method of integrated conservation planning aimed at producing self-sustaining wildlife populations by managing all populations of a species (*in-situ* and *ex-situ*) as one, involving stakeholders and pooling all of the available resources (Fig. 12.10).

The One Plan Approach used to manage the Western swamp tortoise (*Pseudemydura umbrina*) has been described by Jakob-Hoff et al. (2015). The project involves exchanges between wild populations and captive-bred individuals at Perth Zoo in Western Australia, and is summarised in Fig. 12.11.

Most zoos and aquariums do not yet have an integrated plan to guide zoological institutions in species selection and conservation actions. The development of Integrated Collection Planning and Assessment (ICPA) workshops that bring together *in-situ* and *ex-situ* conservation communities has been described by Traylor-Holzer et al. (2019). The first such workshop was held in 2016 for 43 canids and hyaenids working with IUCN Specialist Groups and regional zoo and aquarium associations.

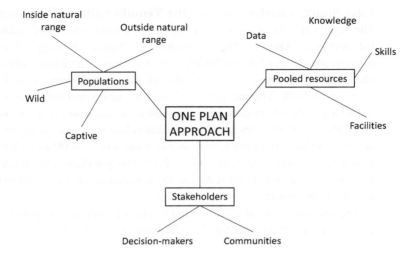

Fig. 12.10 The One Plan Approach.

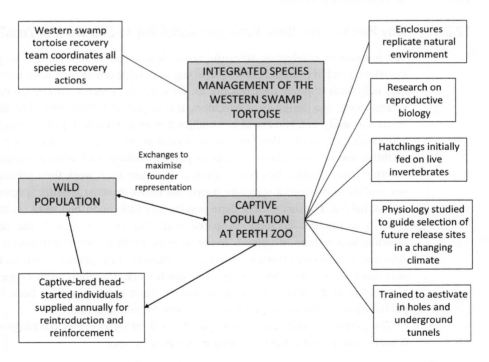

Fig. 12.11 The One Plan Approach has been used to manage the Western swamp tortoise (*Pseudemydura umbrina*).

In addition to the involvement of zoos and NGOs, Kitchener (2020) has suggested that the IUCN's Conservation Specialist Planning Group One Plan Approach makes it important for zoos and museums to work together to enhance conservation, especially in threatened small carnivorans.

Case Study: Chester Zoo and the Tequila Splitfin (*Zoogoneticus tequila*)

The tequila splitfin (*Zoogoneticus tequila*) is a small species of endangered goodeid fish native to Mexico. The Laboratory of Aquatic Biology of the Universidad Michoacana de San Nicolás Hidalgo, Mexico, has established captive breeding colonies whose founders were five pairs of fish from Chester Zoo, England, that it received in 1998. The university released 40 males and 40 females into 6,000 m^2 artificial ponds, where they were exposed to natural conditions, including predators and parasites. Four years later the population had increased to approximately 10,000 and these animals have been the source of reintroductions (Maceda-Veiga et al., 2016; Domínguez-Domínguez et al., 2018). From this population, hundreds of individuals have been released into the wild and the population has expanded from release sites into the river system.

The important role of aquarium hobbyists in maintaining populations of some rare fish species has been discussed by Maceda-Veiga et al. (2016).

12.3   *In-Situ* Conservation

12.3.1   Why Has so Little Been Published About the *In-Situ* Work of Zoos?

The purpose of conducting scientific research in a zoo or aquarium is to gain new knowledge and to publish it, and in so doing to provide others with information which may help them improve their practice as, for example, keepers, curators, nutritionists, veterinarians, zoo designers or educators. It is important to remember that, although a behaviour study in a zoo could be conducted over just a few days by a single person and result in a publication, *in-situ* conservation projects may take many years to establish and involve a large number of people working with several organisations.

When zoos become involved with *in-situ* conservation work their primary aim is not usually to produce a scientific paper, but rather to assist wildlife species in their natural habitat. For this reason, there is relatively little published work on the role of zoos in *in-situ* conservation compared with that concerned with the behaviour, breeding and welfare of animals living in zoos. Furthermore, collaboration between zoos and *in-situ* conservation projects is a relatively new phenomenon, so there has only been a short time for a body of research work describing such cooperation to accumulate in the academic literature. Indeed, there appears to have been very little published about the *in-situ* work of zoos prior to around 2010.

The purpose of this section is to provide a short overview of the range of *in-situ* conservation projects that have been undertaken by zoos.

12.3.2   *In-Situ* Conservation Projects Supported by Zoos

Zoos support a wide range of *in-situ* projects around the world with their staff and expertise providing, among other things, local staff training, community education projects, field equipment and financial resources.

Prior to the establishment of the red panda European Endangered Species Programme (EEP) – now the EAZA Ex-Situ Programme – in 1985 the zoo population of the species in Europe was decreasing. The population began to increase thereafter, and by 31 December 2019 there were 407 red pandas (*Ailurus fulgens*) in the EEP housed at 182 institutions. This success in establishing a stable captive population of red pandas from what was a failing zoo population has been described by Kappelhof and Weerman (2020). Following a decision made in 2012, the Global Species Management Plan (GSMP) for the red panda has been partnered with the Red Panda Network (formerly the Red Panda Project) which, in 2006, became the first community-based project monitoring the species in Nepal. As part of this collaboration, EEP holders of red pandas provide financial support for the Red Panda Network's conservation activities.

Lahore Zoo in Pakistan has worked with the Sindh Wildlife Department and the Punjab Wildlife Department and Parks Department on an *in-situ* project to save the endangered Indus river dolphin (*Platanista minor*) (Javed and Khan, 2005). The project involved the establishment of a rescue unit operated by local people to reduce the mortality of river dolphins stranded in irrigation canals and a river-based rehabilitation centre for dolphins requiring veterinary attention prior to release. A programme of education was devised to inform local fishers of the ecological importance of the species and the Lahore Zoo Education Programme provided awareness sessions for local school children.

Chester Zoo in England has kept Asian elephants (*Elephas maximus*) since 1941 (Rees, 2021). In their range states these elephants are increasingly threatened as a result of human encroachment and habitat fragmentation. The North of England Zoological Society (which operates Chester Zoo) began cooperating with the Assam-based NGO EcoSystems-India in 2004 and created the Assam Haathi Project for human–elephant conflict mitigation. A community-based approach was used to mitigate human–elephant conflict and protect the livelihoods of the local people by integrating elephant research and monitoring. The mitigation measures developed included the use of trip-wire alarms, watch towers, electric fences, searchlights and chilli (which can be burned to produce an unpleasant smoke that acts as a deterrent). This project has produced a handbook for the community entitled *Living with Elephants in Assam*, and provided conservation education workshops and assistance with government compensation schemes. The project received funding from the Darwin Initiative, which is administered by the Department for the Environment, Food and Rural Affairs (DEFRA) as part of the UK's obligation to assist developing countries with biodiversity conservation, as required by the UN Convention on Biological Diversity. The zoo staff involved with this project included conservation officers, research assistants, a nutritionist and a specialist elephant keeper, and it has resulted in the publication of a number of research papers (Zimmermann et al., 2009; Chartier et al., 2011; Davies, 2011; Wilson, 2015) and an educational project at a school local to Chester Zoo (Esson et al., 2014).

The role of North Carolina Zoo in the United States in using emerging technologies to advance the *in-situ* conservation of African elephants (*Loxodonta africana*) over a period of 20 years has included the use of novel anaesthesia techniques to fit elephants with satellite tracking collars in West and Central Africa (Wilson et al., 2019). Data from these collars are used to redirect roaming animals back to protected areas and to avoid human–elephant conflict. The zoo has supported the development of spatial monitoring and reporting tool (SMART) conservation software and its implementation in 14 protected areas in five African countries so that rangers can monitor and record elephant movements using smartphones.

Bairrão Ruivo and Wormell (2012) have discussed an international conservation programme for the white-footed tamarin (*Saguinus leucopus*), which is endemic in Colombia. The programme involves a partnership between a consortium of 21 European zoos, Asociación Colombiana de Parques Zoológicos y Acuarios (ACOPAZOA), local organisations (including nine Colombian zoos), central and regional conservation public authorities, NGOs (especially the Wildlife Conservation Society) and Colombian universities and researchers. It combines *in-situ* conservation and education projects and *ex-situ* in-country reproduction, education and research.

Kleiman and Mallinson (1998) – of the National Zoo, Washington, DC and the Jersey Wildlife Preservation Trust, respectively – have summarised the history, organisation, structure and functioning of the four international oversight recovery and management committees that act as official technical advisers to the Brazilian federal environmental agency – the Instituto Brasileiro do Meio Ambiente e dos Recursos Naturais Renováveis – with respect to the conservation and management of the *in-situ* and *ex-situ* populations of lion tamarins (*Leontopithecus*). The authors considered that the benefits of the recovery process for this species included the establishment of collaborative partnerships between the Brazilian government and a wide range of individuals and organisations, along with the use of multidisciplinary, semi-autonomous teams to implement conservation activities.

A breeding programme for the yellow-breasted capuchin (*Cebus xanthosternos*) began in 1980 at the Rio de Janeiro Primate Centre (Centro de Primatologia do Rio de Janeiro (CPRJ)), Brazil, and was then extended to European facilities with experience of breeding New World monkeys, initially at Mulhouse Zoo in France in 1990 (Lernould et al., 2012). By the end of 2010 there were 140 individuals of the species at 21 European zoos. Cooperation between these organisations included the monitoring of *in-situ* groups of capuchins in fragmented forests, which have helped in the understanding of their ecology and facilitated the formulation of an action plan for their conservation.

The Species Survival Plan$^{®}$ (SSP) for the Aruba Island rattlesnake (*Crotalus unicolor*) was established in 1982 and is the Association of Zoos and Aquariums' (AZA) longest continually functioning snake conservation project. By 2014, 27 potential founders had been imported for assimilation into the SSP to maintain genetic diversity in the captive population. This resulted in gene diversity in the captive population of over 94 per cent. After the establishment of a successful captive breeding programme the SSP began working in partnership with Arubans in 1986 to aid the *in-situ* conservation of the snake and its ecosystem. The programme has included ecological research, education, training, management recommendations, capacity building, workshops and public relations that have been integrated into a successful holistic long-term project (Odum and Reinert, 2015).

Vienna Zoo, Austria, has worked with Bhawal National Park in Bangladesh to establish a breeding population of the Northern river terrapin (*Batagur baska*) in the park, where 84 juveniles were reared over a period of two years (Weissenbacher et al., 2015). The species had become ecologically extinct in the wild – as a result of harvesting and habitat destruction – having formerly inhabited rivers and estuaries in East India, Bangladesh and Myanmar. The first two captive-bred juveniles hatched at the zoo in 2010.

Ziegler (2015) has reported on the *in-situ* and *ex-situ* reptile projects of the Cologne Zoo, Germany, and considered the implications for research and conservation of South-East Asia's herpetodiversity, with particular reference to Vietnam. The German–Vietnamese research team has undertaken work on reptile taxonomy, autecology, population analyses and public awareness, and the zoo has been involved in the development of *in-situ* rescue centres and breeding stations. The author also discusses the roles of collection planning, species identification and biological research in zoological facilities, particularly in relation to monitor lizards and crocodiles.

Zwartepoorte (2015) has described the recovery programme for the Egyptian tortoise (*Testudo kleinmanni*), which links an *in-situ* project in the northern Sinai desert – supported by the local Sweirki Bedouin tribe and coordinated by Nature Conservation Egypt – with the combined *ex-situ* breeding efforts of EAZA and the European Studbook Foundation. This small tortoise species has experienced severe population decline – as a result of overgrazing by cattle, intensification of agriculture and overcollecting for the international pet trade – and has been brought to the brink of extinction in Egypt.

Cali Zoological Foundation, Colombia, and Zoo Zurich, Switzerland, have contributed to the conservation of threatened amphibians in Valle del Cauca in Colombia by running an *ex-situ* conservation centre along with conservation research and education programmes (Furrer and Corredor, 2008).

Yaacob (1994) reported on a captive breeding and reintroduction project for the milky stork (*Mycteria cinerea*) at Zoo Negara in Malaysia following the acquisition of a young group of birds by chance in 1987. The zoo had to overcome a number of problems, including the loss of chicks due to incorrect feeding, collapse of the nest tree and severe damage to the aviary resulting in the loss of eight birds. In addition, pair bonds among the birds were weakened, possibly as a result of egg removal to encourage double clutching. When individuals were released, six milky storks remained near the release site and joined the zoo's free-flying colony of the related painted stork (*M. leucocephala*).

The use of hand-reared scarlet macaws (*Ara macao*) in reintroduction projects in Peru and Costa Rica has been described by Brightsmith et al. (2005). They reported a first-year survival rate of 74 per cent, and this rose to an annual rate of 96 per cent post-first year. Survival rates were higher in birds from large releases than those from smaller releases. Supplemental feeding was important in keeping the released birds near the release sight and facilitating social interactions. Hand-reared birds released in Peru mated with wild conspecifics and successfully fledged young. The authors of the study concluded that hand-reared birds could be successfully released to the wild and would breed, but ex-pets were not suitable for release.

Chaiyarat et al. (2012) reported successful attempts to captive breed oriental pied hornbills (*Anthracoceros albirostris*) in Khao Kheow Open Zoo, Thailand, using artificial nests, and to subsequently reintroduce them to the wild. The first reintroduced pair used an artificial nest to lay and hatch their eggs successfully.

Staff at Chester Zoo, England, have supported work at Cikananga Wildlife Center, Java, for the conservation breeding of several threatened passerines from founder individuals obtained from local private bird keepers (Owen et al., 2014). Breeding success has been achieved for the black-winged starling (*Sturnus melanopterus*), Sumatran laughingthrush (*Garrulax bicolor*) and Javan green magpie (*Cissa thalassina*), with a view to reintroducing these species to the wild.

12.3.3   *In-Situ* Projects and Animal Welfare

Some *in-situ* work involving zoos has had an animal welfare focus in addition to conservation objectives.

San Diego Zoo Global in the United States, and *Free the Bears* – an Australian organisation that works with local communities and governments in Asia to rescue captive bears – have worked together to design surveys aimed at improving the understanding of attitudes, public knowledge and behaviours towards bears and bear-part consumption in South-East Asia (Crudge et al., 2016). Preliminary results from 1,500 surveys completed in Cambodia and the Lao People's Democratic Republic indicate significant differences between Lao and Western responses, attitudes, beliefs and behaviours towards wildlife and the wildlife trade. This work should assist in identifying key messages and the key groups at which they should be targeted to promote sustainable behaviours and to ensure stable wild bear populations in the region.

Guidelines for *in-situ* gibbon rescue, rehabilitation and reintroduction have been produced by staff from Perth Zoo and the Zoology Department of Oxford University based on information available in the literature and experience gained at the Kalaweit Gibbon Rehabilitation Project in Kalimantan and the Javan Gibbon Centre in Java, in Indonesia (Cheyne et al., 2012). In formulating their proposals, the authors paid particular attention to the limited resources available in developing countries.

12.3.4   Evaluating Zoos' Contribution to *In-Situ* Conservation

When they evaluated the contribution of the world zoo and aquarium community to *in-situ* conservation in 2010, Gusset and Dick (2010) concluded that the community had made an appreciable contribution to global biodiversity conservation. The results of a questionnaire survey that evaluated 113 projects showed that they were 'helping to improve the conservation status of high-profile threatened species and habitats in biodiversity-rich regions of the world'. However, the authors suggested that zoos and

Fig. 12.12 The sand lizard (*Lacerta agilis*) is one of the indigenous species whose conservation is being supported by zoos in the UK.

aquariums could make an even greater contribution by allocating more resources to *in-situ* conservation, thereby significantly increasing the projects' conservation impacts. They also advised that zoological institutions should pool their resources. It is worth noting that both of the authors worked for the World Association of Zoos and Aquariums (WAZA) at the time this work was published.

A more recent analysis of the contribution of North American zoos and aquariums to endangered species recovery listed under the Endangered Species Act (ESA) in the United States examined data from federal recovery plans and AZA annual surveys. Che-Castaldo et al. (2018) found that zoos frequently conduct conservation research and monitoring and assessment in the field. However, cooperatively managed programmes in American zoos tend to focus on species that are not listed on the ESA and the authors of the study suggest that it would be beneficial for zoos to manage more native threatened species.

In the UK a group of nine 'large charitable zoos and aquariums' has outlined details of 76 native species it was helping to restore in the UK, its Crown Dependencies and British Overseas Territories, and listed 12 nature reserves and other protected areas in its care (BIAZA, 2021) (Fig. 12.12).

12.3.5 *In-Situ* Conservation and Training for Overseas Conservationists

Modern zoos can play an important role in training the staff of zoos and conservation NGOs in developing countries. This may range from teaching species-specific husbandry techniques to providing training in the use of GPS equipment for monitoring animals in the field.

The zoo biology training programme for developing countries, inaugurated in 1987 by the National Zoo in the United States, has been described by Wemmer et al. (1990). It ran for 2–4 weeks and was targeted mainly at mid-level zoo managers in zoos located in tropical countries. Its purpose was to improve the care and management of animals in zoos through *in-situ* training. The scheme operated through a host zoo in a foreign country which was responsible for recruiting trainees from other zoos in that country or region. Among other things, trainees created an animal inventory for the host zoo, and identified endemic species for studbook compilation and captive management projects.

The Zoological Parks Board of New South Wales (comprising Taronga Zoo and Western Plains Zoo) in Australia has provided specialist training for zoo staff and wildlife managers in Africa, Indo-China, South-East Asia and the South Pacific (Woodside and Kelly, 1995).

Durrell Wildlife Conservation Trust

Some zoos have established strong links with *in-situ* conservation projects. The work of one zoo-based organisation in this respect is exceptional. The Durrell Wildlife Conservation Trust developed from the work of Gerald Durrell and the conservation efforts of the zoo he founded in Jersey in the Channel Islands (Fig. 12.13). The field staff at *Durrell* were operating 50 projects in 18 countries in 2019, focusing on islands and taxa experiencing severe declines, such as primates and amphibians (Table 12.4) (*Durrell*, 2019). In Mauritius, *Durrell* works in partnership with the Mauritius Wildlife Foundation on projects to save bird and reptile species from extinction. The Trust's largest programme region is Madagascar, with 40 conservationists working at eight field sites.

Conservation Research at *Durrell*

Between 1980 and 2019, staff of the Durrell Wildlife Conservation Trust authored or co-authored 338 papers published in peer-reviewed journals, most (98.5 per cent) of which were published after 1996. The subjects of these papers included work on captive management of species (e.g. the Mauritius kestrel (*Falco punctatus*) (Jones, 1984); pink pigeon (*Columba mayeri*) (Swinnerton et al., 2004)), the development of new field methods (e.g. for capturing arboreal geckos (N. C. Cole, 2004)) and work on historical records (e.g. faunal archives in an Asian biodiversity hotspot, Hainan Island, China (Turvey et al., 2019)). *Durrell* has also been involved in research training by facilitating doctoral studies that resulted in the completion of 53 PhDs between 1991 and 2017 (Fig. 12.14).

In-situ projects by *Durrell* staff have examined the impact of weeding and rat control on land snails in Mauritius (Florens et al., 1998), the decline of lemur species in south-east Madagascar (Lehman et al., 2006), the disease status of amphibians on Montserrat (Garcia et al., 2007), monitoring amphibian chytrid fungus in Madagascar (Weldon et al., 2013), the effects of the supplementary feeding of wild pink pigeons (Edmunds et al., 2008), the use of non-indigenous tortoises as a restoration tool to

Fig. 12.13 The Durrell Conservation Academy, Jersey, trains conservationists from many developing countries.

replace extinct ecosystem engineers (Griffiths et al., 2010), a range-wide survey of Livingstone's fruit bat (*Pteropus livingstonii*) (Daniel et al., 2017) and an analysis of deforestation patterns in Madagascar (Zinner et al., 2014).

Conservation Training at *Durrell*

The IUCN Red List Index (RLI) shows trends in overall extinction risks for species. It does this by showing trends in the status of groups of species based on changes in status that are of sufficient magnitude to merit moving species to a more threatened or a less threatened Red List Category. The value of the RLI ranges from 1.0 (all species in a group qualifying as Least Concern) to 0 (all species listed as Extinct). The index is widely used by governments to track their progress towards biodiversity targets.

The potential of the RLI as a useful indicator of the long-term impact of individual conservation agencies on the extinction risk of species has been examined by Young

Table 12.4 Conservation actions of the Durrell Wildlife Conservation Trust (2019).

Action	Number of species
Conservation research	46
Monitoring impacts	35
Managing species	36
Conserving habitats	35
Planning actions	37
Empowering local people	26
Strengthening local partners	43

Source: www.durrell.org/wildlife/conservation/our-approach/in-the-field.

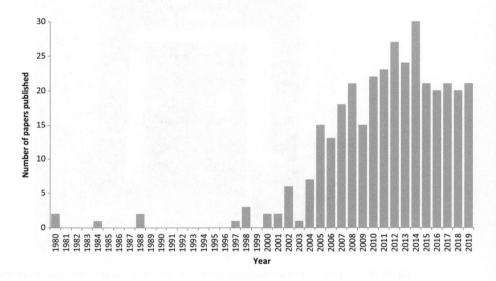

Fig. 12.14 Peer-reviewed papers authored or co-authored by *Durrell* staff 1980–2019 (source: based on information supplied by Dr Tim Wright and Rachael Gerrie, *Durrell*).

et al. (2014) using the work of the Durrell Wildlife Conservation Trust as an example. The trust has run conservation programmes for a number of globally threatened terrestrial vertebrate species for several decades. They found that of the 17 target amphibian, bird and mammal species, eight showed improvements in Red List Category (i.e. reductions in extinction risk) as a result of conservation efforts. Between 1988 and 2012 this resulted in a 67 per cent increase in the value of the RLI, in contrast with a 23 per cent decline in a counterfactual RLI showing projected trends if conservation efforts had been withdrawn in 1988. Young et al. concluded that the RLI is a useful tool for measuring the impact of conservation organisations that target species with circumscribed (restricted) ranges.

The early work of the Jersey (Durrell) Wildlife Preservation Trust's International Training Programme has been discussed by Waugh (1988).

Funding *In-Situ* Conservation

Very little has been published on the contribution of zoos to the funding of *in-situ* conservation projects. The annual reports of zoos should be a useful source of information about this activity. However, the reports of some zoos make no specific reference to funds spent on conservation or research, and those of other zoos concatenate the two categories so that it is not possible to determine how much is spent on each. Inconsistencies in accounting practices between zoos make meaningful comparisons of the amounts of money spent on their various activities impossible.

A bibliography of 470 publications concerned with 'Zoos and Conservation' produced by the North of England Zoological Society (Wilson and Zimmerman, 2005) contained only three documents that referred to funding in the title (Bettinger and Quinn, 2000; Hutchins and Ballentine, 2001; Hutchins and Souza, 2001).

A study examining zoo-based fundraising for *in-situ* wildlife conservation concluded that although it was possible to quantify the zoos' contribution to projects for particular species – for example, tigers (*Panthera tigris*) and amur leopards (*P. pardus orientalis*) – poor data capture made it impossible to determine their global role (Christie, 2007). The author found that zoos supplied or channelled only about 16 per cent of the total NGO budget for *in-situ* tiger conservation between 1998 and 2002, and this included grants obtained by zoos from other sources. However, over the same period zoos supplied or channelled 54 per cent of the funding provided specifically for Sumatran tiger (*P. t. sumatrae*) conservation (Christie, 2007).

A global study of the effect of the management of zoo animal populations on visitor attendance and the *in-situ* conservation contributions of zoos found that zoos with greater visitor attendance made greater *in-situ* conservation contributions (Mooney et al., 2020). Zoos that held more conservation-threatened species invested more in conservation projects *in-situ*. The authors of the study concluded that zoos do not need to compromise their economic viability and entertainment value in order to have a significant value to conservation. They suggested that the absence of large vertebrates – which are very popular with visitors and expensive to keep – may not necessarily result in reduced *in-situ* project activity provided that attendance can be maintained. This could be achieved, for example, by maintaining a collection with a large total number of animals with high species richness that is dissimilar to that of other zoos.

Although many zoos contribute directly or indirectly to the funding of *in-situ* conservation projects to protect species and restore habitats, they are not the main source of funding for these efforts when considered on a global scale. Most organisations that raise money for conservation do not use zoos as a vehicle for this, and many conservation projects are funded by governments.

In their Annual Report 2020–2021, the Royal Society for the Protection of Birds (RSPB) declared an income of £142.4 million, of which £38.8 million was spent on its nature reserves and a further £30.6 million on research and policy/advisory work. In addition, £16.6 million was spent on education and 'inspiring support' (RSPB, 2021). The WWF-UK Annual Report Summary 2020–2021 reported an investment of £21.7 million in restoring threatened habitats and species (WWF-UK, 2021). The 2021

WWF-US Annual Report declared total programme expenses of US$289,203,979, comprising US$180,592,986 spent on conservation field and policy programmes and US$108,610,993 on public education (WWF-US, 2021). Fauna & Flora International (FFI) reported a total expenditure of £24.0 million in 2020, of which 92 per cent was spent on conservation and the remainder was spent on fundraising (FFI, 2020).

The Association of Zoos and Aquariums represents 238 accredited members in 13 countries, but mostly in the United States, that funded field conservation projects in 115 countries and spent US$208 million on conservation initiatives in 2020 (AZA, 2022). However, most of their activity is focused on caring for some 730,000 animals from 8,500 species.

Although the amount of money spent by AZA on *in-situ* conservation appears large, the Detroit Zoological Society alone reported total expenditure of US$26,707,545, including US$7,675,681 spent on animal care and US$7,165,129 spent on 'maintenance and park operations' in 2020 (DZS, 2020). It is not possible to compare the costs of operating individual zoos or the direct financial contribution they make to *in-situ* conservation from their annual reports because there is no common format for these reports, and some contain more detail than others. The 2019 Annual Report of the Zoological Society of Ireland (operator of Dublin Zoo and Fota Wildlife Park) identified operating costs of €20,730,000 (US$22,813,000) – including administrative expenses – €6,389,000 (US$7,031,000) of which related to staff wages and salaries (ZSI, 2019). No other costs relating to animal care were identified.

12.4   Conclusion

Although captive breeding of threatened animals in zoos and aquariums may be important, perhaps even essential, for a number of species it is disingenuous for zoos to suggest that they play a major role in the reintroduction of animals to the wild. Many, probably most, successful releases are achieved by state agencies and wildlife NGOs operating within the range states of the species concerned, with little or no input from zoos.

Nevertheless, zoos have played an important role in the recovery of some species by breeding animals for release or providing expertise and other resources to facilitate reintroductions. Many zoos have made significant contributions to *in-situ* projects to protect threatened species and habitats. While this is laudable, most conservation NGOs fund *in-situ* projects without the need to keep animals in zoos. It is inevitable that most of the income received by zoos will be spent on the very expensive activity of operating a zoo 24 hours a day for 365 days of the year.

13 Animal Nutrition and Conservation Medicine

13.1 Introduction

The profession of zoo veterinarian is of relatively recent origin. Up until just a few decades ago even some quite large and well-known zoos in the UK did not employ their own vet and either relied on those in a local practice – that had developed a certain amount of expertise in treating exotic species by virtue of their proximity to the zoo – or called on the services of specialist companies that could provide suitable vets, sometimes seconding particular staff to a zoo for several days each week.

Nowadays, large professionally run zoos employ veterinarians, veterinary nurses, endocrinologists, nutritionists and other specialist staff, sometimes organised into a well-equipped Conservation Medicine Department. Notwithstanding these advances, such zoos may still need to call upon the services of university veterinary schools and, occasionally, surgeons that normally treat humans, for example to perform procedures on great apes.

As the zoo community has refocused its activities – putting greater emphasis on biodiversity conservation – the zoo vet's work has expanded from concern for the welfare of individual animals to include not only the diversity and sustainability of zoo populations but also the health of individuals and populations in the wild (Braverman, 2018). In an examination of the role of veterinarians working in public aquariums, Braverman (2019) identified the importance of balancing the challenges of managing aquatic taxa – especially marine mammals – with animal rights and welfare considerations, and evolving responsibilities for ocean conservation.

13.2 One Health

In the United States, the Centers for Disease Control and Prevention (CDC) defines the concept of 'One Health' as

A collaborative, multisectoral, and transdisciplinary approach – working at the local, regional, national, and global levels – with the goal of achieving optimal health outcomes recognizing the interconnection between people, animals, plants, and their shared environment. (CDC, 2021)

This is not a new concept, but it has become of greater significance in recent times because of the increased exposure of humans to wild animal diseases as indigenous

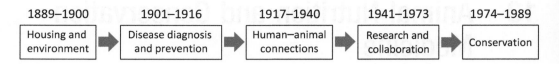

1889–1900	1901–1916	1917–1940	1941–1973	1974–1989
Housing and environment	Disease diagnosis and prevention	Human–animal connections	Research and collaboration	Conservation

Fig. 13.1 The five historical periods identified for the first 100 years of the history of the Smithsonian's National Zoo, based on the way in which animal health was described, treated and understood (based on Gutierrez et al., 2021).

peoples and their livestock come in contact with wild animals as a result of encroachment into, and destruction of, forests and other natural habitats. Furthermore, there are increased movements of animal and plant products around the world, the expansion of the human population into new areas and changes in the global distribution of organisms as a result of climate change.

Conservation medicine emerged as a distinct field of study in the 1990s, combining elements of conservation biology with ecology, the health sciences and the social sciences. It may be regarded as the One Health approach to the conservation of biodiversity.

Many diseases and health problems found in zoo animals are linked to human health and the veterinary treatment these animals receive often incorporates tools and knowledge used in human health care. The intertwined history of human medicine and non-primate health at the Smithsonian's National Zoo and Conservation Biology Institute has been examined by Gutierrez et al. (2021) for the 100 years from the opening of the zoo in 1891. From the Smithsonian's annual reports they were able to distinguish five historical time periods based on how animal health was described, treated and understood (Fig. 13.1). This led to the One Health approach used today.

13.3   Nutrition and Health

Providing appropriate nutrition for a wide range of species is a major challenge for the zoo community. Modern zoos have access to a great deal of research on animal diets and nutrition and many now employ their own nutritionists. This was, of course, not always the case and in the past it was common to find the public feeding the animals in a zoo. Apart from making it impossible for keepers to monitor what the animals were eating, visitors sometimes threw dangerous objects into animal enclosures.

13.3.1   Public Feeding and the Ingestion of Foreign Objects

Feeding of animals by visitors is prohibited in most modern zoos except where the zoo itself provides the food, for example, grain for wildfowl. Feeding by the public makes it impossible for zoo staff to know what foods and how much food animals have consumed, and clearly compromises their health.

In 1967 the Council of the Zoological Society of London (ZSL) decided to ban all feeding by visitors at London Zoo and Whipsnade Park from the beginning of 1968, in the interests of the health and welfare of the animals. Elephants were particularly prone to taking objects from the public as evidenced by these comments made in 1967:

During the same year elephants at London seized fourteen coats, twelve handbags, ten cameras, eight gloves and six return tickets to Leicester, damaging them all beyond repair ... The problem of educating the public to resist throwing buns, nuts and other food to the animals is considerable. (Nature, 2017)

In the nineteenth century London Zoo stopped visitors feeding the African elephant *Jumbo* with their own food when they discovered that some people were inserting pins and other objects into the food (Chambers, 2007) (Fig. 13.2). Having said that, the diet provided by the zoo was far from ideal, and included 'beer and hard liquor, oysters, cakes and candy' (Charles River Editors, 2016).

Even today animals living in zoos are at risk of ingesting foreign objects. Murray et al. (1997) reported examining a Celebes ape (*Macaca nigra*) – also known as a Sulawesi crested macaque – exhibiting zinc toxicosis using radiography at the Smithsonian's National Zoological Park (or National Zoo), Washington, DC. The

A GREAT CHRISTMAS PARTY AT THE ZOOLOGICAL GARDENS.

Fig. 13.2 Feeding a bun to the elephant at London Zoo at Christmas (source: A great Christmas Party at the Zoological Gardens. *The Penny Illustrated Paper*, 9 January 1869, p. 25).

animal recovered after four coins were removed from his gut. The same condition was discovered in a female striped hyena (*Hyena hyena*) by Agnew et al. (1999). Her gut contained 20 coins and 10 coin fragments.

13.3.2   The History of Zoo Nutrition

In modern zoos the formulation of diets suitable for exotic animals has evolved from reliance on the anecdotes of experienced animal keepers to the use of sophisticated computer packages such as Zootrition^{TM} (developed by St Louis Zoo) and FAUNA^{TM} (designed by members of the EAZA Nutrition Group and the Nutrition Advisory Group of AZA) (Fidgett and Webster, 2011)

Abraham Bartlett, one-time Superintendent of London Zoo, listed the following as suitable diets for giraffes and polar bears (Bartlett, 1899):

GIRAFFE

The food of the giraffe in captivity must be as dry as possible, such as good old English clover-hay, crushed oats, beans, bran, crushed Indian corn, chaff with straw; roots, such as mangold,[1] carrots, and particularly onions, are good for them, and in summer a little green tares.[2]

POLAR BEAR (THALASSARCTOS)

This beast is the most strictly fish-eating of all the bears, but in captivity it can be partly fed upon bread, biscuit, etc. At the same time, it requires a strong oily or fat food. Passing a great part of its time in the water, it takes much exercise, and must be well fed, upon fish, fat (horse fat will answer), and now and then the common fish-oil (as it is called), that is, seal or whale oil.

The first specially formulated foods for zoo animals were created by Dr Ellen Corsen-White at the Philadelphia Zoo's Penrose Laboratory in the 1920s. This 'Zoo Cake' was a vitamin-rich mixture of cornmeal, cooked meat, fresh vegetables, eggs, melted fat, salt and baking powder. In 1948 the bird staff at the zoo began adding carrot juice to the diets of their flamingos to replace the carotenoids missing from their zoo diet, resulting in the return of their natural pink colour.

Zoos have traditionally borrowed knowledge from the livestock industry in formulating diets for their animals, but farm animal diets are aimed at maximising economic gain in a relatively small number of species. Kawata (2008) has argued that there is an urgent need to make increased use of the knowledge of the food habits of wild animals acquired by field biologists to improve zoo animal husbandry.

While this is a laudable aim, free-foraging Asian elephants (*Elephas maximus*) at the Seblat Elephant Conservation Center in Sumatra were recorded eating at least 273 plant species belonging to 69 families (Sitompul et al., 2013). It is difficult to see how any zoo located outside South-East Asia could hope to offer the variety of foods provided by nature for this species.

The history of zoo nutrition has been recounted by Crissey (2001).

---

[1] Beet.   [2] Vetch.

13.3.3 Dietary Research

The dietary requirements of many species are difficult to satisfy in a zoo environment, especially in specialist feeders. Diets for the same taxa may vary between institutions and it may not be possible to determine which diet is optimal for a particular type of animal.

Early experiments on diet were very basic and not terribly scientific. Mather (1878) published an article entitled 'Feeding of fishes in confinement' in which he reported:

As another instance of the effect of feeding ... I fed two small crabs on fish, and two on beef for four months; the former cast their shells and grew, while the latter did not. This, however, was not continued long enough to be perfectly satisfactory, and the results require confirming.

Modern studies of diet and nutrition in zoo-housed animals focus largely on providing a suitable diet in terms of macro- and micronutrients, including the development of foods for hand-reared animals. In this respect it is useful to compare the quality of foods utilised in the wild with those provided by zoos.

Studies of wild and captive brown bears (*Ursus arctos*) and polar bears (*U. maritimus*) have shown that they voluntarily select macronutrient proportions resulting in much lower dietary protein and higher fat or digestible carbohydrate than provided by most zoos. These lower protein concentrations maximised growth rates and the efficiency of energy utilisation in brown bears and may help to reduce diseases of the kidneys, liver and cardiovascular system in both species. Robbins et al. (2022) developed a new kibble – a dry compound food pellet – with a higher fat content and lower protein content than is typical in captive diets that was readily consumed by both species that they believed would result in long-term health benefits.

Pangolins are specialist ant feeders. Cabana et al. (2017) analysed the nutrient content of diets used by institutions that were successfully keeping Asian pangolins. They also analysed five different wild food items as a proxy for a wild diet. They found two types of captive diet: one composed mostly or completely of insects and the other high in commercial feeds or animal meat. The large nutrient range found was consistent with a domestic dog nutritional model, but the uppermost nutrient intake data favoured the feline nutrient recommendations. Cabana et al. were unable to determine which nutritional model was most appropriate from their analyses.

Colony composition and nutrient analysis of *Polyrhachis dives* ants, a natural prey of the Chinese pangolin (*Manis pentadactyla*), have been examined by Xu et al. (2022). They found that colony and adult ants differed in chemical composition and concluded that it might be inappropriate to feed this species with adults only. The study provided useful information for the development of artificial diets for Chinese pangolins

Vitamin E deficiency in reptiles, birds and ungulates living in zoos has been discussed by Dierenfeld (1989) and the nutritional deficiencies and toxicities found in captive New World primates have been reviewed by Crissey and Pribyl (2000). In some cases dietary supplements may be important in maintaining animal health.

For example, Durge et al. (2022) have shown that dietary lutein supplementation alleviates stress and improves the immunity and antioxidant status of captive Indian leopards (*Panthera pardus fusca*), whose diet is normally deficient in carotenoids.

Iron storage disorders are common in captive wild mammals (Class and Paglia, 2012). They occur in a wide range of taxa, including rhinoceroses, tapirs, fruit bats, marmosets, lemurs and some other primates, hyraxes, dolphins and pinnipeds. Class and Paglia emphasise the importance of reducing dietary iron levels to the maintenance requirements of individual species as a preventive measure.

Rearing rejected and prematurely born individuals is a major challenge for zoos. When a Nile hippopotamus (*Hippopotamus amphibius*) calf was born prematurely and rejected by its mother at the Cincinnati Zoo and Botanical Gardens, Ohio, staff decided that the calf should be hand-reared even though little published information was available on hand-rearing this species. Samples of the mother's milk were assayed for sugar, protein, fat, mineral and water content, and validated against data held by the National Zoo's Nutrition Science Laboratory for the milk of other species of mammals (Henry et al., 2022). Using this information staff at the zoo were able to formulate a successful milk replacement that allowed the calf to be raised through weaning to maturity.

13.3.4 Obesity, Diabetes and Dietary Restriction

It is important to monitor weight in animals living in zoos because obesity is a risk factor in many diseases, including heart disease, diabetes and some cancers. In some cases it may affect fertility.

Obesity is a problem in some animals living in zoos (e.g. rhinoceroses (Clauss and Hatt, 2006a), elephants (Clauss and Hatt, 2006b), felids (Vester et al., 2008) and chimpanzees (Videan et al., 2007)), and it can occur for a number of reasons, including excessive feeding and lack of exercise. Dominant animals in some species may acquire a disproportionate amount of the food provided by virtue of their position in the hierarchy.

Obesity is associated with decreased activity and increased health risks (e.g. cardiovascular disease, diabetes and high blood pressure in monkeys and apes) and poor reproductive performance (e.g. in elephants (Morfeld and Brown, 2014)). Elephants kept in zoos are prone to obesity due to lack of exercise and excessive calorie intake. Harris et al. (2008) found that 17.1 per cent of 76 elephants living in zoos in the UK were obese in the opinion of their keepers, although no measurements were made.

Cabana and Nekaris (2015) have demonstrated that diets high in fruits and low in gum exudates promote the occurrence and development of dental disease in the pygmy slow loris (*Nycticebus pygmaeus*). A survey of Association of Zoos and Aquariums (AZA) facilities holding primates found that nearly 30 per cent of responding institutions had at least one diabetic primate (Kuhar et al., 2013). All major taxonomic groups were represented, but the majority of cases occurred in Old World monkeys (51 per cent). Almost 80 per cent of all cases occurred in females. The authors of this study concluded that diabetes should be considered in decisions relating to diet, weight and activity levels in zoo-housed primates.

Obesity in chimpanzees has been assessed using a range of morphometric measurements: skinfolds (mm), body mass index (BMI), waist-to-hip ratio (WHR) and total body weight (kg) (Videan et al., 2007). Abdominal skinfold measurements appeared to be an accurate measure of obesity, and thus potential cardiovascular risk, in female chimpanzees, but not in males. Weight was reduced in several obese female chimpanzees by 5–7 per cent over the course of a single year when staff at the Primate Foundation of Arizona monitored and reduced calorific intake and monitored BMI and skinfold measurements.

When measurements from wild individuals are available it may be useful to compare them with data collected from their captive conspecifics. A weight-based definition of obesity was used by Schwitzer and Kaumanns (2001) in their study of obesity in both subspecies of ruffed lemurs (*Varecia variegata variagata* and *V. v. rubra*), comparing the weights of 43 individuals kept in European zoos with those of wild individuals. They found no significant differences in body weight between the two subspecies or between sexes, and no correlation between body weight and age. However, the mean weight of captive individuals was significantly higher than either of two different samples of wild *V. v. variegata*, and the authors concluded that 46.5 per cent of the individuals in the sample from zoos were obese.

Caloric restriction is the reduction in the intake of calories in the diet. Colman et al. (2009) studied rhesus macaques (*Macaca mulatta*) at the Washington National Primate Research Center for 20 years. They found that caloric restriction delayed the onset of age-associated pathologies. Compared with control-fed animals, those with a restricted diet benefited from an extended life and experienced a reduced incidence of diabetes, cancer, cardiovascular disease and brain atrophy.

13.3.5 Food Preference

The food preferences of captive animals are necessarily constrained by the fact that they are completely under human control. Carnivores must obviously eat what keepers provide, and in most zoos this means pieces of meat or carcasses, which may be used as enrichment especially in felids (e.g. McPhee, 2002). In some countries live feeding of vertebrates is not illegal.

Food preference studies have been conducted with a range of species living in zoos, including giant anteaters (*Myrmecophaga tridactyla*) (Redford, 1985), golden lion tamarins (*Leontopithecus rosalia*) (Benz et al., 1992) and stick insects (*Carausius morosus*) (Cassidy, 1978). Such studies may be performed by offering animals different foods in choice experiments or by direct observation of the foods selected when a choice is available, for example different food plants available within a habitat. For example, Barbiers (1985) examined orangutans' colour preference for food items.

Food preferences in the wild are not necessarily reflected in captivity when identical food species are available. When eight species of mound-building termites were offered to giant anteaters in Brasilia Zoo they exhibited marked preferences for some species over others, and these preferences did not correlate with those shown by wild conspecifics (Redford, 1985). For example, termites from the genus *Velocitermes* were most preferred in the wild but least preferred in captivity. This was explained by the fact that there were differences between the manner in which anteaters encountered prey in captivity compared with the way they encountered prey in the wild. Redford and Dorea (1984) studied the nutritional value of termites, ants and other species of invertebrates and concluded that most mammals that eat invertebrates choose their prey based on availability and other aspects of prey biology rather than gross nutritional factors.

The young of some species learn to forage from adults. Ueno and Matsuzawa (2005) found that infant chimpanzees (*Pan troglodytes*) referred to their mothers for some kind of cue before attempting to ingest novel food items. Wild fledgling keas (*Nestor notabilis*) discover little food for themselves (Diamond and Bond, 1991); most new food sources are excavated by adults. Social factors are important in the acquisition of foraging expertise in different ways at different stages of development.

Foods such as fruits and vegetables are often chopped into small pieces in zoos and spread around in enclosures as scatter feeds (Fig. 13.3). Brereton (2020b) has noted that the chopping and placement of food can affect food contamination, nutrient breakdown and desiccation rates.

Fig. 13.3 Animal food in zoos is frequently chopped into small pieces, often so that it can be scattered in enclosures. This takes up a great deal of keepers' time and also deprives animals such as primates of the enrichment value of the activities involved in breaking up the food for themselves.

13.4 Applied Epidemiology and Disease Surveillance

Infectious diseases in animals living in zoos are a risk to captive conspecifics and heterospecifics, wild populations, farm animals and human health. Disease surveillance is essential to monitor the prevalence of diseases in animal populations and their geography. In some cases a disease may be detected in a country for the first time within an animal living in one of its zoos or some other captive environment. For example, the first reported case of Pacheco's disease (*Psittacid herpesvirus*) in Hungary was from wild-caught Patagonian conures (*Cyanoliseus patagonus*) imported to a zoo that subsequently infected three bearded barbets (*Lybius dubius*), despite 61 days of quarantine (Bistyák et al., 2007). The first documented case of elephant endotheliotropic herpesvirus 4 (EEHV4) in the Asian elephant (*Elephas maximus*) in Asia was reported from a captive-born individual in Thailand (Sripiboon et al., 2013).

Disease may be transmitted to animals living in zoos from sympatric wild animal populations. Avian malaria caused by *Plasmodium relictum* in Magellanic penguins (*Spheniscus magellanicus*) has been reported from São Paulo Zoo, Brazil, and has caused mortalities (Bueno et al., 2010). The authors believed that the disease may have been transmitted to the penguins by mosquitoes that had previously bitten infected wild birds as the zoo is located in a remnant of Atlantic forest where the vectors (mosquitoes of the genera *Aedes* and *Culex*) occur and migratory bird species regularly entered bird enclosures in winter. Normally the penguins were housed indoors at night and therefore protected from mosquitoes. The infection occurred following a period when they were kept outside at night after presenting with pododermatitis (bumble foot).

Opportunistic blood samples collected from 14 tigers (*Panthera tigris*) (6 male, 8 female) from a zoo in Oklahoma along with 34 tigers (15 male, 19 female) and 8 African lions (*P. leo*; 6 male, 2 female) from a sanctuary in Tennessee were tested for the presence of a range of vector-borne infections by Cerrata et al. (2022). None of the tigers from the Oklahoma facility was positive for vector-borne organisms, but among tigers at the Tennessee facility animals tested positive for *Cytauxzoon felis* (11.8 per cent), '*Candidatus* Mycoplasma haemominutum'[3] (5.9 per cent) and *Ehrlichia ewingii* (2.9 per cent). This study identified the presence of tick-borne diseases in big cats in the south-eastern United States and identified a need to practice ectoparasite control measures to protect captive carnivores.

During an outbreak of Eastern equine encephalitis virus (EEEV) in Michigan in 2019, two two-month old Mexican wolf pups (*Canis lupus baileyi*) at Binder Park Zoo, Michigan, contracted the disease, probably as a result of a mosquito bite (Thompson et al., 2021).

Although the tapeworms of the genus *Echinococcus* usually live in the small intestine of carnivores, especially canids, they may be passed to humans and other primates.

---

[3] The use of quotation marks and the word '*Candidatus*' indicates that the bacterium has been well characterised by DNA analysis but cannot be maintained or isolated in culture.

Denk et al. (2016) have described fatal cases of cystic echinococcosis in two captive ring-tailed lemurs (*Lemur catta*) and one captive red-ruffed lemur (*Varecia variegata rubra*) in the UK. They noted that the disease is of global public health concern and emphasised the need for enhanced awareness and increased surveillance efforts.

The ZSL is one of the partner organisations responsible for the UK Cetacean Strandings Investigation Programme, which conducts research into mass stranding events around the coast of the UK, and is funded by the Department for Environment, Food and Rural Affairs (DEFRA). One such study involved the necropsy of 26 common dolphins (*Delphinus delphis*) stranded in and around a small tidal tributary of the Fal Estuary in Cornwall in 2008 (Jepson and Deaville, 2008).

The contribution of zoo-based wildlife hospitals to the national surveillance of emerging infectious diseases in Australian wildlife, whereby disease information from free-ranging wildlife populations is incorporated into the national wildlife health information system, has been discussed by Cox-Witton et al. (2014). They argue that this collaborative approach provides a model for surveillance programmes and enhances the capacity for the early detection of emerging diseases.

13.5   Quarantine

The use of quarantine and preshipment testing have long been cornerstones of the management of animal transfers between collections. An increase in our knowledge of animal diseases along with improved health care and disease surveillance in zoos and a greatly reduced dependence on wild populations as a source of animals has led to a re-examination of zoos' disease control practices (Pye et al., 2018).

Recent studies have questioned the necessity for all animals entering a collection to be quarantined and for comprehensive preshipment testing programmes, and some authors advocate a move towards a risk-based approach that relies on the identification of pathogens and the assessment and mitigation of the risk posed by each disease organism. Current practices are time-consuming, expensive and may have negative implications for animal welfare.

McLean et al. (2021) found that birds entering Disney's Animal Kingdom in Florida 'straight to collection' had lower morbidity rates within three months of acquisition than those entering with either 'standard quarantine' or 'risk-based standard quarantine'. Mortality within three months of acquisition showed no significant difference when compared between straight-to-collection and standard quarantine or risk-based standard quarantine methods, and no transmissible pathogens of concern were introduced from the acquired birds using either method. The study was conducted during the period 2013–2018 and McLean et al. concluded that the use of a risk-based approach to the management of animal transfers did not pose a greater threat to morbidity or mortality and still protected collections from the introduction of disease.

Quarantine data for animals arriving at San Diego Zoo in California were analysed by Wallace et al. (2016) to determine if a risk-based approach to quarantine could be used to manage transfers of animals into the zoo. The data related to animals arriving at the zoo from 81 AZA-accredited zoos and 124 other sources (including non-AZA-accredited institutions, government bodies and private dealers and breeders) over the period 2009–2013. No mammal or herptile failed quarantine due to a transmissible disease of concern, and although 2.5 per cent of birds failed quarantine for this reason, all 14 birds originated from non-accredited sources (confiscation and private breeders). Wallace et al. concluded that quarantine could be minimised or eliminated by using a risk-based approach where animals are transferred from institutions with comprehensive disease surveillance programmes and/or where preshipment testing is practiced.

Marinkovich et al. (2016) have suggested that a risk-based animal and institution-specific approach to transmissible disease preshipment testing should replace the industry standard of dogmatic preshipment testing because it is more cost-effective and provides better animal welfare. Such an approach would be based on a comprehensive surveillance programme including necropsy and preventive medicine examination testing and data. The authors analysed data collected over five years for animals shipped from San Diego Zoo and San Diego Zoo Safari Park, California, to 116 AZA-accredited and 29 non-AZA-accredited institutions. These animals consisted of 341 mammals, 607 birds and 704 reptiles and amphibians. The analysis found no evidence of the specific diseases tested for during the preshipment exam being present within the San Diego Zoo animal collection.

A survey of quarantine practices in aquariums ($n = 42$, mostly in North America) found considerable variation between institutions (Hadfield and Clayton, 2011) (Table 13.1). Only 25 per cent of institutions employed specialist quarantine staff, but 64 per cent used isolated areas for at least some of their fish quarantine. Separate quarantine protocols were used by most aquariums for freshwater teleosts, marine teleosts and elasmobranchs. In closed systems prophylactic treatments were common, especially formalin immersion for teleosts, freshwater dips and copper sulphate immersion for marine teleosts, and praziquantel immersion for marine teleosts and elasmobranchs. Food medicated with fenbendazole and praziquantel were used commonly in teleosts, but dosages varied greatly.

Table 13.1 Variation in quarantine practices in aquariums.

Procedure	Percentage of institutions
Use of visual health assessments	100
Necropsies on all mortalities	100
Minimum quarantine period of at least 30 days	95
Routine hands-on diagnostics carried out on some fish	54
Histopathology performed on almost all fresh mortalities	15

Source: data from Hadfield and Clayton, 2011.

13.6   Anaesthesia

13.6.1   Anaesthesia and Pain Relief

The routine use of anaesthesia in the treatment of captive animals is a relatively recent development in the history of zoo animal management. The former curator of Toronto Zoo, Canada, made the following statement in the *Canadian Journal of Comparative Medicine* in 1950 (Campbell, 1950):

> when I was appointed Veterinarian and Curator of the Toronto Zoological Collection, a definite ruling was made that no animal in our collection would be permitted to suffer any pain through medical and surgical treatment, and thus wherever possible, anaesthetics were to be used.

Campbell went on to describe the procedure used to subdue dangerous animals – such as lions and tigers – at Toronto Zoo when he was employed there as veterinarian and curator:

> To subdue the animal, instead of the four husky caretakers, we now have on hand four to six one-pound containers of chloroform. After closing all openings in the front of the den with tarpaulins to confine the fumes, chloroform is withdrawn from the can by means of a syringe and quietly squirted on the floor of the den in close proximity to the animal. There is no fuss and commotion; usually, after the fourth can has been used, the patient becomes drowsy, wobbles its head about, and then slumps on the floor. At this moment, a large handful of absorbent cotton is placed in a piece of cloth and tied to the end of a light stick about five feet long. The cotton is saturated with chloroform and held close to the animal's nose. Often, the animal tried to assist in anaesthetizing itself, as it were, by biting the end of the saturated cotton, and in a few minutes, it loses consciousness completely. To be sure there is no feeling, it is advisable to poke the body with a stick.

Seventy years later, although anaesthesia is much more sophisticated, it still carries risks for the patients.

13.6.2   Recent Developments in Anaesthesia Research

Veterinarians have recently turned their attention to the anaesthesia of lower vertebrates and invertebrates. A study of the pharmacokinetics of a single dose of intramuscular and oral meloxicam – a nonsteroidal anti-inflammatory drug that can reduce pain – in 17 healthy yellow stingrays (*Urobatis jamaicensis*) was undertaken at the John G. Shedd Aquarium, Chicago, Illinois (Kane et al., 2022). The results obtained led the authors of the study to recommend a dosage of 2 mg/kg orally once daily. They established that the drug was rapidly eliminated when administered intramuscularly and concluded that it would be difficult to maintain clinically relevant plasma concentrations using this route.

The effects of the anaesthetic drug tricaine methanesulfonate (MS-222) in a managed collection of moon jellyfish (*Aurelia aurita*) maintained by the North Carolina Museum of Natural Sciences, Raleigh, North Carolina, were examined by Gorges et al. (2022). The purpose of the study was to examine the effects of the drug on movement and response to stimuli. Clinically healthy jellyfish were assigned to one of three groups of eight for trials of 0.3 g/L MS-222, 0.6 g/L MS-222 and a saltwater control. Movement and response to stimuli were both decreased when jellyfish were exposed to MS-222 at 0.3 and 0.6 g/L. All of the animals used in the trials recovered uneventfully and there were no mortalities.

13.6.3 Anaesthesia and Mortality

The risk of mortality among great apes resulting from anaesthesia has been studied by Masters et al. (2007) using 1,182 records relating to individuals from 16 zoos in the UK and Ireland that underwent anaesthesia between 1 January 1990 and 30 June 2005. Fifteen deaths were recorded; 20 per cent of these occurred during maintenance and the remaining 80 per cent occurred within 7 days post-anaesthetic. The authors concluded that great ape anaesthesia carried a high risk of mortality: at least five deaths appeared to be anaesthetic-related, resulting in a mortality risk of at least 0.42 per cent (5/1,182). An increased risk of mortality was identified in sick and older animals (over 30 years).

13.7 *In-Situ* Wildlife Disease Management

Zoo-housed animals have been used to study the biology of disease vectors. The distances flown by mosquitoes following a blood meal were studied by Greenberg et al. (2012). Four species of mosquitoes were trapped in the Rio Grande Zoo in Albuquerque, New Mexico, and the source of the blood found in blood-gorged individuals was determined by using blood meal analysis. The authors found that mosquitoes trapped in the zoo flew an average of 106.7 m after feeding and never further than 170 m.

Martínez-de la Puente et al. (2020) sampled mosquitoes in Barcelona Zoo in Spain and captured three species: *Culex pipiens sensu lato*,[4] *Culiseta longiareolata* and *Aedes albopictus*. Birds dominated the diets of the first two species and *Aedes* fed only on humans. Mosquitoes had a mean flight distance of 95.67 m after feeding on blood, with a maximum of 168.51 m (remarkably similar to the distances measured in Rio Grande Zoo). The authors found evidence that *C. pipiens* s.l. was involved in the local transmission of avian *Plasmodium*, potentially facilitating circulation been wildlife and zoo-housed animals. They concluded that zoos need to be vigilant in monitoring potential mosquito breeding sites within their grounds.

Zoo collections are useful in identifying the range of hosts utilised by parasites and determining the extent to which they occur in non-native species. For example, in the UK Boufana et al. (2012) identified – using molecular techniques – the tapeworms *Echinococcus granulosus* (G1 genotype) in a guenon monkey (*Cercopithecus* sp.) and a Philippine spotted deer (*Rusa alfredi*); *E. equinus* in a zebra and a lemur; *E. ortleppi*

---

[4] In this context '*sensu lato*' (in the broad sense) means that the preceding scientific name should be taken as including all subordinate taxa and/or other taxa that may have been considered as distinct at other times.

in a Philippine spotted deer; *E. multilocularis* in a macaque monkey (*Macaca* sp.) and *Taenia polyacantha* in jumping rats (*Hypogeomys antimena*).

13.8 Infectious and Zoonotic Diseases

An infectious disease is an illness that is caused by an organism such as a bacterium, virus, protozoan, fungus or helminth. A zoonotic disease is one that can be transmitted between animals and humans. The zoonoses of greatest concern within the United States have been identified by the CDC (2017). They are zoonotic influenza, salmonellosis, West Nile virus, plague, rabies, brucellosis, Lyme disease and emerging coronaviruses (e.g. severe acute respiratory syndrome (SARS) and Middle East respiratory syndrome (MERS)).

13.8.1 Salmonellosis

Jang et al. (2008) determined the rate of *Salmonella* spp. infection in animals at Seoul Grand Park, a zoo in South Korea. They took anal swabs from 294 animals and found that reptiles were at particular risk, with 14 of the 46 (30.4 per cent) individuals examined testing positive. Only 1 of the 15 birds (6.7 per cent) and 2 of the 233 mammals (0.9 per cent) were infected with *Salmonella*. Jang et al. drew attention to the need for vigilance and education about personal hygiene in zoos where visitors are allowed to handle animals.

Contact with contaminated surfaces may be an important source of *Salmonella* infection. In 1996, 39 children became infected with *Salmonella* after visiting a Komodo dragon (*Varanus komodoensis*) exhibit at Denver Zoo in Colorado (Fig. 13.4). The visitors were only separated from the dragons by a wooden fence and the reptiles were allowed to wander freely behind this. None of the children touched the Komodo dragons, suggesting that they became infected by contact with the infected wooden barrier (Friedman et al., 1998).

In September 2009 a petting farm in Surrey, England, was temporarily closed when a number of children were hospitalised after an outbreak of *E. coli* 0157:H7, which is potentially fatal (Rees, 2011).

13.8.2 COVID-19

In 2020 the SARS-CoV-2 (COVID-19) pandemic led to millions of human deaths and the closure of many zoos and other visitor attractions around the world in an attempt to reduce community transmission.

Gollakner and Capua (2020) have discussed the possibility that this pandemic could evolve into a panzootic because transmission from humans to dogs (two cases), domestic cats (two cases), tigers (four cases) and lions (three cases) had been recorded and they suggested that non-human primates, pigs and ferrets may also be susceptible. Infection and transmission in ferrets has since been reported (Kim et al., 2020). The

Fig. 13.4 Komodo dragon (*Varanus komodoensis*). In 1996, 39 children became infected with *Salmonella* after visiting a Komodo dragon exhibit at Denver Zoo in Colorado.

infected tigers and lions were housed at the Bronx Zoo, New York, and these cases were the first reports of infection in non-domestic species in the world (McAloose et al., 2020). Genomic analysis indicated that the virus had been transmitted to the tigers from a keeper. No clear transmission route was identified for the lions, but it seemed likely that they were also infected by the keeper(s). Subsequently, Fernández-Bellon et al. (2021) reported SARS-CoV-2 infection in four lions (*Panthera leo*) and three caretakers at Barcelona Zoo, Spain. Again, genomic analysis supported human-to-lion transmission as the origin of the infection.

Gollakner and Capua (2020) propose a One Health approach to surveillance, intervention and management of the disease – that is to say, a decompartmentalisation of human, animal and ecosystem health to protect both human and wildlife populations. A more comprehensive review of the susceptibility of domestic and wild animals to SARS-CoV-2 has been provided by Kumar et al. (2020).

Ten et al. (2021) have published seven disease control measures that they believe should be introduced in Malaysian zoos to prevent the spread of SARS-CoV-2 and other zooanthroponoses in animals. These included restricting human–animal interactive activities and animal performances involving susceptible species, and disease screening for SARS-CoV-2 in the pre-release protocol for any Malayan tigers (*Panthera tigris jacksoni*) that might be released into the wild.

13.8.3   Other Zoonoses

In a survey of workers in North American zoos who worked with mammals (including primates) for antibodies to simian foamy viruses (SFV), 4 out of 133 (3 per cent) were seropositive, primarily with chimp-like viruses (Sandstrom et al., 2000). Simian foamy virus antibodies were detected in three persons who worked in zoos by Switzer et al. (2004). European bat lyssaviruses (ELB types 1 and 2) have been identified as emerging zoonoses (Fooks et al., 2003). European bat lyssavirus was isolated in Denmark from a colony of Egyptian flying foxes (*Rousettus aegyptiacus*) originating from a Dutch zoo (Van der Poel et al., 2000).

Tuberculosis is a threat to a wide range of species kept in zoos, including rhinoceroses (Miller et al., 2017) and elephants (Mikota et al., 2000), and infected animals may pose a health risk to keepers. Bovine tuberculosis was identified in six out of eight antelopes tested at a zoo in Poland (Krajewska et al., 2015a) and in giraffe (*Giraffa camelopardalis*) (Krajewska-Wędzina et al., 2018). Avian TB was identified in a cassowary (*Casuarius casuarius*) at a Polish zoo in 2010 after transfer from a Dutch zoo (Krajewska et al., 2015b). In a zoo in the Netherlands, six keepers tested positive for TB (*Mycobacterium pinnipedii*) after exposure to infected sea lions, probably as a result of nebulisation resulting from cleaning their enclosure (Kiers et al., 2008).

13.8.4   Elephants and Herpesviruses

Herpesviruses (EEHVs) have emerged as a significant threat to the survival of elephants of both species in zoos (Montali et al., 1998; Rickman et al., 1999). An analysis of deaths from EEHV-HD in Asian elephants (*Elephas maximus*) born in European zoos between 1985 and 2017 found that, of those that survived for more than one day and tracked until the age of eight years, or until death, 57 per cent ($n = 25/44$) died as a result of the disease (Perrin et al., 2021). The median age of death was 2.6 years, with no difference between the sexes. Exposure to new elephants did not affect the risk of EEHV death, but deaths were more likely in institutions with a previous history of EEHV death.

The high incidence of EEHVs in zoos has provided important opportunities for the disease to be investigated. Dastjerdi et al. (2016) investigated the use of anti-herpesviral therapy and reported on two cases at Whipsnade Zoo, England.

A longitudinal study of the effects of between- and within-herd movements on EEHV recrudescence was conducted at Whipsnade Zoo (Titus et al., 2022). Between-herd moves consisted of relocating two bulls out of the collection and housing a new bull in an adjacent paddock with limited trunk touching possible; within-herd moves consisted of unrestricted full contact between the new bull and the females, including active mating. A period of social stability acted as a control. The study found that all management-derived social changes promoted recrudescence, but within-herd movements posed the most significant increase in the probability of EEHV reactivation.

The first report of the use of the Zelnate DNA immunostimulant and recombinant human interferon alpha (rhIFNα) in the successful treatment of EEHV1A-HD in an Asian elephant at Chester Zoo, England was provided by Drake et al. (2020). At the time of writing Chester Zoo announced that it was embarking on a trial of a vaccine for EEHV in collaboration with scientists at the University of Surrey (Gill, 2022).

13.9   Animal Injuries

Congenital and traumatic injuries in animals living in zoos are occasionally reported in academic journals and often emphasise the lack of available information on the physiology and development of exotic species.

Silva et al. (2022) described fatal congenital and traumatic cervical spine injuries in a newborn plains zebra (*Equus quagga*) born at Zoo da Maia in Portugal. Sayer et al. (2007) discussed locomotor, postural and manipulative behaviour in a juvenile male gibbon who had one of his arms amputated as a result of an untreatable injury. He found unique solutions to foraging and locomotion, often using his feet and teeth, and was capable of one-armed brachiation.

Diseases of the urinogenital tract or endocrine disorders associated with the presence of foreign bodies in the vagina have been reported in primates living in zoos by Lamglait et al. (2022). The animals concerned were four Japanese macaques (*Macaca fuscata*) from the same group, a Wolf's guenon (*Cercopithecus wolfi*) and a Western lowland gorilla (*Gorilla g. gorilla*). All of these animals suffered from vaginitis and a variety of other conditions (Table 13.2). The authors emphasised the need to investigate this abnormal behaviour to determine the underlying cause, especially when encountering vaginal bleeding.

Injuries caused to animals living in zoos by enrichment devices are discussed in Section 10.14.

13.10   Gait Analysis

The analysis of an animal's gait can provide useful information about the mechanics of its locomotion and, in some cases, may help to diagnose and alleviate skeletal and other problems.

Table 13.2 Diseases and disorders associated with vaginal foreign bodies in non-human primates.

Species	Individual	Age (yrs)	Disease/ disorder
Japanese macaque	1	20	Cavernous uterine haemangioma
(*Macaca fuscata*)	2	21	Diffuse endometritis
	3	24	*In-situ* endometrial carcinoma
	4	24	Chronic cystitis and chronic renal disease
Wolf's guenon (*Cercopithecus wolfi*)	5	12	History of hypothyroidism with irregular reproductive cycles
Western lowland gorilla (*Gorilla g. gorilla*).	6	27	Endometritis and an ovarian cyst-like structure

Source: based on information in Lamglait et al., 2022.

The gait characteristics of polar bears (*Ursus maritimus*) during normal non-repetitive locomotion and during pacing were compared by Cless et al. (2015). They analysed high-speed video recordings of 11 zoo-housed polar bears and concluded that pacing is quantitatively different from non-repetitive locomotion and may reflect disengagement with the environment, incorporating reduced sensory feedback and cognitive input. The bears exhibited lower variability of gait characteristics while pacing. Step cycle duration was shorter while pacing than during normal locomotion and head height was higher. Both step cycle duration and head height displayed lower variation while pacing.

Locomotor kinematic data from more than 2,400 strides for 46 Asian elephants (*Elephas maximus*) and 14 African elephants (*Loxodonta africana*) were used to establish the normal kinematics of elephants (Hutchinson et al., 2006). The elephants used were kept in facilities in the United States (California and Indiana), Germany, the UK and Thailand, and the results of the work could be used to identify gait abnormalities indicative of musculoskeletal pathologies.

The use of a pressure-sensitive walkway to analyse gait in Humboldt penguins (*Spheniscus humboldti*) has been described by Sheldon et al. (2020). They analysed gait in 21 adult penguins, including five abnormal individuals who were suffering from either right-sided or historical lameness causing disease. Step and stride distances and velocities, maximum force, impulse and peak pressure were calculated for each foot of each penguin. In normal penguins there were no significant differences between feet or sex, but abnormal penguins had a shorter left step width than normal penguins. The method should be useful in assessing the causes of lameness, including pododermatitis (bumble foot) and osteoarthritis.

Kinetic gait analysis using a pressure plate system has been performed on a female Indian rhinoceros (*Rhinoceros unicornis*) at Zoo Vienna in Austria as a clinical tool for the detection of early chronic foot disease (Pfistermüller et al., 2011). The condition is characterised by lesions between the central toe and the pad that frequently become infected.

13.11   Causes of Death and Post-Mortem Records

Animals in zoos are frequently found dead, often having shown no signs of disease, so it is essential that a post mortem is performed to ascertain the cause of death to protect other animals and humans from the possibility of exposure to infectious disease. In a study of mortality in Eurasian lynx (*Lynx lynx*) in zoos in the UK between 2000 and 2015, of 44 animals for which data were available, 13 (30 per cent) were found dead and the remainder were euthanised (Heaver and Waters, 2019).

Zoos have a long history of recording the causes of death of their animals. Table 13.3 shows part of a record of the causes of deaths of animals in 1913 at Victoria Zoo, Melbourne, Australia as determined by the staff of the Veterinary School of Melbourne University.

Table 13.3 Facsimile of part of a page of a report of the causes of death of animals at Victoria Zoo, Melbourne, Australia (January 1913).

37
Report of Causes of Death
during the year 1913
From the VETERINARY SCHOOL, MELBOURNE UNIVERSITY

Jan. 6 – Carpet Snake: intestinal catarrh, blood parasites.
    ” 7 – Black Wallaby: congestion of lungs.
    ” 9 – Carpet Snake: gastro-intestinal catarrh, parasites.
 ” 21 – Black snake: gastro-intestinal catarrh.
 ” 21 – Boobook Owl: intestinal catarrh.
 ” 21 – Ring-tail Lemur: haemorrhage of the pleural cavity.
 ” 24 – Masked Owl: blood parasites.
 ” 28 – Ring necked Pheasant: wound on wing.
 ” 28 – Ring Dove: intestinal catarrh.
 ” 29 – Skink Lizard: Emaciated.

Source: Victoria Zoo Annual Report c.1913, Belle Vue Zoo Archive, Chetham's Library, Manchester, England.

The deaths that occurred in the ZSL's gardens during the years 1939–1940 were reported by Hamerton (1941). This included reports on the animals euthanised during the Second World War: 38 mammals, 35 birds and 98 'dangerous reptiles'. Those individuals found not to be diseased or carrying parasites were fed to carnivores at the zoo.

Analysis of the causes of death of animals living in zoos can provide important information about the prevalence of disease in particular taxa and the history of animal welfare in zoos. An analysis of mortality in Australian marsupials kept at London Zoo between 1872 and 1972 found that 60 per cent of the deaths occurred between 1992 and 1931, often shortly after arrival due to poor transport conditions (Canfield and Cunningham, 1993). Increased mortality often occurred after periods of cold weather, especially from pneumonia. Marsupials commonly suffered from pneumonia, gastroenteritis, gastric ulceration and lumpy jaw (necro-bacillosis). In addition, liver disease, gastric neoplasia and intestinal obstruction occurred. During the time that records were kept, the range of diseases found in marsupials at the zoo were similar to those found in other zoos. They had not changed greatly over time, but the prevalence of the diseases was influenced by the changing environment.

Post-mortem findings for over 12,000 specimens from more than 2,000 species and subspecies that died at San Diego Zoo and San Diego Wild Animal Park in California between 1964 and 1978 have been described by Griner (1983). A study of the causes of death of 353 animals that died at the Ljubljana Zoo, Slovenia, between 2005 and 2015 found that almost 40 per cent died of infectious diseases (Cigler et al., 2020). Other causes of death are indicated in Fig. 13.5. Of the animals examined, 69 per cent were mammals, 24 per cent were birds and 7 per cent were reptiles.

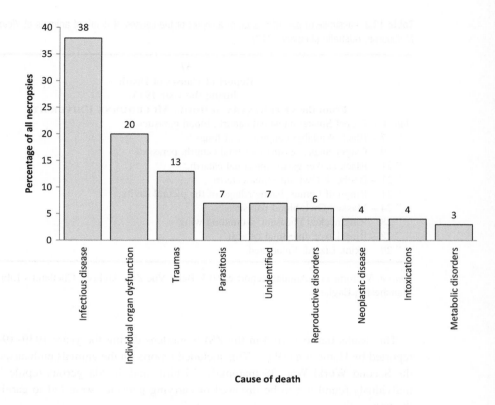

Fig. 13.5 Causes of death of animals in Ljubljana Zoo, Slovenia, 2005–2015.

A study of causes of death among Eurasian lynx (*Lynx lynx*) in zoos in the UK (2000–2015) found that culling as part of population management was the commonest cause of death (21 per cent) as many individuals in the population had high inbreeding coefficients (Heaver and Waters, 2019). It should be noted that the European studbook for this species was created in 2002 and all of the neonatal culls occurred before this. The next most common causes of death were neoplastic (16 per cent), circulatory (11 per cent), neurological (11 per cent) and genitourinary (11 per cent) disease.

A retrospective study of the causes of mortality in tufted puffins (*Fratercula cirrhata*) at Point Defiance Zoo and Aquarium, Tacoma, Washington, between 1982 and 2017 found that the most common pathological finding across all age classes was aspergillosis, particularly in adults (Heinz et al., 2022). Other frequent causes of death were haemoparasitism, predation and trauma. Among neonates, mortality was primarily caused by omphalitis, yolk sac disease and bacterial septicaemia, and most cultures detected *Escherichia coli*.

Baker et al. (2022) conducted a retrospective evaluation of cases of obstructive and incidental urolithiasis in 21 individuals representing five Asian species of colobine monkeys (*Trachypithecus* spp. and *Pygathrix nemaeus*) from eight institutions, and found that 86 per cent of cases occurred in males and all cases of obstructive urolithiasis were found in males. Death or euthanasia secondary to obstructive urolithiasis occurred in 52.4 per cent of cases and the authors concluded that urinary obstruction secondary to urolithiasis appears to be an important cause of morbidity and mortality in these monkeys.

13.12 Morphological Effects of Captivity

Distinctive cranial differences have been reported between wild and captive lions (*Panthera leo*) from the same locality in Kenya by Hollister (1917) probably in part due to nutritional inadequacies in zoo diets at the time. More recently, Duckler (1998) reported malformed external occipital protuberances in zoo specimens of tigers (*P. tigris*) that were not found in wild-caught individuals. The condition was most likely the result of increased lateral rotation of the head and neck combined with reduced jaw activities, which may have been the consequences of non-natural diets and the increased grooming behaviours fostered in captive environments.

A study of the flight musculoskeletal system of Rodrigues fruit bats (*Pteropus rodricensis*) found no difference in wild-caught founders and captive-bred bats adapted to captivity (Fig. 13.6). Founders and captive-bred individuals deposited increasing quantities of subcutaneous fat with age and no significant difference was

Fig. 13.6 Rodrigues fruit bats (*Pteropus rodricensis*).

identified in the dimensions of their limb bones or the dry weights of their principal flight muscles.

The effects of captivity on the morphology of mammals has been reviewed by O'Regan and Kitchener (2005).

13.13   Skeletal and Dental Problems

13.13.1   Skeletal Problems

Animals held in zoos are prone to accidents causing skeletal problems. For example, macropods are prone to fractures of the cervical vertebrae as a result of collisions (Kragness et al., 2016). Sometimes accidents occur as a result of poor enclosure design. A study of mandibular fractures in giraffes (*Giraffa camelopardalis*) in European zoos found that 14 of 86 responding zoos (16.3 per cent) reported jaw fractures, most of which (71.4 per cent) required surgery (Remport et al., 2022). Seven (50 per cent) of these 14 cases were associated with hay racks where the muzzle could fit between the grid of the feeding contraptions. Most fractures of the mandible occurred in young giraffes (mean age 3.4 years) and the authors of the study recommended that facilities keeping this species should evaluate their hay feeders for entrapment risk, especially if they house young individuals.

Da Costa et al. (2022) have described the surgical management of upward fixation of the patella via medial patellar ligament desmotomy in a lowland tapir (*Tapirus terrestris*) housed at the Quinzinho de Barros Zoological Park in Brazil. As part of this work an anatomical study was made of the limbs of a different tapir that had been killed in a car accident. The patient was returned to her enclosure on the same day as the operation and normal locomotion was restored.

Spinal spondylosis and acute intervertebral disc prolapse have been reported from a 22-year-old male European brown bear (*Ursus a. arctos*) at Johannesburg Zoo in South Africa (Wagner et al., 2005). The prognosis for this animal was poor so he was euthanised and the diagnosis was confirmed on necropsy.

Degenerative spinal disease is an important problem in big cats living in zoos. Kolmstetter et al. (2000) conducted a retrospective study of the medical records of big cats that died or were euthanised at Knoxville Zoo, Tennessee, between 1976 and 1996. The records related to 13 lions (*Panthera leo*), 16 tigers (*P. tigris*), 4 leopards (*P. pardus*), 3 jaguars (*P. onca*) and 1 snow leopard (*Uncia uncia*). Of these, 8 cats (3 lions, 4 tigers and 1 leopard) were diagnosed with degenerative spinal disease. The median age of onset of clinical signs was 18 years.

13.13.2   Dental Problems

Dental problems experienced by animals living in AZA zoos have been surveyed by Glatt et al. (2008). In the 58 institutions examined, the commonest problems encountered were malocclusions, dental fractures and periodontal disease.

Wenker et al. (1999) compared the teeth from 63 brown bears (*Ursus arctos*) that had lived in the Bernese Bear Pit in Switzerland between 1850 and 1995 (zoo bears) with teeth from 14 skulls from free-ranging Alaskan brown bears (*U. a. horribilis*). Severe enamel and dentinal attrition was found in the canine teeth of zoo bears over 10 years old, along with exposed pulp and proximal lesions in the molars. Deposits of calcified dental calculus were widely distributed in the zoo bears and increased with age. A much lower degree of calculus deposition was present in the Alaskan bears. These bears had a higher frequency of caries with large individual variation. The authors concluded that stereotypic chewing was the likely cause of canine tooth and secondary alveolar lesions in the zoo bears and that extensive calculus formation in these individuals was the result of an inappropriate diet and inadequate natural tooth cleaning opportunities.

The repair of tusk fractures using composite materials has been described by Sim et al. (2017). A Kevlar–fibreglass composite was used to cap both tusks of a seven-year-old male Asian elephant who had fractured the distal end of each tusk, which prevented further damage. In addition, a 34-year-old male African elephant received a carbon fibre–fibreglass composite circumferential wrap to potentially stabilise a longitudinal crack.

A study of two spectacled bears (*Tremarctos ornatus*) at Wroclaw Zoo in Poland found that stereotypic behaviour in the more stereotypic individual of the two reduced significantly and foraging increased following dental treatment involving multiple extractions (Maslak et al., 2013).

Tooth development may provide a record of stressful events in the life of some animals. Cipriano (2002) showed that incremental lines in dental cementum can record exposure to periods of cold weather in orangutans (*Pongo* sp.). Four out of five animals examined showed marked irregularities in terms of hypomineralised bands that could be dated to the year 1963. These animals had been kept in a zoo in the northern hemisphere, where 1963 had had an extremely cold winter. Cold stress consumes calcium and the lack of available calcium could have been responsible for the hypomineralised bands. A record of stressful life events was recorded in the dental development of a juvenile gorilla by Schwartz et al. (2006).

Other publications concerned with dental problems in animals living in zoos have included work on the bacteriology of dental plaque (Dent, 1979), dental occlusions in cheetahs (*Acinonyx jubatus*) (Fitch and Fagan, 1982), the relationship between diet and dental disease in the pygmy slow loris (*Nycticebus pygmaeus*) (Cabana and Nekaris, 2015), dental resorptive lesions in leopards and lions (Berger et al., 1996) and the prevention of dental calculus formation in lemurs (*Lemur catta* and *Eulemur fulvus*) and baboons (*Papio cynocephalus*) (Willis et al., 1999). Other studies have reported the findings of dental examinations of Bolivian squirrel monkeys (*Saimiri boliviensis*), black-tufted marmosets (*Callithrix pencillata*) (Žagar et al., 2021) and meerkats (*Suricata suricatta*) (Kvapil et al., 2006), and dental emergence norms in orangutans (Fooden and Izor, 1983). Dental care for a killer whale (*Orcinus orca*) has been described by Graham and Dow (1990)

An up-to-date account of zoo and wild animal dentistry has been produced by Emily and Eisner (2021).

13.13.3 Problems of Old Age

Individuals of many species live longer in zoos than do their wild conspecifics, developing diseases of old age that would make it difficult for them to survive without human intervention. These animals are problematic for zoo managers because they take up valuable space needed for breeding individuals, although some may play an important social role, such as post-reproductive cow elephants which may act as allomothers to the calves of younger animals (Rees, 2021)

Dental and skeletal pathologies are prevalent in captive bears in old age (Kitchener, 2004). In individuals aged 15 years and above a wide range of problems were identified, including fused vertebrae in around 55 per cent of animals and broken canine teeth in over 70 per cent.

Necropsies performed on mammals following euthanasia have shown that pathological changes were already at an advanced stage in many individuals, causing pain, a reduced quality of life and compromised welfare. In order to develop a decision-making framework for the euthanasia of geriatric mammals, a scoring system to evaluate physical condition and quality of life has been developed by Föllmi et al. (2007). This system was based on investigations of 70 geriatric mammals in five European zoos. The subjects were from 24 species, including bears, big cats, giraffids, antelopes, zebra, wolves, monkeys, camels, tapirs, rhinoceroses and an African elephant.

13.14 Diagnostic Predictors

Bliss et al. (2022) made a comparison of diagnostic predictors of neonatal survivability in non-domestic Caprinae using animals kept at San Diego Zoo and San Diego Zoo Safari Park in California over a period of 29 years. They tested 184 neonates from 10 non-domestic Caprinae species within seven days of birth. The results of the study found that glucose and fibrinogen levels were better predictors of neonate survival than the other tests reviewed: serum gamma glutamyltransferase, glutaraldehyde coagulation test and sodium sulphite precipitation test. The authors of the study noted that neonates who were known to have nursed had a lower chance of requiring major medical intervention.

A study of cholelithiasis – gallstones and related disease – in the mountain chicken frog (*Leptodactylus fallax*) involved a retrospective evaluation of available data from 133 postmetamorphic individuals kept at Chester Zoo, *Durrell* and the ZSL's zoos (Martinez et al., 2022) (Fig. 13.7). This comprised 139 ultrasonographic images, 156 radiographic images and 32 histopathology samples obtained between 2014 and 2020. The authors described the use of an ultrasound score to standardise monitoring of changes in the gall bladder in this species.

The use of allostatic load indices together with data on serum cholesterol and triglycerides to predict disease and mortality risk in zoo-housed Western lowland gorillas (*Gorilla g. gorilla*) has been discussed by Edes et al. (2020). The development of a quantitative immunoassay for serum haptoglobin as a putative disease marker in

Fig. 13.7 Mountain chicken frog (*Leptodactylus fallax*).

the Southern white rhinoceros (*Ceratotherium simum simum*) has been described by Petersen et al. (2022).

13.15 Veterinary Care of Invertebrates

The veterinary care of invertebrates is a relatively new area of research, but it is growing in importance as zoos increasingly add a wider range of invertebrates to their collections. There is also increasing welfare concern about the use of live invertebrates as food for other animals (see Section 6.3.5).

Relatively little research has been conducted on baseline haematological values and biochemical parameters in invertebrates compared with that conducted on vertebrates. Griffioen et al. (2022) determined baseline haemocyte concentrations and biochemistry values for Japanese spider crabs (*Macrocheira kaempferi*) housed at six public aquariums (Fig. 13.8). They identified distinct haemocyte types including hyaline cells (the predominant type), semigranulocytes and granulocytes. Crabs with exoskeletal lesions were found to have higher absolute semigranulocyte counts and triglyceride levels than those without lesions. Higher total protein, cholesterol, triglyceride and amylase levels were found in non-moulting crabs compared with those that have previously moulted since acquisition. When crabs kept in aquariums that used ozone

Fig. 13.8 Baseline haematological values and biochemical parameters have been determined for Japanese spider crabs (*Macrocheira kaempferi*) housed in aquariums.

sterilisation systems were compared with those that were not, the authors found lower relative and absolute granulocyte counts and higher albumin concentrations in individuals housed in aquariums that used ozone.

Researchers from the University of Illinois, University of Wisconsin-Madison and Lincoln Park Zoo, Illinois, used diagnostic imaging to improve baseline knowledge of the radiology and radiographic anatomy of six terrestrial invertebrate species: Madagascar hissing cockroach (*Gromphadorhina portentosa*), desert millipede (*Orthoporus* sp.), emperor scorpion (*Pandinus imperator*), Chilean rose tarantula (*Grammostola spatulata*), Mexican fireleg tarantula (*Brachypelma boehmei*) and Mexican redknee tarantula (*Brachypelma smithi*) (Davis et al., 2008). Individuals were fed radiographic contrast media and contrast-containing food items in order to produce visualisations of the gastrointestinal anatomy of each of these species. Anaesthesia of Chilean rose tarantulas has been examined by Dombrowski et al. (2013).

The emergency care and first aid for common injuries found in invertebrates has been discussed by Pellet and Bushell (2015) and a more comprehensive account of invertebrate medicine has been published by Lewbart (2022).

13.16   Conclusion

Catering for the many and varied diets of animals living in zoos and aquariums presents a significant challenge for these institutions, but wild animal nutrition now has a solid foundation in science. As the body of knowledge and expertise in animal nutrition and veterinary medicine expands we can expect to see ongoing improvements in not only the health and welfare of animals living in zoos and aquariums, but also that of animals living wild. Improving our understanding of the transmission of zoonotic diseases is essential if we are to prevent, or at least contain, the next global pandemic that originates from a population of wild animals. Zoos can make an important contribution to the One Health approach to dealing with disease.

14 The Past and Future of Zoos

14.1 Introduction

A chapter on 'the future of zoos' presupposes that zoos have a future. Many animal welfare organisations have actively worked to remove particular individual animals (e.g. the Free Morgan Foundation established to free a killer whale) or taxa (e.g. elephants, cetaceans) from zoos, and others challenge the value of zoo conservation and education (e.g. *ZooCheck*).

Many historically important zoos have closed. In the UK, Belle Vue Zoological Gardens in Manchester closed in 1979, after more than 140 years, Glasgow Zoo closed in 2003, Windsor Safari Park closed in 1992, Lambton Lion Park in County Durham closed in 1980 and Marineland in Morecambe, Lancashire closed in 1990. In 1991, London Zoo – perhaps the most famous zoo in the world – almost closed due to financial difficulties and increased competition from other visitor attractions (Rees, 2011). But new zoos have emerged and many existing zoos are investing in new facilities. In 2022, Bristol Zoo closed its city site after 186 years and moved to a new, larger, out-of-town site five miles away.

Some zoos of great historical importance cannot avoid acknowledging that they have legacy issues that need to be addressed (Fig. 14.1). Many are associated with the colonial history of European countries and were established with animals taken from the wild in a time when this was considered perfectly acceptable. Others have been associated with circuses and animal cruelty. Critics of zoos constantly remind us that, regardless of their current good intentions, many old zoos have a chequered history.

Twenty-three years ago, in 1999, Turley noted that, in spite of increasing competition in the day-visit market and mounting concerns about the keeping of wild animals in captivity for human amusement, visitor numbers at zoos in the UK were likely to remain steady 'but at a lower level', and that while some would falter there was 'scope for success and continued existence'. This conclusion was based on a national survey of zoo visitors (Turley, 1999). The COVID-19 pandemic demonstrated to the zoo community just how vulnerable it is to global disasters.

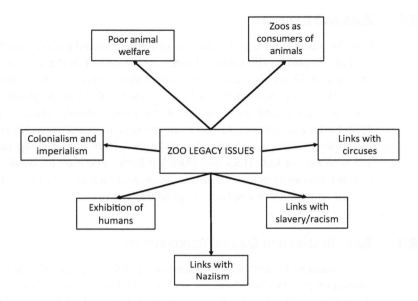

Fig. 14.1 Legacy issues for zoos.

14.2 The Politically Correct Zoo?

14.2.1 Heroes or Villains?

Recent developments in the recognition of injustice in society, especially in relation to racism and the past actions and attitudes of the colonial powers, have led to the renaming of buildings and removal of statues associated with slavery and colonialism (Burch-Brown, 2017). I cannot help wondering if this societal concern might affect zoos in the near future.

Many of the world's zoos were founded using animals acquired (stolen?) by the European colonial powers from their colonies. Many of their descendants still live in zoos. Some zoos display statues of former animal dealers who filled the world's zoos with animals at a time when such activity was acceptable and routine. They were active at a time when the world's resources, including its wildlife, were perceived to be inexhaustible and the international animal trade was unregulated.

In their youth, some of our best-known conservationists, including Gerald Durrell and Sir David Attenborough, took animals from wild places and transported them to zoos. Should we now vilify them for past actions that were perfectly acceptable (and legal) at the time, or praise them for their overall contributions to conservation?

In the 1970s drive-through safari parks were created in England with the help of the famous circus owner Jimmy Chipperfield. Should we now boycott these parks because of their past links with circuses that exhibited big cats, bears and elephants, and kept them in what would now be considered completely unacceptable conditions?

14.2.2   Zoos and Naziism

Ludwig (Lutz) Heck was the Director of Berlin Zoo and a Nazi. Lutz Heck succeeded his father – also named Ludwig Heck – as Director of Berlin Zoo in 1932. He joined the SS in 1933 and the National Socialist German Workers' Party (the Nazi Party) in 1937. During the Second World War he was involved in the plundering of Warsaw Zoo, in Poland, and the removal of the most valuable animals to Germany. In 1938 Berlin Zoo removed Jewish members from its board, forced Jewish shareholders to sell their shares at a loss and later banned Jews from visiting the zoo (Haaretz, 2013). A bust of Lutz Heck was placed in Berlin Zoo after his death, alongside those of other famous directors. Berlin Zoo now acknowledges its role in the war and has created an impressive display to explain this to visitors.

14.2.3   Zoos, Racism and Cultural Appropriation

In the nineteenth and twentieth centuries public displays of people from different ethnic groups, often in naturalistic settings with their traditional houses, were common (Blanchard et al., 2008b). They were often displayed with their animals and sometimes in zoos (see Section 6.5). Such displays would be unacceptable now, but are part of the history of some zoos, especially in Europe.

Between 1852 and 1969 Belle Vue Zoo in Manchester, England, held spectacular fireworks displays, the theme of which changed annually. Most were re-enactments of historical events, but in 1925 the display was entitled 'Cannibals' and featured black Africans depicted as savages (Nicholls, 1992).

Even in more recent times some zoos have shown a lack of sensitivity towards the racist implications of some of their activities. In 2005, Augsburg Zoo in Germany held a four-day African Festival. It featured craft sellers, drummers, music groups, storytellers and traditional food from around the African continent. This event was not well received by some representatives of Germany's black community and some academics argued that setting the festival in a zoo was racist. They claimed that using this location implied that non-whites were not really part of German society (BBC, 2005).

It is important that zoos recognise the historical role that they have played in what would now be considered racist activities and take a much greater account of the diversity of human life in their activities. This must include those involved in *in-situ* conservation activities that impinge on the livelihoods of indigenous peoples.

Many modern zoo exhibits are immersive and contain elements that relate to the people and culture of the areas where threatened animals live. Zoos need to take great care to avoid accusations of cultural appropriation in their exhibits and to avoid making stereotypical references to particular cultures. I was once present at a keeper talk at a tiger enclosure where an inexperienced young keeper made reference to the decline in tiger populations as a result of the demand for tiger parts for use in the traditional medicines of Chinese people. She failed to notice the presence of a family of Chinese heritage in the audience at the time!

14.2.4 Zoos and International Relations

The Chinese government loans giant pandas (*Ailuropoda melanoleuca*) to a number of zoos around the world, including Edinburgh Zoo (Scotland), the Smithsonian's National Zoological Park (or National Zoo, Washington, DC) and Berlin Zoo (Germany). These arrangements are really rentals as zoos pay for temporarily holding the animals and any cubs born in zoos outside China become the property of the Chinese government.

Pandas are popular with the public and attract visitors to zoos. However, those concerned about the poor human rights record of the Chinese government might argue that other countries should not be collaborating with it in a project of dubious conservation value that has produced relatively few offspring. Collard (2013) has argued that China's agreement to loan two pandas to Canada in 2012 was 'emblematic of animals' simultaneous material-symbolic inclusion and exclusion in contemporary politics' as they drew a connection between China's 'gift' and the promise of Chinese access to Canadian resources.

14.2.5 Renaming Species?

Many species are named after historical figures who were linked to what is now considered immoral and even criminal behaviour. John James Audubon – the author of *The Birds of America*, printed between 1827 and 1838 – was a slaveholder and a grave robber, and hunted birds and sold their skins (Nobles, 2017). Should we now use alternative vernacular names for Audubon's oriole (*Icterus graduacauda*), Audubon's warbler (*Setophaga auduboni*) and Audubon's shearwater (*Puffinus lherminieri*)? Bendire's thrasher (*Toxostoma bendirei*) is named after Major Charles Bendire of the US Army, who fought many battles aimed at eradicating Native Americans from the western United States, and the British explorer Captain James Cook – commemorated in the name of Cook's petrel (*Pterodroma cookii*) – collected artefacts from the indigenous peoples who inhabit the islands of the Pacific Ocean, Australia and New Zealand during his voyages, and is seen by many as an enabler of colonisation. Should we now purge our zoos of any species whose names make reference to explorers and others whose behaviour we now perceive as immoral?

14.3 The Compassionate Zoo

Organisations that work to remove particular taxa from zoos (such as elephants, cetaceans and great apes) or close them completely are unlikely to moderate or reverse their views in the future. It seems much more likely that, as more scientific information about the adverse effects of captivity on animals becomes available, this movement will gather pace and zoos will be compelled to adjust their activities appropriately. This will probably involve zoos making better arguments for keeping threatened species, for example by keeping only those taxa where captive breeding can make a significant contribution to their conservation beyond what can be achieved *in-situ*, and phasing out common species and those that adapt poorly to captive environments.

In 2018, Dr Marc Bekoff – an advocate of compassionate conservation and animal protection – and *In Defense of Animals* declared 9 February as World Zoothanasia Day to raise awareness of the number of animals whose lives they say are 'needlessly terminated by zoos' (IDA, 2022). *BornFree* is working to phase out the keeping of elephants in zoos in the UK. This may happen by attrition and as a result of limited breeding of the elephant populations here. At 14 June 1995 there were 18 zoos holding 4.33 *Elephas* and 6.27 *Loxodonta*, with 0.2 *Elephas* in Dublin (Ireland) and 0.1 in Dyfed (at a Hindu Temple in Wales): a total of 70 elephants in British zoos. At the beginning of 2022 there were just 51 elephants in Great Britain (excluding those in Dublin, Ireland). If the Howletts translocation of its 14 elephants to Kenya goes ahead there could soon be fewer than 40.

As societal attitudes towards animal change it seems likely that more countries will pass legislation that improves animal welfare, recognises animal sentience and acknowledges that many species – not just the more intelligent social species – are worthy of ethical consideration.

14.4   The Technological Zoo

The use of virtual reality (VR) in a zoo context has been explored by studying the reactions of a small number of people to a VR experience ($n = 12$) and a short video ($n = 12$) at Edinburgh Zoo in Scotland, both of which were concerned with the biology and conservation of African hunting dogs (*Lycaon pictus*) (Lugosi and Lee, 2021). Younger participants (13–18 years old) emphasised that VR allowed close and personal access to the animals, while older participants (19 years old and above) highlighted the educational value of the VR experience enabling visitors to see the hunting dogs in their natural environment. The authors concluded that the use of VR is potentially valuable as an educational tool and that it could offer a 'welfare positive addition to the visitor experience'. A similar study of the use of VR to provide visitors with a virtual visit to a little penguin (*Eudyptula minor*) enclosure and a 'behind-the-scenes' experience of food preparation with narration by a zookeeper found that visitors valued this experience as an addition to seeing live animals (Carter et al., 2020).

Virtual reality experiences may not only add to the enjoyment of zoo visitors in the future by offering immersive experiences, but may also have the potential to enhance the welfare of the animals by reducing their exposure to visitors. However, some developments in virtual and augmented reality may signal the end of live animal shows and zoos in a world where children now use technology from an early age and their urban existence deprives them of any significant contact with nature. Advances in hologram technology have allowed the creation of animal experiences without the use of live animals. Holograms have been used to produce 'live' performances from deceased human artists such as Whitney Houston, Roy Orbison and Maria Callas, and are also being used as an alternative to presenting live animals in circuses (Katz, 2019).

Scientists and veterinarians will undoubtedly continue to develop assisted reproductive technologies in an attempt to increase the numbers of individuals of rare species kept in zoos as part of cooperative breeding programmes. The development of these technologies is an investment in the future of some species but will not be needed for all. Some believe that cloning is the future for zoos. Daniel Wright of the University of Central Lancashire, England, has suggested that cloning could support future tourism markets by supplying animals for food tourism (luxury species dining), sport hunting and safari-zoo tourism (Wright, 2018). Although this is an interesting idea, it is entirely fanciful to imagine that zoos will be filled with cloned animals any time soon, considering the current state of cloning technology and the scale on which it would need to take place.

14.5 The Financially Secure Zoo

Tourist attractions, including zoos and aquariums, tend to generate an income that is seasonal. In fine weather zoo visitor numbers increase whereas aquarium visitor numbers tend to fall. In bad weather aquarium attendance increases and zoo attendance falls. To some extent institutions can plan for these variations in visitor numbers. Disease outbreaks pose a more serious problem. Zoos may be forced to close during a foot-and-mouth disease outbreak to prevent the disease being introduced into zoo animal populations from agricultural animals and vice versa. The COVID-19 outbreak introduced zoos to a new risk: zoonoses entering the human population from wild animals.

The COVID-19 outbreak had a devastating effect on the finances of zoos and aquariums all around the world as they were initially forced to close to prevent large numbers of people congregating, and then allowed to reopen with very restricted visitor numbers. The consequent fall in income led to zoos restricting their core activities such as education – due to the absence of school visits – and reducing their investment in conservation (see Section 2.12.2).

14.6 The COVID-Secure Zoo

Once global death rates from COVID-19 had subsided, an outbreak of monkeypox (Mpox) began spreading around the world. The World Health Organization (WHO) reported 3,413 confirmed cases of Mpox from 50 countries/territories between 1 January and 22 June 2022 (WHO, 2022). On 14 December 2022, WHO reported a global total of 82,809 cases. While the human death rate from this disease is low, its sudden expansion out of Africa highlights the needs for continued vigilance and the importance of surveillance in the control of zoonotic diseases. It is impossible to predict which zoonosis is likely to reach pandemic proportions next, but the lesson of the COVID pandemic for zoos is that zoo staff and visitors should not be unnecessarily exposed to zoonotic risks by exposure to animals. Glass has been widely used to separate animals from visitors – especially primates and dangerous animals like big cats – but it seems likely that new enclosures will increasingly use glass as a visitor barrier to reduce the likelihood of disease transmission. During the COVID pandemic a number of cases of transmission of the virus between animals and zoo staff were recorded (see Section 13.8.2).

14.7   The Homogenised Zoo

Over two decades ago Beardsworth and Bryman (2001) argued that zoos were undergoing crucial changes in their legitimating narratives and that they were exhibiting a tendency towards Disneyisation, entailing the interlinked features of theming, dedifferentiation of consumption, merchandising and emotional labour. Many modern zoos now have large themed exhibits reminiscent of attractions at Disney Resorts, such as *Elephant Odyssey* at San Diego Zoo, *Congo Gorilla Forest* at the Bronx Zoo and *Masoala Rainforest* at Zurich Zoo.

In parallel with the process of Disneyisation, I would argue that zoos are also undergoing a process of McDonaldisation. Ritzer (2008) has defined McDonaldisation as

the process by which the principles of the fast-food restaurant are coming to dominate more and more sectors of American society as well as the rest of the world.

This process is well known in businesses and many other aspects of society. In relation to biodiversity conservation, the term 'McDonaldisation' has been used to describe the global homogenisation of island faunas and floras as a result of invasions by alien species (Holmes, 1998), and, more recently, farm nature conservation projects (Morris and Reed, 2007). McDonaldisation has already been identified in theme parks (Bryman, 1999) and it now appears to be affecting zoos.

Ritzer identifies four dimensions that define McDonaldisation: efficiency, calculability, predictability and control. These may be simply defined as follows:

1. Efficiency refers to the optimum method of completing a task, and results in the loss of individuality.
2. Calculability is the assessment of outcomes based on quantifiable rather than subjective criteria.
3. Predictability is the guarantee of uniformity of product and standardised outcomes and services.
4. Control is achieved by the deskilling of the workforce and the introduction of automation.

To substantiate a claim of McDonaldisation in zoos the presence of all of these dimensions within the zoo community must be demonstrated. A consideration of each of Ritzer's four dimensions should help us to establish whether or not zoos may be considered to be McDonaldised institutions.

14.7.1 Efficiency

Some zoos have devised means of rapidly and efficiently moving visitors through popular exhibits. When Edinburgh Zoo obtained its first giant pandas, visitors had to book a time slot to visit the exhibit in advance. Aquariums with underwater tunnels are frequently equipped with moving floorways to control the flow of visitors. Walk-through enclosures such as aviaries and those containing lemurs or bats often have a linear structure with a single entrance and a separate exit. Drive-through safari parks usually have a single route that visitors must follow so that enclosures are viewed in a fixed sequence. On busy days pressure from following vehicles forces drivers to keep moving.

14.7.2 Calculability

Many zoos have built new exhibits that are defined by their size: the largest marine tank, the longest underwater tunnel, the largest elephant house or free-flight bat exhibit:

Howletts is home to the only herd of African elephants in Kent and the herd is the largest in the UK (Howletts, 2022)

World's largest indoor rainforest. (Henry Doorly Zoo, 2008)

Sometimes animal groups themselves are defined in terms of their size rather than, for example, their breeding success. Size alone may be an important reason why a species is kept in a zoo; visitors to zoos generally find large species more attractive than small species (Ward et al., 1998). Zoos publish attendance figures and data on the number of children participating in educational visits. Although this quantifies some aspects of the economic activity of a zoo, it does not measure its performance in terms of education or conservation.

14.7.3 Predictability

In 1959, London Zoo kept 290 different mammal species (Jarvis and Morris, 1960). By 2006 this had fallen to just 62 species (Fisken, 2007). Over the same period mammal species fell from 133 to 65 at Paignton Zoo, England, and from 120 to 66 at Chester Zoo, England. In addition to this reduction in species numbers, the composition of animal collections is becoming homogenised as zoos focus their efforts on keeping those species that can be maintained by captive breeding programmes.

The design of enclosures is becoming predictable as zoos decide the most appropriate ways of housing animals: great apes are kept behind glass; lemurs often feature in walk-through exhibits; enclosures for red pandas (*Ailurus fulgens*) contain bamboo and several small 'huts' located off the ground and linked by suspended wooden walkways (Fig. 14.2). The behaviour of visitors affects the behaviour of zoo animals

Fig. 14.2 Zoo exhibits are becoming homogenised. Red panda (*Ailurus fulgens*) exhibits are tending to look almost identical in zoos. (a) Whipsnade Zoo, England; (b) Barcelona Zoo, Spain; (c) Yorkshire Wildlife Park, England.

(Davey, 2007b) and as zoos increasingly take this into consideration in their exhibits further homogenisation will occur, for example, by restricting the viewing of animals using screens or small windows.

14.7.4   Control

There is some evidence of control in zoo environments. Some modern zoos have introduced electronic aids into their interpretation, pandering to the ubiquitous desire of young people to interact with computers. In some zoos personalised 'keeper talks' given by experienced keepers have been replaced by scripted – controlled – lectures by zoo educators using sound systems hard-wired into the exhibits.

Some zoo management activities are increasingly becoming computerised. Zoos use computer databases to manage their breeding programmes. This is essential to maintain genetic diversity in zoo populations. Studbook keepers decide which animals will be used for breeding and therefore exert a considerable control over which zoos keep which species, and the movements of individuals between institutions.

Some elements of McDonaldisation are inevitable in modern zoos. They may act to standardise the experience of zoo visitors, but they are an inevitable consequence of zoos attempting to find a legitimate function in the modern world where biodiversity is under threat and individual animals have the legal right to high welfare standards. If zoos are to maintain a captive breeding function for at least some species, some elements of McDonaldisation are not only inevitable, they are essential.

This process is being driven by a body of scientific evidence that animal welfare and reproduction can be improved by changes to husbandry and improvements in enclosure design. Complementary work on the behaviour of humans in zoos is also beginning to influence zoo design.

In addition to these scientific drivers, there are also legislative drivers. There is a global trend towards requiring higher standards of welfare for zoo animals, for example the Zoos Directive in the European Union. Other laws limit the availability of species to zoos, especially the Convention on International Trade in Endangered Species of Wild Fauna and Flora (CITES) 1973.

These laws, together with the conservation priorities determined by the International Union for Conservation of Nature (IUCN) and zoo associations such as the Association of Zoos and Aquariums (AZA) and the European Association of Zoos and Aquaria (EAZA), have a significant effect on the range of species exhibited by zoos.

14.8   Conclusion

Almost 40 years ago the zoologist Dr Jeremy Cherfas published a book entitled *Zoo 2000: A Look Beyond the Bars* to accompany a series of documentaries about zoos broadcast by the BBC (Cherfas, 1984). In his concluding remarks he wrote:

What zoos offer, which nothing else can, is the simple pleasure of contact. For that alone, they are worth it.

The future of zoos may be affected by issues relating to their legacy, animal welfare, the long-term viability of captive populations and their financial viability. They are becoming homogenised in a world that increasingly values diversity. Many keep animals that probably should not be in zoos because of their complex welfare requirements. If they can overcome these challenges the very best of the world's zoos have a future and an important contribution to make towards the conservation of biodiversity. For small children they will continue to provide an exciting day out filled with the sounds, smells and sometimes the touch of animals from the four corners of the Earth. A small proportion of those children will be inspired to follow a career studying or conserving wildlife, teaching biology or working to protect the environment.

Bibliography

Baratay, E., and Hardouin-Fugier, E. (2002). *Zoo: A History of Zoological Gardens in the West.* Reaktion Books Ltd, London.

Bell, C. E. (ed.) (2001). *Encyclopedia of the World's Zoos* (3 volumes). Fitzroy Dearborn Publishers, London.

Bostock, S. St C. (1993). *Zoos and Animal Rights: The Ethics of Keeping Animals.* Routledge, New York.

Braverman, I. (2012). *Zooland: The Institution of Captivity.* Stanford University Press, Stanford.

Brunner, B. (2005). *The Ocean at Home: An Illustrated History of the Aquarium.* Princeton Architectural Press, Hudson.

Crandall, L. S. (1964). *Management of Wild Mammals in Captivity.* University of Chicago Press, Chicago.

Donahue, J. C. (ed.) (2017). *Increasing Legal Rights for Zoo Animals: Justice on the Ark.* Lexington Books, Lanham.

Donahue, J. C. and Trump, E. K. (2006). *The Politics of Zoos: Exotic Animals and Their Protectors.* Northern Illinois University Press, DeKalb.

Donahue, J. C. and Trump, E. K. (2010). *American Zoos During the Depression: A New Deal for Animals.* McFarland & Co., Jefferson.

Emily, P. P. and Eisner, E. R. (2021). *Zoo and Wild Animal Dentistry.* Wiley-Blackwell, Chichester.

Fa, J. E., Funk, S. M. and O'Connell, D. (2011). *Zoo Conservation Biology.* Cambridge University Press, Cambridge.

Fraser, J. and Switzer, T. (2021). *The Social Value of Zoos.* Cambridge University Press, Cambridge.

Frost, W. (ed.) (2010). *Zoos and Tourism: Conservation, Education, Entertainment.* Channel View Publications, Bristol.

Gray, J. (2017). *Zoo Ethics: The Challenges of Compassionate Conservation.* Cornell University Press, Ithaca.

Grigson, C. (2016). *Menagerie: The History of Exotic Animals in England, 1100–1837.* Oxford University Press, Oxford.

Gullard, F. M. D., Dierauf, L. A. and Whitman, K. L. (2018). *Marine Mammal Medicine* (3rd ed.). CRC Press, Abingdon.

Hahn, A. (2019). *Zoo and Wild Mammal Formulary.* Wiley-Blackwell, Hoboken.

Hancocks, D. (2001). *A Different Nature: The Paradoxical World of Zoos and Their Uncertain Future.* University of California Press, Berkeley.

Harrison, E. (2002). *Animal Attractions: Nature on Display in American Zoos.* Princeton University Press, Princeton.

Hediger, H. (1969). *Man and Animal in the Zoo: Zoo Biology*. Delacorte, New York.

Hediger, H. (1969). *The Psychology and Behaviour of Animals in Zoos and Circuses*. Dover Publications, Mineola.

Hoage, R. J. and Deiss, W. A. (eds.) (1996). *New Worlds, New Animals: From Menagerie to Zoological Park in the Nineteenth Century*. Johns Hopkins University Press, Baltimore.

Hosey, G., Melfi, V. and Pankhurst, S. (2013). *Zoo Animals: Behaviour, Management, and Welfare* (2nd ed.). Oxford University Press, Oxford.

Irwin, M. D., Stoner, J. B. and Cobaugh, A. M. (2013). *Zookeeping: An Introduction to the Science and Technology*. University of Chicago Press, Chicago.

Johnson, W. (1994). *The Rose-Tinted Menagerie*. Heretic Books Ltd., Farnham.

Kaufman, A. B., Bashaw, M. J. and Maple, T. L. (eds.) (2019). *Scientific Foundations of Zoos and Aquariums: Their Role in Conservation and Research*. Cambridge University Press, Cambridge.

Kisling. V. N. (ed.) (2000). *Zoo and Aquarium History: Ancient Animal Collections to Zoological Gardens*. CRC Press, Boca Raton.

Kleiman, D. G., Thompson, K. V. and Baer, C. K. (2012). *Wild Mammals in Captivity: Principles & Techniques for Zoo Management* (2nd ed.). University of Chicago Press, Chicago.

Lewbart, G. A. (ed.) (2022). *Invertebrate Medicine*. Wiley-Blackwell, Chichester.

Malamud, R. (1998). *Reading Zoos: Representations of Animals and Captivity*. New York University Press, New York.

Maple, T. L. and Perdue, B. M. (2013). *Zoo Animal Welfare*. Springer, Berlin.

Markowitz, H. (1982). *Behavioral Enrichment in the Zoo*. Van Nostrand Reinhold Company, New York.

Markowitz, H. (2011). *Enriching Animal Lives*. Mauka Press, USA.

Melfi, V. A., Dorey, N. and Ward, S. J. (eds.) (2020). *Zoo Animal Learning and Training*. Wiley & Sons, Chichester.

Meuser, N. (2018). *Zoo Buildings: Construction and Design Manual*. DOM Publishers, Berlin.

Miller, R. E., Lamberski, N. and Calle, P. (eds.) (2018). *Fowler's Zoo and Wild Animal Medicine: Current Therapy*. Volume 9. Saunders, San Diego.

Minteer, B. A., Maienschein, J. and Collins, J. P. (2018). *The Ark and Beyond: The Evolution of Zoo and Aquarium Conservation*. University of Chicago Press, Chicago.

Mullan, B. and Marwin, G. (1997). *Zoo Culture* (2nd ed.). University of Illinois Press, Urbana.

Murphy, J. B. (2007). *Herpetological History of the Zoo and Aquarium World*. Krieger Publishing Company, Malabar.

Norton, B. G., Hutchins, M., Stevens, E. F. and Maple, T. L. (eds.) (1995). *Ethics on the Ark: Zoos, Animal Welfare, and Wildlife Conservation*. Smithsonian Institution Press, Washington.

Olney, P. J. S., Mace, G. M. and Feistner, A. T. C. (eds.) (1993). *Creative Conservation: Interactive Management of Wild and Captive Animals*. Chapman & Hall, New York.

Patrick, P. G. and Tunnicliffe, S. D. (2013). *Zoo Talk*. Springer, Berlin.

Rees, P. A. (2011). *An Introduction to Zoo Biology and Management*. Wiley-Blackwell, Chichester.

Rees, P. A. (2013). *Dictionary of Zoo Biology and Animal Management*. Wiley-Blackwell, Chichester.

Rees, P. A. (2015). *Studying Captive Animals: A Workbook of Methods in Behaviour, Welfare and Ecology*. Wiley-Blackwell, Chichester.

Rees, P. A. (2021). *Key Questions in Zoo and Aquarium Science: A Study and Revision Guide.* CABI, Wallingford.

Scott, D. (2020). *Raptor Medicine, Surgery and Rehabilitation.* CABI, Wallingford.

Shepherdson, D. J., Mellen, J. D. and Hutchins, M. (1998). *Second Nature: Environmental Enrichment for Captive Animals.* Smithsonian Institution Press, Washington.

Stott, R. (2003). *Theatres of Glass: The Woman Who Brought the Sea to the City.* Short Books, London.

Tait, P. (2016). *Fighting Nature: Travelling Menageries, Animal Acts and War Shows.* Sydney University Press, Sydney.

Terio, K. A., McAloose, D. and St. Leger, J. (eds.) (2018). *Pathology of Wildlife and Zoo Animals.* Academic Press, San Diego.

Tudge, C. (1991). *Last Animals at the Zoo.* Hutchinson Radius, London.

Tyson, E. (2021). *Licensing Laws and Animal Welfare: The Legal Protection of Wild Animals.* Palgrave Macmillan, London.

West, G., Heard, D. and Caulkett, N. (eds.) (2014). *Zoo Animal and Wildlife Immobilization and Anesthesia* (2nd ed.). Wiley-Blackwell, Chichester.

Yew, W. (1991). *Noah's Art: Zoo, Aquarium, Aviary and Wildlife Park Graphics.* Quon Editions, Singapore.

York, T. (2015). *The End of Captivity? A Primate's Reflections on Zoos, Conservation and Christian Ethics.* Cascade Books, Eugene.

Young, R. J. (2003). *Environmental Enrichment for Captive Animals.* Wiley-Blackwell, Chichester.

Zimmerman, A., Hatchwell, M., Dickie, L. and West, C. (eds.) (2007). *Zoos in the 21st Century: Catalysts for Conservation?* Cambridge University Press, Cambridge.

Zuckerman, S. (1980). *Great Zoos of the World: Their Origins and Significance.* Routledge, London.

References

Abbott, M. and Tan-Kantor, A. (2022). Accounting for zoo animals: it is a jungle out there. *Australian Accounting Review*. https://doi.org/10.1111/auar.12362.

Acampora, R. R. (1998). Extinction by exhibition: looking at and in the zoo. *Human Ecology Review*, 5, 1–4.

Acevedo, M. A. and Villanueva-Rivera, L. J. (2006). From the field: using automated digital recording systems as effective tools for the monitoring of birds and amphibians. *Wildlife Society Bulletin*, 34, 211–214.

Ackers, J. S. and Schildkraut, D. S. (1985). Regurgitation/reingestion and coprophagy in captive gorillas. *Zoo Biology*, 4, 99–109.

Adelman, L. M., Falk, J. F. and James, S. (2010). Impact of national aquarium in Baltimore on visitors' conservation attitudes, behavior, and knowledge. *Curator: The Museum Journal*, 43, 33–41.

Adetola, B. O. and Adedire, O. P. (2018). Visitors' motivation and willingness to pay for conservation in selected zoos in Southwest Nigeria. *Journal of Applied Sciences and Environmental Management*, 22, 531–537.

Agnew, D. W., Barbiers, R. B., Poppenga, R. H. and Watson, G. L. (1999). Zinc toxicosis in a captive striped hyena (*Hyena hyena*). *Journal of Zoo and Wildlife Medicine*, 30, 431–434.

Agnew, M. K., Asa, C. S., Franklin, A. D., McDonald, M. M. and Cowl, V. B. (2021). Deslorelin (Suprelorin®) use in North American and European zoos and aquariums: taxonomic scope, dosing and efficacy. *Journal of Zoo and Wildlife Medicine*, 52, 427–436.

Agrillo, C., Gori, S. and Beran, M. J. (2015). Do rhesus monkeys (*Macaca mulatta*) perceive illusory motion? *Animal Cognition,* 18, 895–910.

Akiyama, J., Sakagami, T., Uchiyama, H. and Ohta, M. (2021). The health benefits of visiting a zoo, park and aquarium for older Japanese. *Anthrozoös*, 34, 463–473.

Alberts, A. C. (1994). Dominance hierarchies in male lizards: implications for zoo management programs. *Zoo Biology*, 13, 479–490.

Allard, S., Fuller, G., Torgerson-White, L., Starking, M. and Yoder-Nowak, T. (2019). Personality in zoo-hatched Blanding's turtles affects behavior and survival after reintroduction into the wild. *Frontiers in Psychology*, 10, 2324. DOI: 10.3389/fpsyg.2019.02324.

Altmann, J. (1974). Observational studies of behavior: sampling methods. *Behaviour*, 49, 227–266.

Altman, J. D. (1998). Animal activity and visitor learning at the zoo. *Anthrozoös*, 11, 12–21.

Altman, J. D., Gross, K. L. and Lowry, S. R. (2005). Nutritional and behavioral effects of gorge and fast feeding in captive lions. *Journal of Applied Animal Welfare Science*, 8, 47–57.

Altschul, D. M., Wallace, E. K., Sonnweber, R., Tomonaga, M. and Weiss, A. (2017). Chimpanzee intellect: personality, performance and motivation with touchscreen tasks. *Royal Society Open Science*, 4(5), 170169.

Alward, L. (2008). Why circuses are unsuited to elephants. In: Wemmer, C. and Christen, C. A. (eds.). *Elephants and Ethics: Towards a Morality of Coexistence.* Johns Hopkins University Press, Baltimore, pp. 205–224.

Amrein, M., Heistermann, M. and Weingrill, T. (2014). The effect of fission–fusion zoo housing on hormonal and behavioural indicators of stress in Bornean orangutans (*Pongo pygmaeus*). *International Journal of Primatology*, 35, 509–528.

Andersen, K. F. (1992). Size, design and interspecific interactions as restrictors of natural behaviour in multi-species exhibits: 3. Interspecific interactions of plains zebra (*Equus burchelli*) and eland (*Taurotragus oryx*). *Applied Animal Behaviour Science*, 34, 273–284.

Andersen, L. L. (2003). Zoo education: from formal school programmes to exhibit design and interpretation. *International Zoo Yearbook*, 38, 75–81

Andersen, M. J., Arnold, M., Barclay, M., et al. (2010). *Oregon Silverspot Butterfly Husbandry Manual.* Oregon Zoo, Portland.

Anderson, D., Lawson, B. and Mayer-Smith, J. (2006). Investigating the impact of a practicum experience in an aquarium on pre-service teachers. *Teaching Education*, 17, 341–353.

Anderson, K. (1995). Culture and nature at the Adelaide Zoo: at the frontiers of 'human' geography. *Transactions of the Institute of British Geographers*, 20, 275–294.

Anderson, P. A., Berzins, I. K., Fogarty, F., Hamlin, H. J. and Guillette, Jr., L. J. (2011). Sound, stress and seahorses: the consequences of a noisy environment to animal health. *Aquaculture*, 311, 129–138.

Anderson, R. C. (2005). How smart are octopuses? *Coral*, 2, 44–48.

Anderson, R. C. and Wood, J. B. (2001). Enrichment for giant Pacific octopuses: happy as a clam? *Journal of Applied Animal Welfare Science*, 4, 157–168.

Anderson, U. S., Benne, M., Bloomsmith, M. A. and Maple, T. L. (2002). Retreat space and human visitor density moderate undesirable behavior in petting zoo animals. *Journal of Applied Animal Welfare Science*, 5, 125–137.

Anderson, U. S., Kelling, A. S., Pressley-Keough, R., Bloomsmith, M. A. and Maple, T. L. (2003). Enhancing the zoo visitor's experience by public animal training and oral interpretation at an otter exhibit. *Environment and Behavior*, 35, 826–841.

Anderson, U. S., Maple, T. L. and Bloomsmith, M. A. (2004). A close keeper–nonhuman animal distance does not reduce undesirable behavior in contact yard goats and sheep. *Journal of Applied Animal Welfare Science*, 7, 59–69.

Anderson, U. S., Kelling, A. S. and Maple, T. L. (2008). Twenty-five years of *Zoo Biology*: a publication analysis. *Zoo Biology*, 27, 444–457.

Anderson, U. S., Maple, T. L. and Bloomsmith, M. A. (2010). Factors facilitating research: a survey of zoo and aquarium professionals. *Zoo Biology*, 29, 663–675.

Ang, M. Y., Shender, M. A. and Ross, S. R. (2017). Assessment of behavior and space use before and after forelimb amputation in a zoo-housed chimpanzee (*Pan troglodytes*). *Zoo Biology*, 36, 5–10.

Auckland Zoo (2014). Zoo achieves record visitation for third year. Media release, 10 July. www.aucklandzoo.co.nz/sites/news/media-releases/Zoo-achieves-record-visitation-for-third-year (accessed 13 August 2017).

Audigé, L., Wilson, P. R. and Morris, R. S. (1998). A body condition score system and its use for farmed red deer hinds. *New Zealand Journal of Agricultural Research*, 41, 545–553.

Axelsson, T. and May, S. (2008). Constructed landscapes in zoos and heritage. *International Journal of Heritage Studies*, 14, 43–59.

Aylen, J., Albertson, K. and Cavan, G. (2014). The impact of weather and climate on tourist demand: the case of Chester Zoo. *Climate Change*, 127, 183–197.

AZA (2009). Behaviour Advisory Group. www.aza.org/behavior-advisory-group (accessed 15 June 2009).

AZA (2020). Visitor demographics. www.aza.org/partnerships-visitor-demographics?locale=en (accessed 14 June 2022).

AZA (2021). Association of Zoos and Aquariums. www.aza.org/species-survival-plan-pro grams?locale=en (accessed 19 January 2021).

AZA (2022). Zoo and aquarium statistics. www.aza.org/zoo-and-aquarium-statistics?locale=en (accessed 22 March 2022).

Azevedo, C. S. de and Young, R. J. (2021). Animal personality and conservation: basics for inspiring new research. *Animals*, 11, 1019. https://doi.org/10.3390/ani11041019.

Azevedo, C. S. de, Young, R. J. and Rodrigues, M. (2011). Role of Brazilian zoos in ex-situ bird conservation: from 1981 to 2005. *Zoo Biology*, 30, 655–671.

Azevedo, C. S. de, Young, R. J. and Rodriges, M. (2012). Failure of captive-born greater rheas (*Rhea americana*, Rheidae, Aves) to discriminate between predator and nonpredator models. *Acta Ethologica*, 15, 179–185.

Azevedo, C. S. de, Caldeira, J. R., Faggioloi, A. B. and Cipreste, C. F. (2016). Effects of different enrichment items on the behavior of the endangered Lear's Macaw (*Anodorhyncus leari*, Psittacidae) at Belo Horizonte Zoo, Brazil. *Revista Brasiliera de Ornitologia*, 24, 204–210.

Bååth, R., Seno, T. and Kitaoka, A. (2014). Cats and illusory motion. *Psychology,* 5, 1131–1134.

Bacon, H., Vigors, B., Shaw, D. J., et al. (2021). Zookeepers: the most important animal in the zoo? *Journal of Applied Animal Welfare Science*. https://doi.org/10.1080/10888705.2021 .2012784.

Bader, H. (1983). Electroejaculation in chimpanzees and gorillas and artificial insemination in chimpanzees. *Zoo Biology*, 2, 307–314.

Baechler, B., Granek, E. F., Carlin-Morgan, K. A., Smith, T. E. and Nielsen-Pincus, M. (2021). Aquarium visitor engagement with an ocean plastics exhibit: effects on self-reported intended single-use plastic reductions and plastic-related environmental stewardship actions. *Journal of Interpretation Research*. https://doi.org/10.1177/10925872211021183.

Baechli, J., Bellis, L. M., García Capocasa, M. C. and Busso, J. M. (2021). Activity budget of zoo-housed *Dolichotis patagonum* mates. *Journal of Zoo and Aquarium Research*, 9, 14–19.

Bagaria, A. and Sharma, A. K. (2014). A knowledge and practices study of health hazards among animal handlers in zoological gardens. *International Journal of Occupational Health*, 4(1). https://doi.org/10.3126/ijosh.v4i1.9146.

Bairrão Ruivo, E. and Wormell, D. (2012). The international conservation programme for the white-footed tamarin *Saguinus leucopus* in Colombia. *International Zoo Yearbook*, 46, 46–55.

Bajomi, B., Pullin, A. S., Stewart, G. B. and Takács-Sánta, A. (2010). Bias and dispersal in the animal reintroduction literature. *Oryx*, 44, 358–365.

Baker, A. (1994). Variation in the parental care systems of mammals and the impact on zoo breeding programs. *Zoo Biology*, 13, 413–421.

Baker, K. C., Georoff, T. A., Ialeggio, D. M., et al. (2022). Retrospective review of urolithiasis-related morbidity and mortality in Asian colobine monkeys. *Journal of Zoo and Wildlife Medicine*, 53, 1–10.

Ball, V. (1886). Observations on lion-breeding in the gardens of the Royal Zoological Society of Ireland. *The Transactions of the Royal Irish Academy*, 28, 723–758.

Ballou. J. D. (1993). Assessing the risks of infectious diseases in captive breeding and reintroduction programs. *Journal of Zoo and Wildlife Medicine*, 24, 327–335.

Ballouard, J.-M., Brischoux, F. and Bonnet, X. (2011). Children prioritize virtual exotic biodiversity over local biodiversity. *PLoS ONE*, 6(8), e23152. https://doi.org/10.1371/journal.pone.0023152

Balmford, A., Leader-Williams, N., Mace, G. M., et al. (2007). Message received? Quantifying the impact of informal conservation education on adults visiting zoos. In: Zimmerman, A., Hatchwell, M., Dickie, L. and West, C. (eds.), *Zoos in the 21st Century: Catalysts for Conservation*. Cambridge University Press, Cambridge, pp. 120–136.

Bandoli, F. and Cavicchio, P. (2021). The COVID-19 pandemic and the fragile balance of a small zoo: the case of Pistoia Zoo in Italy. *Journal of Applied Animal Ethics Research*, 3, 57–73.

Banks, P. B., Norrdahl, K. and Korpimäki, E. (2002). Mobility decisions and the predation risks of reintroduction. *Biological Conservation*, 103, 133–138.

Baratay, E. and Hardouin-Fugier, E. (2002). *Zoo: A History of Zoological Gardens in the West*. Reaktion Books Ltd, London.

Barbiers, R. B. (1985). Orangutans' color preference for food items. *Zoo Biology*, 4, 287–290.

Barnett, R., Yamaguchi, N., Barnes, I. and Cooper, A. (2006). Lost populations and preserving genetic diversity in the lion *Panthera leo*: implications for its ex situ conservation. *Conservation Genetics*, 7, 507–514.

Barongi, R., Fisken, F. A., Parker, M. and Gusset, M. (eds.) (2015). *Committing to Conservation: The World Zoo and Aquarium Conservation Strategy*. WAZA Executive Office, Gland.

Barr, D. (2005). Zoo and aquarium libraries: an overview and update. *Science & Technology Libraries*, 25, 71–87.

Barreiros, J. P. and Haddad, Jr, V. (2016). Zoo animals and humans killed because of human negligent behavior. *Journal of Coastal Life Medicine*, 4, 1008.

Barrett, L. P. and Benson-Amram, S. (2020). Can Asian elephants use water as a tool in the floating object task. *Animal Behavior and Cognition*, 7, 310–326.

Bartlett, A. D. (1899). *Wild Animals in Captivity: Being an Account of the Habits, Food, Management and Treatment of the Beasts and Birds at the 'Zoo' with Reminisces and Anecdotes* (compiled and edited by Edward Bartlett). Chapman and Hall, London.

Bashaw, M. J. and Maple, T. L. (2001). Signs fail to increase zoo visitors' ability to see tigers. *Curator: The Museum Journal*, 44, 297–304.

Bashaw, M. J., Tarou, L. R., Maki, T. S. and Maple, T. L. (2001). A survey assessment of variables related to stereotypy in captive giraffe and okapi. *Applied Animal Behaviour Science*, 73, 235–247.

Bashaw, M. J., Bloomsmith, M. A., Marr, M. J. and Maple, T. L. (2003). To hunt or not to hunt? A feeding enrichment experiment with captive large felids. *Zoo Biology*, 22, 189–198.

Bateson, M. and Martin, P. (2021). *Measuring Behaviour. An Introductory Guide* (4th ed.). Cambridge University Press, Cambridge.

Bayne, K., Dexter, S., Mainzer, H., et al. (1992). The use of artificial turf as a foraging substrate for individually housed rhesus monkeys (*Macaca mulatta*). *Animal Welfare*, 1, 39–54.

BBC (1955). *Zoo Quest to West Africa*. BBC.

BBC (2000). Zoo kills endangered antelopes. BBC News, 17 December. http://news.bbc.co.uk/2/hi/uk_news/scotland/1075140.stm.

BBC (2005). Row over German zoo's Africa show. BBC News, 8 June. http://news.bbc.co.uk/go/pr/fr/-/1/hi/world/africa/4070816stm.

BBC (2021). Somerset Noah's Ark Zoo elephant M'Changa dies in attack. BBC News, 23 June. www.bbc.co.uk/news/uk-england-somerset-57578702.amp.

Beardsworth, A. and Bryman, A. (2001). The wild animal in late modernity: the case of the Disneyization of zoos. *Tourist Studies*, 1, 83–104.

Bechert, U. S., Brown, J. L., Dierenfeld, E. S., et al. (2019). Zoo elephant research: contributions to conservation of captive and free-ranging species. *International Zoo Yearbook*, 53, 89–115.

Beck, B. B., Kleiman, D. G., Dietz, J. M., et al. (1991). Losses and reproduction in reintroduced golden lion tamarins, *Leontopithecus rosalia*. *Dodo, Journal of the Jersey Wildlife Preservation Trust*, 27, 50–61.

Beck, B. B., Rapaport, L. G., Stanley-Price, M. and Wilson, A. C. (1994). Reintroduction of captive-born animals. In: Olney, P. J. S., Mace, B. M. and Feistner, A. T. C. (eds.), *Creative Conservation: Interactive Management of Wild and Captive Animals*. Springer, Dordrecht, pp. 265–286.

Behr, B., Rath, D., Mueller, P., et al. (2009). Feasibility of sex-sorting sperm from the white and the black rhinoceros (*Ceratotherium simum, Diceros bicornis*). *Theriogenology*, 72, 353–364.

Behringer, V., Stevens, J. M. G., Deschner, T. and Hohmann, G. (2018). Getting closer: contributions of zoo studies to research on the physiology and development of bonobos (*Pan paniscus*), chimpanzees (*P. troglodytes*) and other primates. *International Zoo Yearbook*, 52, 34–47.

Bell, E., Price, E., Balthes, S., Cordon, M. and Wormell, D. (2019). Flight patterns in zoo-housed fruit bats (*Pteropus* spp.). *Zoo Biology*, 38, 248–257.

Benedict, F. G. (1936). *The Physiology of the Elephant*. Carnegie Institution, Washington.

Bengston, S. E., Pruitt, J. N. and Riechert, S. E. (2014). Differences in environmental enrichment generate contrasting behavioural syndromes in a basal spider lineage. *Animal Behaviour*, 93, 105–110.

Benirschke, K. and Roocroft, A. (1992). Elephant inflicted injuries. *Internationalen Symposiums uber die Erkrankungender Zootiere*, 34, 239–247.

Benz, J. J., Leger, D. W. and French, J. A. (1992). Relation between food preference and food-elicited vocalizations in golden lion tamarins (*Leontopithecus rosalia*). *Journal of Comparative Psychology*, 106, 142–149.

Bercovitch, F. B., Bashaw, M. J., del Castillo, S. M. (2006). Sociosexual behaviour, male mating tactics, and the reproductive cycle of giraffe *Giraffa camelopardalis*. *Hormones and Behavior*, 50, 314–321.

Berg, W., Jolly, A., Rambeloarivony, H., Andrianome, V. and Rasamimanana, H. (2009). A scoring system for coat and tail condition in ringtailed lemurs, *Lemur catta*. *American Journal of Primatology: Official Journal of the American Society of Primatologists*, 71, 183–190.

Berger, M., Schawalder, P., Stich, H. and Lussi, A. (1996). Feline dental resorptive lesions in captive and wild leopards and lions. *Journal of Veterinary Dentistry*, 13, 13–21.

Berger-Tal, O., Blumstein, D. T. and Swaisgood, R. R. (2020). Conservation translocations: a review of common difficulties and promising direction. *Animal Conservation*, 23, 121–131.

Bernstein-Kurtycz, L. M., Hopper, L. M., Ross, S. R. and Tennie, C. (2020). Zoo-housed chimpanzees can spontaneously use tool sets but perseverate on previously successful tool-use methods. *Animal Behavior and Cognition*, 7, 288–309.

Bettinger, T. and Quinn, H. (2000). Conservation funds: how do zoos and aquariums decide which projects to fund? *AZA American Zoo and Aquarium Association Annual Conference Proceedings*. American Zoo and Aquarium Association, St Louis, pp. 52–54.

Bexell, S., Jarrett, O. S., Lan, L., et al. (2007). Observing panda play: implications for zoo programming and conservation efforts. *Curator: The Museum Journal*, 50, 287–297.

Biasetti, P., Florio, D., Gili, C. and de Mori, B. (2020). The ethical assessment of touch pools in aquariums by means of the Ethical Matrix. *Journal of Agricultural & Environmental Ethics*, 33, 337–353.

BIAZA (2021). *In Our Hands*. British and Irish Association of Zoos and Aquariums, London.

Biggins, D. E., Godbey, J. L., Hanebury, L. R., et al. (1998). The effect of rearing methods on survival of reintroduced black-footed ferrets. *Journal of Wildlife Management*, 62, 643–653.

Bildstein, K. L., Golden, C. B., McCraith, B. J., Bohmke, B. W. and Seibels, R. E. (1993). Feeding behavior, aggression, and the conservation biology of flamingos: integrating studies of captive and free-ranging birds. *American Zoologist*, 33, 117–125.

Birch, J. (2017). Animal sentience and the precautionary principle. *Animal Sentience*, 2(16), 1.

Birke, L. (2002). Effects of browse, human visitors and noise on the behaviour of captive orang utans. *Animal Welfare*, 11, 189–202.

Birke, L., Hosey, G. and Melfi, V. (2019). 'You can't really hug a tiger': zoo keepers and their bonds with animals. *Anthrozoös*, 32, 597–612.

Birkett, L. P. and Newton-Fisher, N. E. (2011). How abnormal is the behaviour of captive, zoo-living chimpanzees? *PLoS ONE*, 6(6), e20101. https://doi.org/10.1371/journal.pone.0020101

Bistyák, A., Kecskeméti, S., Glávits, R., et al. (2007). Pacheco's disease in a Hungarian zoo bird population: a case report. *Acta Veterinaria Hungarica*, 55, 213–218.

Bitgood, S. (1987). Selected bibliography on exhibit design and evaluation in zoos. *Visitor Behaviour*, 2, 8.

Bitgood, S. (1988). Problems in visitor orientation and circulation. In: Bitgood, S., Roper, J. and Benefield, A. (eds.), *Visitor Studies: 1988 – Theory, Research, and Practice*. Center for Social Design, Jacksonville, pp. 155–170.

Bitgood, S. (2006). An analysis of visitor circulation movement patterns and the general value principle. *Curator: The Museum Journal*, 49, 463–475.

Bitgood, S. and Patterson, D. (1993). The effects of gallery changes on visitor reading and object viewing. *Environment and Behavior*, 25, 761–781.

Bitgood, S. and Richardson, K. (1986). Wayfinding at the Birmingham Zoo. *Visitor Behavior*, 1, 9.

Bitgood, S., Benefield, A., Patterson, D., Lewis, D. and Landers, A. (1985). Zoo visitors: can we make them behave? *Annual Proceedings of the 1985 American Association of Zoological Parks and Aquariums*. Columbus.

Bitgood, S., Patterson, D. and Benefield, A. (1988a). Exhibit design and visitor behavior: empirical relationships. *Environment and Behavior*, 20, 474–491.

Bitgood, S., Carnes, J., Nabors, A. and Patterson, D. (1988b). Controlling public feeding of zoo animals. *Visitor Behavior*, 2, 6.

Bitgood, S., Benefield, A. and Patterson, D. (1989). The importance of label placement: a neglected factor in exhibit design. *Current Trends in Audience Research*, 4, 49–52.

Bitgood, S., Benefield, A., Patterson, D. and Litwak, H. (1990). Influencing visitor attention: effects of life-size silhouettes on visitor behavior. In Bitgood, S., Benefield, A. and Patterson, D. (eds.), *Visitor Studies: Theory, Research, and Practice*, Vol. 3. Center for Social Design, Jacksonville, pp. 221–230.

Bjerke, T., Ødegårdstuen, T. S. and Kaltenborn, B. P. (1998). Attitudes toward animals among Norwegian children and adolescents: species preferences. *Anthrozoös*, 11, 227–235

Bjornvad, C. R., Nielsen, D. H., Armstrong, P. J., et al. (2011). Evaluation of a nine-point body condition scoring system in physically inactive pet cats. *American Journal of Veterinary Research*, 72, 433–437.

Black, S., Yamaguchi, N., Harland, A. and Groombridge, J. (2010). Maintaining the genetic health of putative Barbary lions in captivity: an analysis of Moroccan Royal Lions. *European Journal of Wildlife Research*, 56, 21–31.

Blanchard, P., Bancel, N., Boëtsch, G., et al. (eds.) (2008a). *Human Zoos: Science and Spectacle in the Age of Colonial Empires*. Liverpool University Press, Liverpool.

Blanchard, P., Bancel, N., Boëtsch, G., Deroo, E. and Lemaire, S. (2008b). Human zoos: the greatest exotic shows in the West. In: Blanchard, P., Bancel, N., Boëtsch, G., et al. (eds.), *Human Zoos: Science and Spectacle in the Age of Colonial Empires*. Liverpool University Press, Liverpool, pp. 1–49.

Blanchett, M. K. S., Finegan, E. and Atkinson, J. (2020). The effects of increasing visitor and noise levels on birds within a free-flight aviary examined through enclosure use and behavior. *Animal Behavior and Cognition*, 7, 49–69.

Blaney, E. C. and Wells, D. L. (2004). The influence of a camouflage net barrier on the behaviour, welfare and public perceptions of zoo-housed gorillas. *Animal Welfare*, 13, 111–118.

Blattner, C. E. (2019). The recognition of animal sentience by the law. *Journal of Animal Ethics*, 9, 121–136.

Bliss, T. N., Marinkovich, M. J., Burns, R. E., et al. (2022). Comparison of diagnostic predictors of neonatal survivability in nondomestic Caprinae. *Journal of Zoo and Wildlife Medicine*, 53, 31–40.

Bloomfield, R. C., Gillespie, G. R., Kerswell, K. J., Butler, K. L. and Hemsworth, P. H. (2015). Effect of partial covering of the visitor viewing area window on positioning and orientation of zoo orangutans: a preference test. *Zoo Biology*, 34, 223–229.

Bloomsmith, M. A. and Lambeth, S. P. (2000). Videotapes as enrichment for captive chimpanzees (*Pan troglodytes*). *Zoo Biology*, 19, 541–551.

Bloomsmith, M. A., Kuhar, C., Baker, K., et al. (2003). Primiparous chimpanzee mothers: behavior and success in a short-term assessment of infant rearing. *Applied Animal Behaviour Science*, 84, 235–250.

Bollen, K. S. and Novak, M. A. (2000). A survey of abnormal behavior in captive zoo primates. *American Journal of Primatology* 51(Suppl.1), 47 (abstract).

Bolliger, G. (2015). Legal protection of animal dignity in Switzerland: status quo and future perspectives. *Animal Law*, 22, 311–395.

Bonal, B. S., Sharma, S. C., Patnaik, S. K. and Gupta, B. K. (2012). *Guidelines on Minimum Dimensions of Enclosures for Housing Exotic Animals of Different Species*. Central Zoo Authority, Delhi.

Boorer, M. K. (1966). Educational facilities for school parties at London Zoo and Whipsnade Park. *International Zoo News*, 6, 239–242.

Boorer, M. K. (1967). Zoos as teaching aids. *Journal of Biological Education*, 1, 233–238.

Borun, M. and Miller, M. S. (1980). To label or not to label. *Museum News*, 58(4), 64–67.

Bostock, S. St C. (1993). *Zoos and Animal Rights: The Ethics of Keeping Animals*. Routledge, London.

Boufana, B., Stidworthy, M. F., Bell, S., et al. (2012). *Echinococcus* and *Taenia* spp. from captive mammals in the United Kingdom. *Veterinary Parasitology*, 190, 95–103.

Bowler, M. T., Buchanan-Smith, H. M. and Whiten, A. (2012). Assessing public engagement with science in a university primate research centre in a national zoo. *PLoS ONE*, 7(4), e34505.

Bowler, M., Messer, E. J., Claidière, N. and Whiten, A. (2015). Mutual medication in capuchin monkeys :social anointing improves coverage of topically applied anti-parasite medicines. *Scientific Reports*, 5, 1–10.

Bowler, P. J. (1992). *The Fontana History of the Environmental Sciences*. Fontana Press, London.

Boyd, B. S., Colon, F., Doty, J. F. and Sanders, K. C. (2021). Beware of the dragon: a case report of a komodo dragon attack. *Foot & Ankle Orthopaedics*, 6(2). https://doi.org/10.1177/24730114211015623.

Boyd, L. E. (1988). Time budgets of adult Przewalski horses: effects of sex, reproductive status and enclosure. *Applied Animal Behaviour Science*, 21, 19–39.

Brady, M., Rehling, M., Mueller, J. and Lukas, K. (2010). Giant Pacific octopus behavior and enrichment. *International Zoo News*, 57, 134–145.

Braverman, I. (2010). Zoo registrars: a bewildering bureaucracy. *Duke Environmental Law and Policy Forum*, 21, 165.

Braverman, I. (2011). States of exemption: the legal and animal geographies of American zoos. *Environment and Planning A*, 43, 1693–1706.

Braverman, I. (2012). Zooveillance: Foucault goes to the zoo. *Surveillance and Society*, 10, 119.

Braverman, I. (2018). Saving species, one individual at a time: zoo veterinarians between welfare and conservation. *Humanimalia*, 9, 1–27.

Braverman, I. (2019). Fish encounters: aquariums and their veterinarians on a rapidly changing planet. *Humanimalia*, 11, 1–29.

Breed, D., Meyer, L. C. R., Steyl, J. C. A., et al. (2019). Conserving wildlife in a changing world: understanding capture myopathy – a malignant outcome of stress during capture and translocation. *Conservation Physiology*, 7. https://doi.org/10.1093/conphys/coz027.

Bremner-Harrison, S., Prodohl, P. A. and Elwood, R. W. (2004). Behavioural trait assessment as a release criterion: boldness predicts early death in a reintroduction programme of captive-bred swift fox (*Vulpes velox*). *Animal Conservation*, 7, 313–320.

Brent, L. (1992). Woodchip bedding as enrichment for captive chimpanzees in an outdoor enclosure. *Animal Welfare*, 1, 161–170.

Brereton, A. W. (1968). Education: a primary function of the San Diego Zoo. *International Zoo Yearbook*, 8, 173–174.

Brereton, J. E. (2020a). Directions in animal enclosure use studies. *Journal of Zoo and Aquarium Research*, 8, 1–9.

Brereton, J. E. (2020b). Challenges and directions in zoo and aquarium food presentation research: a review. *Journal of Zoological and Botanical Gardens*, 1, 13–23.

Brichieri-Colombi, T. A., Lloyd, N. A., McPherson, M. and Moehrenschlager, A. (2019). Limited contributions of released animals from zoos to North American conservation translocations. *Conservation Biology*, 33, 33–39.

Brightsmith, D., Hilburn, J., del Campo, A., et al. (2005). The use of hand-raised psittacines for reintroduction: a case study of scarlet macaws (*Ara macao*) in Peru and Costa Rica. *Biological Conservation*, 121, 465–472.

Broad, G. (1996). Visitor profile and evaluation of informal educational at Jersey Zoo. *Dodo*, 32, 166–192.

Broad, S. and Smith, L. (2004). Who educates the public about conservation issues? Examining the role of zoos and the media. In: Frost, W., Croy, G. and Beeton, S. (eds.), *International Tourism and Media Conference Proceedings*, Tourism Research Unit, Monash University, Melbourne, pp. 15–23.

Broad, S. and Weiler, B. (1998). Captive animals and interpretation: a tail of two tiger exhibits. *Journal of Tourism Studies*, 9, 14–27.

Broom, D. M. (1986). Indicators of poor welfare. *British Veterinary Journal*, 142, 524–526.

Broom, D. M. (1991). Animal welfare: concepts and measurement. *Journal of Animal Science*, 69, 4167–4175.

Broom, D. M. (2007). Cognitive ability and sentience: which aquatic animals should be protected?. *Diseases of Aquatic Organisms*, 75, 99–108.

Broom, D. M. (2014). *Sentience and Animal Welfare*. CABI, Wallingford.

Brown, C. (2015). Fish intelligence, sentience and ethics. *Animal Cognition*, 18, 1–17.

Brown, T., Ashby, A. and Schwitzer, C. (2011). *An Illustrated History of Bristol Zoo Gardens*. Independent Zoo Enthusiasts Society, Todmorden.

Browning, H. (2018). No room at the zoo: management euthanasia and animal welfare. *Journal of Agricultural and Environmental Ethics*, 31, 483–498.

Browning, H. and Maple, T. L. (2019). Developing a metric of usable space for zoo exhibits. *Frontiers in Psychology*, 10, 791.

Bruni, C. M., Fraser, J. and Schultz, P. W. (2008). The value of zoo experiences for connecting people with nature. *Visitor Studies*, 11, 139–150.

Bryman, A. (1999). Theme parks and McDonaldization. In: Smart, B. (ed.). *Resisting McDonaldization*, Sage Publications, Los Angeles, pp. 101–115.

Buchanan-Smith, H. M., Griciute, J., Daoudi, S., Leonardi, R. and Whiten, A. (2013). Interspecific interactions and welfare implications in mixed species communities of capuchin (*Sapajus apella*) and squirrel monkeys (*Saimiri sciureus*) over 3 years. *Applied Animal Behaviour Science*, 147, 324–333.

Buckley, K. A., Smith, L. D., Crook, D. A., Pillans, R. D. and Kyne, P. M. (2020). Conservation impact scores identify shortfalls in demonstrating the benefits of threatened wildlife displays in zoos and aquaria. *Journal of Sustainable Tourism*, 28, 978–1002.

Bueno, M. G., Lopez, R. P. G., de Menezes, R. M. T., et al. (2010). Identification of *Plasmodium relictum* causing mortality in penguins (*Spheniscus magellanicus*) from São Paulo Zoo, Brazil. *Veterinary Parasitology*, 173, 123–127.

Bullock, N., James, C. and Williams, E. (2021). Using keeper questionnaires to capture zoo-housed tiger (*Panthera tigris*) personality: considerations for animal management. *Journal of Zoological and Botanical Gardens*, 2, 650–663.

Bunderson, J. S. and Thompson, J. A. (2009). The call of the wild: zookeepers, callings, and the double-edged sword of deeply meaningful work. *Administrative Science Quarterly*, 54, 32–57.

Bundesanstalt für Landwirtschaft und Ernährung (2021). *Einheimische Nutztierrassen in Deutschland und Rote Liste gefä hrdeter Nutztierrassen 2021*. Bundesanstalt für Landwirtschaft und Ernährung. Informations – und Koordinationszentrum für Biologische Vielfalt (IBV), Bonn.

Burch-Brown, J. (2017). Is it wrong to topple statues and rename schools? *Journal of Political Theory and Philosophy*, 1, 59–88.

Burden, F. (2012). Practical feeding and condition scoring for donkeys and mules. *Equine Veterinary Education*, 24, 589–596.

Burger, J. and Hemmer, H. (2006). Urgent call for further breeding of the relic zoo population of the critically endangered Barbary lion (*Panthera leo leo* Linnaeus 1758). *European Journal of Wildlife Research*, 52, 54–58.

Burgess, J. and Unwin, D. (1984). Exploring the living planet with David Attenborough. *Journal of Geography in Higher Education*, 8, 93–113.

Bustamante, J. (1996). Population viability analysis of captive and released bearded vulture populations. *Conservation Biology*, 10, 822–831.

Cabana, F. and Nekaris, K. A. I. (2015). Diets high in fruits and low in gum exudates promote the occurrence and development of dental disease in pygmy slow loris (*Nycticebus pygmaeus*). *Zoo Biology*, 6, 547–553.

Cabana, F., Plowman, A., Van Nguyen, T., et al. (2017). Feeding Asian pangolins: an assessment of current diets fed in institutions worldwide. *Zoo Biology*, 36, 298–305.

Caillaud, D., Eckardt, W., Vecellio, V., et al. (2020). Violent encounters between social units hinder the growth of a high-density mountain gorilla population. *Science Advances*, 6(45), eaba0724.

Campbell, J. A. (1950). Use of anaesthesia in treatment of zoo inmates. *Canadian Journal of Comparative Medicine and Veterinary Science*, 14, 39–41.

Campbell, K. H. S., McWhir, J., Ritchie, W. A. and Wilmut, I. (1996). Sheep cloned by nuclear transfer from a cultured cell line. *Nature*, 380 (6569), 64–66.

Campbell-Palmer, R. and Rosell, F. (2015). Captive care and welfare considerations for beavers. *Zoo Biology*, 34, 101–109.

Canfield, P. J. and Cunningham, A. A. (1993). Disease and mortality in Australian marsupials held at London Zoo, 1872–1972. *Journal of Zoo and Wildlife Medicine*, 24, 158–167.

Canino, W. and Powell, D. (2010). Formal behavioral evaluation of enrichment programs on a zookeeper's schedule: a case study with a polar bear (*Ursus maritimus*) at the Bronx Zoo. *Zoo Biology*, 29, 503–508.

Card, W. C., Roberts, D. T. and Odum, R. A. (1998). Does zoo herpetology have a future? *Zoo Biology*, 17, 453–462.

Carlson, A. (2000). *Aesthetics and the Environment*. Routledge, New York.

Carlstead, K. and Brown, J. L. (2005). Relationships between patterns of fecal corticoid secretion and behaviour, reproduction, and environmental factors in captive black (*Diceros bicornis*) and white (*Ceraotherium simum*) rhinoceros. *Zoo Biology*, 24, 215–232.

Carlstead, K. and Seidensticker, J. (1991). Seasonal variation in stereotypic pacing in an American black bear *Ursus americanus*. *Behavioural Processes*, 25, 155–161.

Carlstead, K., Seidensticker, J. and Baldwin, R. (1991). Environmental enrichment for bears. *Zoo Biology*, 10, 3–16.

Carlstead, K., Brown, J. L. and Seidensticker, J. C. (1993). Behavioral and adrenocortical responses to environmental changes in leopard cats (*Felis bengalensis*). *Zoo Biology*, 12, 321–331.

Carlstead, K., Mellen, J. and Kleiman, D. G. (1999). Black rhinoceros (*Diceros bicornis*) in U.S. zoos: I. Individual behavior profiles and their relationship to breeding success. *Zoo Biology*, 18, 17–34.

Carlstead, K., Paris, S. and Brown, J. L. (2018). Good keeper–elephant relationships in North American zoos are mutually beneficial to welfare. *Applied Animal Behaviour Science*, 211, 103–111.

Caro, T. M. (1994). *Cheetahs of the Serengeti Plains: Group Living in an Asocial Species.* University of Chicago Press, Chicago.

Carpenter, J. W., Gabel, R. R. and Goodwin, Jr., J. G. (1991). Captive breeding and reintroduction of the endangered masked bobwhite. *Zoo Biology*, 10, 439–449.

Carr, N. (2016a). Ideal animals and animal traits for zoos: general public perspectives. *Tourism Management*, 57, 37–44.

Carr, N. (2016b). An analysis of zoo visitors' favourite and least favourite animals. *Tourism Management Perspectives*, 20, 70–76.

Carr, N. and Cohen, S. (2015). The public face of zoos: images of entertainment, education and conservation. *Anthrozoös*, 24, 175–189.

Carter, M., Webber, S., Rawson, S., et al. (2020). Virtual reality in the zoo: a qualitative evaluation of a stereoscopic virtual reality video encounter with little penguins (*Eudyptula minor*). *Journal of Zoo and Aquarium Research*, 8, 239–245.

Cassey, P. and Hogg, C. J. (2015). Escaping captivity: the biological invasion risk from vertebrate species in zoos. *Biological Conservation*, 181, 18–26.

Cassidy, M. D. (1978). Development of an induced food plant preference in the Indian stick insect *Carausius morosus*. *Entomologia Experimentalis et Applicata*, 24, 287–293.

Castanheira, M. F., Conceição, L. E. C., Millot, S., et al. (2017). Coping styles in farmed fish: consequences for aquaculture. *Reviews in Aquaculture*, 9, 23–41.

CDC (2017). *Prioritizing Zoonotic Diseases for Multisectoral, One Health Collaboration in the United States. Workshop Summary.* One Health Zoonotic Disease Prioritization Workshop, 5–7 December, Department of Health and Human Services Office of the Assistant Secretary for Preparedness and Response, Washington, DC.

CDC (2021). One Health. www.cdc.gov/onehealth/index.html (accessed 6 December 2021).

Cerreta, A. J., Yang, T. S., Ramsay, E. C., et al. (2022). Detection of vector-borne infections in lions and tigers at two zoos in Tennessee and Oklahoma, USA. *Journal of Zoo and Wildlife Medicine*, 53, 50–59.

Chadwick, C. L. (2014). Social behaviour and personality assessment as a tool for improving the management of captive cheetahs (*Acinonyx jubatus*). PhD thesis, University of Salford.

Chadwick, C. L., Rees, P. A. and Stevens-Wood, B. (2013). Captive-housed male cheetahs (*Acinonyx jubatus soemmeringii*) form naturalistic coalitions: measuring associations and calculating chance encounters. *Zoo Biology*, 32, 518–527.

Chadwick, C. L., Springate, D. A., Rees, P. A., Armitage, R. P. and O'Hara, S. J. (2015). Calculating association indices in captive animals: controlling for enclosure size and shape. *Applied Animal Behaviour Science*, 169, 100–106.

Chadwick, C. L., Williams, E., Asher, L. and Yon, L. (2017). Incorporating stakeholder perspectives into the assessment and provision of captive elephant welfare. *Animal Welfare*, 26, 461–472.

Chaiyarat, R., Kongprom, U., Manathamkamon, D., Wanpradab, S. and Sangarang, S. (2012). Captive breeding and reintroduction of the oriental pied hornbill (*Anthracoceros albirostris*) in Khao Kheow Open Zoo, Thailand. *Zoo Biology*, 31, 683–693.

Chalmers, C., Fergus, P., Curbelo Montanez, C. A., Longmore, S. N. and Wich, S. A. (2021). Video analysis for the detection of animals using convolutional neural networks and consumer-grade drones. *Journal of Unmanned Vehicle Systems*, 9, 112–127.

Chalmin-Pui, L. S. and Perkins, R. R. (2017). How do visitors relate to biodiversity conservation? An analysis of London Zoo's "BUGS" exhibit. *Environmental Education Research*, 23, 1462–1475.

Chambers, P. (2007). *Jumbo: This Being the True Story of the Greatest Elephant in the World.* André Deutsch, London.

Chamove, A. S., Hosey, G. R. and Schaetzel, P. (1988). Visitors excite primates in zoos. *Zoo Biology*, 7, 359–369.

Champoux, M., Hempel, M. and Reinhardt, V. (1987). Environmental enrichment with sticks for singly-caged adult rhesus monkeys. *Laboratory Primate Newsletter*, 26(4), 5–7.

Chandon, P., Morwitz, V. G. and Reinartz, W. J. (2005). Do intentions really predict behavior? Self-generated validity effects in survey research. *Journal of Marketing*, 69, 1–14.

Charles River Editors (2016). *Jumbo the Elephant: The Life and Legacy of History's Most Famous Circus Animal.* Charles River Editors.

Chartier, L., Zimmermann, A. and Ladle, R. (2011). Habitat loss and human–elephant conflict in Assam, India: does a critical threshold exist? *Oryx*, 45, 528–533.

Che-Castaldo, J. P., Grow, S. A. and Faust, L. J. (2018). Evaluating the contribution of North American zoos and aquariums to endangered species recovery. *Scientific Reports*, 8, 9789 https://doi.org/10.1038/s41598-018-27806-2.

Che-Castaldo, J. P., Gray, S. M., Rodriguez-Clark, K. M., Schad Eebes, K. and Faust, L. J. (2021). Expected demographic and genetic declines not found in most zoos and aquarium populations. *Frontiers in Ecology and the Environment*, 19, 435–442.

Chen, Y.-H., Yu, J.-F., Chang, Y.-J., et al. (2020). Novel low-voltage electro-ejaculation approach for sperm collection from zoo captive Lanyu miniature pigs (*Sus barbatus sumatrensis*). *Animals*, 10, 1825. https://doi.org/10.3390/ani10101825.

Cheng, Y. Y., Chen, T. Y., Yu, P. H. and Chi, C. H. (2010). Observations on the female reproductive cycles of captive Asian yellow pond turtles (*Mauremys mutica*) with radiography and ultrasonography. *Zoo Biology*, 29, 50–58.

Cherfas, C. (1984). *Zoo 2000: A Look Beyond the Bars.* BBC, London.

Cheyne, S. M., Campbell, C. O. and Payne, K. L. (2012). Proposed guidelines for in situ gibbon rescue, rehabilitation and reintroduction. *International Zoo Yearbook*, 46, 265–281.

Chiew, S. J., Butler, K. L., Sherwen, S. L., et al. (2020). Effect of covering a visitor viewing area window on the behaviour of zoo-housed little penguins (*Eudyptula minor*). *Animals*, 10, 1224. https://doi.org/10.3390/ani10071224.

Chikwanha, O. C., Halimani, T. E., Chimonyo, M., Dzama, K. and Bhebhe, E. (2007). Seasonal changes in body condition scores of pigs and chemical composition of pig feed resources in a semi-arid smallholder farming area of Zimbabwe. *African Journal of Agricultural Research*, 2, 468–474.

Choo, Y., Todd, P. A. and Li, D. (2011). Visitor effects on zoo orang-utans in two novel, naturalistic enclosures. *Applied Animal Behaviour Science*, 133, 78–86.

Christensen, J. W., Zharkikh, T., Ladewig, J. and Yasinetskaya, N. (2002). Social behaviour in stallion groups (*Equus przewalskii* and *Equus caballus*) kept under natural and domestic conditions. *Applied Animal Behaviour Science*, 76, 11–20.

Christie, S. (2007). Zoo-based fundraising for in situ wildlife conservation. In Zimmerman, A., Hatchwell, M., Dickie, L. and West, C. (eds.), *Zoos in the 21st Century: Catalysts for Conservation.* Cambridge University Press, Cambridge, pp. 257–274.

Ciborowska, P., Michalczuk, M. and Bień, D. (2021). The effect of music on livestock: cattle, poultry and pigs. *Animals*, 11, 3572.

Cigler, P., Kvapil, P., Kastelic, M., et al. (2020). Retrospective study of causes of animal mortality in Ljubljana Zoo 2005–2015. *Journal of Zoo and Wildlife Medicine*, 51, 571–577.

Ciminelli, G., Martin, M. S., Swaisgood, R. R., et al. (2021). Social distancing: high population density increases cub rejection and decreases maternal care in the giant panda. *Applied Animal Behaviour Science*, 243, 105457.

Cipriano, A. (2002). Cold stress in captive great apes recorded in incremental lines of dental cementum. *Folia Primatologica*, 73, 21–31.

Claidière, N., Bowler, M. and Whiten, A. (2012). Evidence for weak or linear conformity but not for hyper-conformity in an everyday social learning context. *PLoS ONE*, 7(2), e30970.

Claidière, N., Bowler, M., Brookes, S., Brown, R. and Whiten, A. (2014). Frequency of behavior witnessed and conformity in an everyday social context. *PLoS ONE*, 9(6), e99874.

Clare, E. L., Economou, C. K., Bennett, F. J., et al. (2022). Measuring biodiversity from DNA in the air. *Current Biology*, 32, 693–700.

Clark, F. E. (2011a). Great ape cognition and captive care: can cognitive challenges enhance well-being. *Applied Animal Behaviour Science*, 135, 1–12.

Clark, F. E. (2011b). Space to choose: network analysis of social preferences in a captive chimpanzee community, and implications for management. *American Journal of Primatology*, 73, 748–757.

Clark, F. E., Gray, S. I., Bennett, P., Mason, L. J. and Burgeaa, K. V. (2019). High-tech and tactile: cognitive enrichment for zoo-housed gorillas. *Frontiers in Psychology*, 10, 1574. https://doi.org/10.3389/fpsyg.2019.01574.

Clauss, M. and Hatt, J. M. (2006a). The feeding of rhinoceros in captivity. *International Zoo Yearbook*, 40, 197–209.

Clauss, M. and Hatt, J. M. (2006b). Feeding Asian and African elephants *Elephas maximus* and *Loxodonta africana* in captivity. *International Zoo Yearbook*, 40, 88–95.

Clauss, M. and Paglia, D. E. (2012). Iron storage disorders in captive wild mammals: the comparative evidence. *Journal of Zoo and Wildlife Medicine*, 43(3s). https://doi.org/10.1638/2011-0152.1.

Clavadetscher, I., Bond, M., Martin, L., et al. (2021). Development of an image-based body condition score for giraffes *Giraffa camelopardalis* and a comparison of zoo-housed and free-ranging individuals. *Journal of Zoo and Aquarium Research*, 9, 170–185.

Clay, A. W., Perdue, B. M., Gaalema, D. E., Dolins, F. L. and Bloomsmith, M. A. (2010). The use of technology to enhance zoological parks. *Zoo Biology*, 30, 487–497.

Clay, Z. and Zuberbühler, K. (2009). Food-associated calling sequences in bonobos. *Animal Behaviour*, 77, 1387–1396.

Clay, Z. and Zuberbühler, K. (2011). Bonobos extract meaning from call sequences. *PLoS ONE*, 6(4), e18786.

Clayton, S., Fraser, J. and Saunders, C. D. (2009). Zoo experiences: conversations, connections, and concern for animals. *Zoo Biology*, 28, 377–397.

Clayton, S., Fraser, J. and Burgess, C. (2011). The role of zoos in fostering environmental identity. *Ecopsychology*, 3, 87–96.

Clayton, S., Prévot, A., Germain, L. and Saint-Jalme, M. (2017). Public support for biodiversity after a zoo visit: environmental concern, conservation knowledge, and self-efficacy. *Curator: The Museum Journal*, 60, 87–100.

Clements, J. and Sanchez, J. N. (2015). Creation and validation of a novel body condition scoring method for the Magellanic penguin (*Spheniscus magellanicus*) in the zoo setting. *Zoo Biology*, 34, 538–546.

Cless, I. T., Voss-Hoynes, H. A., Ritzmann, R. E. and Lukas, K. E. (2015). Defining pacing quantitatively: a comparison of gait characteristics between pacing and non-repetitive locomotion in zoo-housed polar bears. *Applied Animal Welfare Science*, 169, 78–85.

Clubb, R. and Mason, G. J. (2003). Captivity effects on wide-ranging carnivores. *Nature*, 425, 473–474.

Clubb, R. and Mason, G. J. (2007). Natural behavioural biology as a risk factor in carnivore welfare: how analysing species differences could help zoos improve enclosures. *Applied Animal Welfare Science*, 102, 303–328.

Clubb, R., Rowcliffe, M., Lee, P., et al. (2008). Compromised survivorship in zoo elephants. *Science*, 322, 1649.

Clulow, S., Clulow, J., Marcec-Greaves, R., et al. (2022). Common goals, different stages: the state of the ARTs for reptile and amphibian conservation. *Reproduction, Fertility and Development*, 34(5), i–ix.

Clyvia, A., Faggioli, A. B. and Cipreste, C. F. (2015). Effects of environmental enrichment in a captive pair of golden parakeet (*Guaruba guarouba*) with *Revista Brasiliera de Ornitologia*, 23, 309–314.

Coe, J. C. (1995). Zoo animal rotation: new opportunities from home range to habitat theater. In *Proceedings of the American Zoo and Aquarium Association*, Wheeling.

Coe, J. C. (2004). Mixed species rotation exhibits. In: *2004 ARAZPA Conference Proceedings*. www.joncoedesign.com/pub/technical.htm

Coe, J. C., Scott, D. and Lukas, K. E. (2009). Facility design for bachelor gorilla groups. *Zoo Biology*, 28, 144–162.

Cohen, E. and Fennell, D. (2016). The elimination of Marius, the giraffe: humanitarian act or callous management decision? *Tourism Recreation Research*, 41, 168–176.

Cohen, S. C. and Clark, A. (2019). Loss, grief, and bereavement in the context of human–animal relationships. In: Tedeschi, P. and Jenkins, M. A. (eds.), *Transforming Trauma: Resilience and Health Through Our Connections With Animals*. Purdue University Press, West Lafayette, pp. 395–422.

Colahan, H. and Breder, C. (2003). Primate training at Disney's Animal Kingdom. *Journal of Applied Animal Welfare Science*, 6, 235–246.

Cole, N. C. (2004). A novel technique for capturing arboreal geckos. *Herpetological Review*, 35, 358–359.

Cole, P. (2004). *Suspended Animation: An Unauthorised History of Herald and Britains Plastic Figures*. Plastic Warrior Publications, UK.

Coleman, K. and Maier, A. (2010). The use of positive reinforcement training to reduce stereotypic behaviour in rhesus macaques. *Applied Animal Behaviour Science*, 124, 142–148.

Collard, R. C. (2013). Panda politics. *The Canadian Geographer/Le Géographe canadien*, 57, 226–232.

Collinge, N. E. (1989). Mirror reactions in a zoo colony of Cebus monkeys. *Zoo Biology*, 8, 89–98.

Colman, R. J., Anderson, R. M., Johnson, S. C., et al. (2009). Caloric restriction delays disease onset and mortality in rhesus monkeys. *Science*, 325, 201–204.

Comfort, A. (1962). Survival curves of some birds in the London Zoo. *Ibis*, 104, 115–117.

Comizzoli, P., Crosier, A. E., Songsasen, N., et al. (2009). Advances in reproductive science for wild carnivore conservation. *Reproduction in Domestic Animals*, 44, 47–52.

Conde, D. A., Colchero, F., Gusset, M., et al. (2013). Zoos through the lens of the IUCN Red List: a global metapopulation approach to support conservation breeding programs. *PLoS ONE*, 8(12), e80311.

Conde, D. A., Staerk, J., Colchero, F., et al. (2019). Data gaps and opportunities for comparative and conservation biology. *Proceedings of the National Academy of Sciences*, 116(19), 9658–9664.

Conway, W. (1967). The opportunity for zoos to save vanishing species. *Oryx, 9*, 154–160.

Cooke, C. M. and Schillaci, M. A. (2007). Behavioral responses to the zoo environment by white handed gibbons. *Applied Animal Behaviour Science*, 106, 125–133.

Cooper, M. E. (2003). Zoo Legislation. *International Zoo Yearbook*, 38, 81–93.

Cortez, M. V., Valdez, D. J., Navarro, J. L. and Martella, M. B. (2015). Efficiency of antipredator training in captive-bred greater rheas reintroduced into the wild. *Acta Ethologica*, 18, 187–195.

Cortez, M. V., Navarro, J. L. and Martella, M. B. (2018). Effect of antipredator training on spatial behaviour of male and female greater rheas *Rhea americana* reintroduced into the wild. *Acta Ornithologica*, 53, 81–90.

Corwin, A. L. (2012). Training fish and aquatic invertebrates for husbandry and medical behaviors. *Veterinary Clinics: Exotic Animal Practice*, 15, 455–467.

Cottle, L., Tamir, D., Hyseni, M., Bühler, D. and Lindemann-Matthies, P. (2010). Feeding live prey to zoo animals: response of zoo visitors in Switzerland. *Zoo Biology*, 29, 344–350.

Courtenay, J. and Santow, G. (1989). Mortality of wild and captive chimpanzees. *Folia Primatologica* 52, 167–177.

Cox-Witton, K., Reiss, A., Woods, R., et al. (2014). Emerging infectious diseases in free-ranging wildlife: Australian zoo based wildlife hospitals contribute to national surveillance. *PLoS ONE*, 9(5), e95127.

Craig, L. E. and Vick, S. J. (2021). Engaging zoo visitors at chimpanzee (*Pan troglodytes*) exhibits promotes positive attitudes towards chimpanzees and conservation. *Antrozoös*, 34, 1–15.

Crane, A. L. and Mathis, A. (2011). Predator-recognition training: a conservation strategy to increase postrelease survival of hellbenders in head-starting programs. *Zoo Biology*, 30, 611–622.

Cranfield, M. R., Schaffer, N., Bavister, B. D., et al. (1989). Assessment of oocytes retrieved from stimulated and unstimulated ovaries of pig-tailed macaques (*Macaca nemestrina*) as a model to enhance the genetic diversity of captive lion-tailed macaques (*Macaca silenus*). *Zoo Biology*, 8(S1), 33–46.

Crates, R., Langmore, N., Ranjard, L., et al. (2021). Loss of vocal culture and fitness costs in a critically endangered songbird. *Proceedings of the Royal Society B*, 288(1947), 20210225.

Crawford, M. P. (1938). A behavior rating scale for young chimpanzees. *Journal of Comparative Psychology*, 26, 79–92.

Cray, C., Arhwart, K. L., Hunt, M., et al. (2013). Acute phase protein quantification in serum samples from healthy Atlantic bottlenose dolphins (*Tursiops truncatus*). *Journal of Veterinary Diagnostic Investigation*, 25, 107–111.

Cresswell, R. (1883). *Aristotle's History of Animals in Ten Books* (translated by Richard Cresswell). George Bell & Sons, London.

Crissey, S. D. (2001). The history of zoo nutrition. In: Edwards, M., Lisi, K. J., Schlegel, M. L. and Bray, R. E. (eds.), *Proceedings of the Fourth Conference on Zoo and Wildlife Nutrition*, AZA Nutrition Advisory Group, Lake Buena Vista.

Crissey, S. D. and Pribyl, L. (2000). A review of nutritional deficiencies and toxicities in captive New World primates. *International Zoo Yearbook*, 37, 355–360.

Croft, V. F. (2008). Animal health libraries, librarians, and librarianship: a bibliography. http://hdl.handle.net/2376/1469 (accessed 24 February 2021).

Crone, E. E., Pickering, D. and Schultz, C. B. (2007). Can captive rearing promote recovery of endangered butterflies? An assessment in the face of uncertainty. *Biological Conservation*, 139, 103–112.

Cronin, K. A., Bethell, E. J., Jacobson, S. L., et al. (2018). Evaluating mood changes in response to anthropogenic noise with a response-slowing task in three species of primates. *Animal Behavior and Cognition*, 5, 209–221.

Cronin, K. A., Tank, A., Ness, T., Leahy, M. and Ross, S. R. (2020). Sex and season predict wounds in zoo-housed Japanese macaques (*Macaca fuscata*): a multi-institutional study. *Zoo Biology*, 39, 147–155.

Crosier, A. E., Pukazhenthi, B. S., Henghali, J. N., et al. (2006). Cryopreservation of spermatozoa from wild-born Namibian cheetahs (*Acinonyx jubatus*) and influence of glycerol on cryosurvival. *Cryobiology*, 52, 169–181.

Crudge, B., O'Connor, D., Hunt, M., Davis, E. O. and Browne-Nuñez, C. (2016). Groundwork for effective conservation education: an example of in situ and ex situ collaboration in South East Asia. *International Zoo Yearbook*, 50, 34–48.

Curry, E., Skogen, M. and Roth, T. (2021). Evaluation of an odour detection dog for non-invasive pregnancy diagnosis in polar bears (*Ursus maritimus*): considerations for training sniffer dogs for biomedical investigations in wildlife species. *Journal of Zoo and Aquarium Research*, 9, 1–7.

Curtis, D. J. and Zaramody, A. (1999). Social structure and seasonal variation in the behaviour of *Eulemur mongoz*. *Folia Primatologica*, 70, 79–96.

Cushing, A. (2011). Pursuing zoo veterinary work via the USA. *Veterinary Record*, 168, i.

Czekala, N., McGeehan, L., Steinman, K., Xuebing, L. and Gual-Sil, F. (2003). Endocrine monitoring and its application to the management of the giant panda. *Zoo Biology*, 22, 389–400.

da Costa, A. L. M., Appolonio, E. V. P., de Mattos, L. H. L., et al. (2022). Surgical management of upward fixation of the patella via medial patellar ligament desmotomy in a lowland tapir (*Tapirus terrestris*). *Journal of Zoo and Wildlife Medicine*, 53, 222–227.

Dalton, R. and Buchanan-Smith, H. M. (2005). A mixed-species exhibit for Goeldi's monkeys and Pygmy marmosets *Callimico goeldii* and *Callithrix pygmaea* at Edinburgh Zoo. *International Zoo Yearbook*, 39, 176–184.

Dancer, A. M. M. and Burn, C. C. (2019). Visitor effects on zoo-housed Sulawesi crested macaque (*Macaca nigra*) behaviour: can signs with 'watching' eyes requesting quietness help? *Applied Animal Behaviour Science*, 211, 88–94.

Daniel, B. M., Green, K. E., Doulton, H., et al. (2017). A bat on the brink? A range-wide survey of the Critically Endangered Livingstone's fruit bat *Pteropus livingstonii*. *Oryx*, 51, 742–751.

Dastjerdi, A., Seilern-Moy, K., Darpel, K., Steinbech, F. and Molenaar, F. (2016). Surviving and fatal Elephant Endotheliotropic Herpes-1A infections in juvenile Asian elephants: lessons learned and recommendation on anti-herpesviral therapy. *BMC Veterinary Research*, 12, 178. https://doi.org/10.1186/s12917-016-0806-5.

Davey, G. (2006a). An hourly variation in zoo visitor interest: measurement and significance for animal welfare research. *Journal of Applied Animal Welfare Science*, 9, 249–256.

Davey, G. (2006b). Relationships between exhibit naturalism, animal visibility and visitor interest in a Chinese zoo. *Applied Animal Behaviour Science*, 96, 93–102.

Davey, G. (2007a). An analysis of country, socio-economic and time factors on worldwide zoo attendance during a 40 year period. *International Zoo Yearbook*, 41, 217–225.

Davey, G. (2007b). Visitors' effects on the welfare of animals in the zoo: a review. *Journal of Applied Animal Welfare Science*, 10, 169–183.

Davies, J., Foxall, G. R. and Pallister, J. (2002). Beyond the intention–behaviour mythology: an integrated model of recycling. *Marketing Theory*, 2, 29 https://doi.org/10.1177/1470593102002001645.

Davies, T. E., Wilson, S., Hazarika, N., et al. (2011). Effectiveness of intervention methods against crop-raiding elephants. *Conservation Letters*, 4, 346–354.

Davis, M. R., Gamble, K. C. and Matheson, J. S. (2008). Diagnostic imaging in terrestrial invertebrates: Madagascar hissing cockroach (*Gromphadorhina portentosa*), desert millipede (*Orthoporus* sp.), emperor scorpion (*Pandinus imperator*), Chilean rosehair tarantula (*Grammostola spatulata*), Mexican fireleg tarantula (*Brachypelma boehmei*), and Mexican redknee tarantula (*Brachypelma smithi*). *Zoo Biology*, 27, 109–125.

Davis, N., Schaffner, C. M. and Wehnelt, S. (2009). Patterns of injury in zoo-housed spider monkeys: a problem with males? *Applied Animal Behaviour Science*, 116, 250–259.

De Faria, C. M., de Sousa Sá, F., Costa, D. D. L., et al. (2018). Captive-born collard peccary (*Pecari tajacu*, Tayassuidae) fails to discriminate between predator and non-predator models. *Acta Ethologica*, 21, 175–184.

de Waal, F. (1996). *Good Natured: The Origins of Right and Wrong in Humans and Other Animals*. Harvard University Press, Cambridge, MA.

de White, T. G. and Jacobson, S. K. (1994). Evaluating conservation education programs at a South American zoo. *The Journal of Environmental Education*, 25, 18–22.

de Wit, J. J. (1995). Mortality of rheas caused by *Synchamus trachea* infection. *Veterinary Quarterly*, 17, 39–40.

Deans, C., Martin, J., Neon, K., Nuesea, B. and O'Reilly, J. (1987). *A zoo for who? A pilot study in zoo design for children. The Reid Park Zoo*. Center for Social Design, Jacksonville.

D'Eath, R. B. and Keeling, L. D. (2003). Social discrimination and aggression by laying hens in large groups: from peck order to social intolerance. *Applied Animal Behaviour Science*, 84, 197–212.

Debevec, K. and Kernan, J. B. (1984). More evidence on the effects of a presenter's attractiveness: some cognitive, affective, and behavioural consequences. *Advances in Consumer Research*, 11, 127–132.

DEFRA (2007). Reports received by Defra of escapes of non-native cats in the UK 1975 to present day. www.defra.gov.uk/wildlife-countryside/vertebrates/reports/exotic-cat-escapes .pdf (accessed 1 January 2008).

DEFRA (2021). *Our Action Plan for Animal Welfare*. DEFRA, London.

DeGrazia, D. (2002). *Animal Rights: A Very Short Introduction*. Oxford University Press, Oxford.

Deleu, R., Veenhuizen, R. and Nelissen, M. (2003). Evaluation of the mixed-species exhibit of African elephants and Hamadryas baboons in Safari Beekse Bergen, the Netherlands. *Primate Report*, 65, 5–19.

DeMatteo, K. E., Porton, I. J., Kleiman, D. G. and Asa, C. S. (2006). The effect of the male bush dog (*Speothos venaticus*) on the female reproductive cycle. *Journal of Mammalogy*, 87, 723–732.

Dembiec, D. P., Snider, R. J. and Zanella, A. J. (2004). The effects of transport stress on tiger physiology and behavior. *Zoo Biology*, 23, 335–346.

Demetriou, D. (2017). Japanese zoo culls 57 snow monkeys with 'invasive alien' genes by lethal injection. *The Telegraph*. www.telegraph.co.uk/news/2017/02/22/japanese-zoo-culls-57-snow-monkeys-invasive-alien-genes-lethal/ (accessed 14 August 2017).

Denk, D., Boufana, B., Masters, N. J. and Stidworthy, M. F. (2016). Fatal echinococcosis in three lemurs in the United Kingdom: a case series. *Veterinary Parasitology*, 218, 10–14.

Dent, V. E. (1979). The bacteriology of dental plaque from a variety of zoo-maintained mammalian species. *Archives of Oral Biology*, 24, 277–282.

DeSmet, A. and Ogle, B. (2022). The influence of welfare and bonds with animals on the job satisfaction of felid keepers in North America. *Zoo Biology.* https://doi.org/10.1002/zoo.21667.

DeVault, M. L. (2000). Producing family time: practices of leisure activity beyond the home. *Qualitative Sociology*, 23, 485–503.

Diamond, J. and Bond, A. B. (1991). Social behavior and the ontogeny of foraging in the kea (*Nestor notabilis*). *Ethology*, 88, 128–144.

Dickel, L., Boal, J. G. and Budelmann, B. U. (2000). The effect of early experience on learning and memory in cuttlefish. *Developmental Psychobiology*, 36, 101–110.

Dickens, M. J., Earle, K. A. and Romero, L. M. (2009). Initial transference of wild birds to captivity alters stress physiology. *General and Comparative Endocrinology*, 160, 76–83.

Dierenfeld, E. S. (1989). Vitamin E deficiency in zoo reptiles, birds, and ungulates. *Journal of Zoo and Wildlife Medicine*, 20, 3–11.

Dishman, D. L., Thomson, D. M. and Karnovsky, N. J. (2009). Does simple feeding enrichment raise activity levels of captive ring-tailed lemurs (*Lemur catta*)? *Applied Animal Behaviour Science*, 116, 88–95.

Dixon, L. M., Hardiman, J. R. and Cooper, J. J. (2010). The effects of spatial restriction on the behavior of rabbits (*Oryctolagus cuniculus*). *Journal of Veterinary Behavior: Clinical Applications and Research*, 5, 302–308.

Dombrowski, D. S., De Voe, R. S. and Lewbart, G. A. (2013). Comparison of isoflurane and carbon dioxide anesthesia in Chilean rose tarantulas (*Grammostola rosea*). *Zoo Biology*, 32, 101–103.

Domínguez-Domínguez, O., Morales, R. H., Nava, M. M., et al. (2018). Progress in the reintroduction program of the tequila splitfin in the springs of Teuchitlán, Jalisco, Mexico. In: Soorae. P. S. (ed.), *Global Reintroduction Perspectives: 2018. Case Studies from Around the Globe*. IUCN, Gland, pp. 38–42.

Donahue, J. (2017). Beyond Personhood: Legal Rights for Zoo Animals. Donahue, J. (ed.), *Increasing Legal Rights for Zoo Animals: Justice on the Ark*. Lexington Books, Rowman & Littlefield, Lanham, pp. 147–154.

Donlan, J., Berger, C. J., Bock, C. E., et al. (2006). Pleistocene rewilding: an optimistic agenda for the twenty-first century. *The American Naturalist*, 168, 660–681.

Doolittle, R. L. and Grand, T. I. (1995). Benefits of the zoological park to the teaching of comparative vertebrate anatomy. *Zoo Biology*, 14, 453–462.

Dorsten, C. M. and Cooper, D. M. (2004). Use of body condition scoring to manage body weight in dogs. *Journal of the American Association for Laboratory Animal Science*, 43, 34–37.

Douglas-Hamilton, I. and Douglas-Hamilton, O. (1975). *Among the Elephants*. Collins, Glasgow.

Dove, T. and Byrne, J. (2014). Do zoo visitors need zoology knowledge to understand conservation messages? An exploration of the public understanding of animal biology and of the conservation of biodiversity in a zoo setting. *International Journal of Science Education, Part B*, 4, 323–342.

Doyle, C. (2014). Captive elephants. In: Gruen, L. (ed.), *The Ethics of Captivity*. Oxford University Press, Oxford, pp. 38–56.

Draganova, I. G. and Gurnell, J. (2004). The behaviour of Przewalski horses (*Equus przewalskii*) during formation of bachelor groups. In *Proceedings of the British Society of Animal Science*. Cambridge University Press, Cambridge, p. 141.

Drake, G. J., Haycock, J., Dastjerdi, A., Davies, H. and Lopez, F. J. (2020). Use of immunostimulants in the successful treatment of a clinical EEHV1A infection in an Asian elephant (*Elephas maximus*). *VetRecord*, 8. https://doi.org/10.1136/vetreccr-2020-001158.

Draper, C. and Harris, S. (2012). The assessment of animal welfare in British zoos by government-appointed inspectors. *Animals*, 2, 507–528.

Draper, C., Browne, W. and Harris, S. (2013). Do formal inspections ensure that British zoos meet and improve on minimum animal welfare standards? *Animals*, 3, 1058–1072.

Dröscher, I. and Waitt, C. D. (2012). Social housing of surplus males of Javan langurs (*Trachypithecus auratus*): compatibility of intact and castrated males in different social settings. *Applied Animal Behaviour Science*, 141, 184–190.

Duckler, G. L. (1998). An unusual osteological formation in the posterior skulls of captive tigers (*Panthera tigris*). *Zoo Biology*, 17, 135–142.

Duffy, D. C., Todd, F. S. and Siegfried, W. R. (1987). Submarine foraging behavior of alcids in an artificial environment. *Zoo Biology*, 6, 373–378.

Dufour, V., Sueur, S. Whiten, A. and Buchanan-Smith, H. M. (2011). The impact of moving to a novel environment on social networks, activity and wellbeing in two new world primates. *American Journal of Primatology*, 73, 802–811.

Dugdale, A. H., Grove-White, D., Curtis, G. C., Harris, P. A. and Argo, C. M. (2012). Body condition scoring as a predictor of body fat in horses and ponies. *The Veterinary Journal*, 194, 173–178.

Dukas, R. and Mooers, A. Ø. (2003). Environmental enrichment improves mating success in fruit flies. *Animal Behaviour*, 66, 741–749.

Duncan, I. J. (2006). The changing concept of animal sentience. *Applied Animal Behaviour Science*, 100, 11–19.

Durge, S. M., Das, A., Saha, S. K., et al. (2022). Dietary lutein supplementation improves immunity and antioxidant status of captive Indian leopards (*Panthera fusca*). *Zoo Biology*. https://doi.org/10.1002/zoo.21671.

Durrant, B. S., Millard, S. E., Zimmerman, D. M. and Lindurg, D. G. (2001). Lifetime semen production in a cheetah (*Acinonyx jubatus*). *Zoo Biology*, 20, 359–366.

Durrell (2019). Home page. www.durrell.org (accessed 1 September 2019).

Durrell, G. (1960). *A Zoo in My Luggage.* Rupert Hart-Davis, London.

Dyke, B., Gage, T. B., Alford, P. L., Swenson, B. and Williams-Blangero, S. (1995). Model life table for captive chimpanzees. *American Journal of Primatology*, 37, 25–37.

DZS (2020). *Financial Statements: Years Ended December 31, 2020 and 2019.* The Detroit Zoological Society, Detroit.

Echarte, G. V. and Vasallo, R. O. (2016). Occupational accidents in workers associated with wild animals at the National Zoo Park of Cuba. *Revista Cubana de Salud y Trabajo*, 17(3), 15–20.

Edes, A. N., Edwards, K. L., Wolfe, B. A., Brown, J. L. and Crews, D. E. (2020). Allostatic load indices with cholesterol and triglycerides predict disease and mortality risk in zoo-housed Western lowland gorillas. *Biomarker Insights*, 15, 1177271920914585.

Edmonson, A. J., Lean, I. J., Weaver, L. D., Farver, T. and Webster, G. (1989). A body condition scoring chart for Holstein dairy cows. *Journal of Dairy Science*, 72, 68–78.

Edmunds, K., Bunbury, N., Sawmy, S., Jones, C. G. and Bell, D. J. (2008). Restoring avian island endemics: use of supplementary food by the endangered Pink Pigeon (*Columba mayeri*). *Emu-Austral Ornithology*, 108, 74–80.

Edwards, M. C., Ford, C., Hoy, J. M., FitzGibbon, S. and Murray, P. J. (2021). How to train your wildlife: a review of predator avoidance training. *Applied Animal Behaviour Science*, 234. https://doi.org/10.1016/j.applanim.2020.105170.

Einwiller, S., Viererbl, B. and Himmelreich, S. (2017). Journalists' coverage of online firestorms in German-language news media. *Journalism Practice*, 11, 1178–1197.

ElephantVoices (2019). Elephants in zoos. https://elephantvoices.org/elephants-in-captivity-7/in-zoos.html (accessed 5 June 2019).

Emily, P. P. and Eisner, E. R. (eds.) (2021). *Zoo and Wild Animal Dentistry*. Wiley-Blackwell, Chichester.

Emmett, C., Digby, M., Pope, J. and Williams, E. (2021). Social behaviour in zoo bachelor groups: a case study of related South American fur seals. *Animals*, 11(9), 2682.

Encinoso, M., Orós, J., Ramírez, G., et al. (2019). Anatomic study of the elbow joint in a Bengal tiger (*Panthera tigris tigris*) using magnetic resonance imaging and gross dissections. *Animals*, 9(12), 1058.

Esson, M. and Moss, A. (2016). The challenges of evaluating conservation education across cultures. *International Zoo Yearbook*, 50, 61–67.

Esson, M., Moss, A. and Pitchford, L. (2014). The 'Thinking Big' Elephant Project. *International Zoo Educators' Journal*, 50, 14–16.

European Commission. (2018). *Evaluation of Council Directive 1999/22/EC of 29 March 1999 relating to the keeping of wild animals in zoos (Zoos Directive).* (SWD (2018) 456 final).

Ezenwa, V. O., Jolles, A. E. and O'Brien, M. P. (2009). A reliable body condition scoring technique for estimating condition in African buffalo. *African Journal of Ecology*, 47, 476–481.

Fàbregas, M. C., Guillén-Salazar, F. and Garcés-Narro, C. (2010). The risk of zoological parks as potential pathways for the introduction of non-indigenous species. *Biological Invasions*, 12, 3627–3636.

Fagot, J. and Vauclair, J. (1988). Handedness and bimanual coordination in the lowland gorilla. *Brain, Behavior and Evolution*, 32, 89–95.

Fair, P. A., Schaefer, A. M., Houser, D. S., et al. (2017). The environment as a driver of immune and endocrine responses in dolphins (*Tursiops truncatus*). *PLoS ONE*, 12, e0176202.

Falk, J. H., Reinhard, E. M., Vernon, C. L., et al. (2007). *Why Zoos & Aquariums Matter: Assessing the Impact of a Visit*. Association of Zoos & Aquariums, Silver Spring

Fant, J. B., Havens, K., Kramer, A. T., et al. (2016). What to do when we can't bank on seeds: what botanic gardens can learn from the zoo community about conserving plants in living collections. *American Journal of Botany*, 103, 1541–1543.

Farhadinia, M. S., Alinezhad, H., Hadipour, E., et al. (2018). Intraspecific killing among leopards (*Panthera pardus*) in Iran (Mammalia: Felidae). *Zoology in the Middle East*, 64, 189–194.

Faria, C., Boaventura, D. and Guilherme, E. (2020). Personal meaning maps as an assessment tool for a Planetarium session: a study with primary school children. *Education 3–13*, 48, 66–75.

Farmer, H. L., Plowman, A. B. and Leaver, L. A. (2011). Role of vocalisations and social housing in breeding in captive howler monkeys (*Alouatta caraya*). *Applied Animal Behaviour Science*, 134, 177–183.

Farr, J. A. (1976). Social facilitation of male sexual behaviour, intrasexual competition, and sexual selection in the guppy, *Poecilia reticulata* (Pisces: Poeciliidae). *Evolution*, 30, 707–717.

Farrell, M. A., Barry, E. and Marples, N. (2000). Breeding behavior in a flock of Chilean flamingos (*Phoenicopterus chilensis*) at Dublin Zoo. *Zoo Biology*, 19, 227–237.

Faust, L. J., Jackson, R., Ford, A., Earnhardt, J. M. and Thompson, S. D. (2004). Models for management of wildlife populations: lessons from spectacled bears in zoos and grizzly bears in Yellowstone. *System Dynamics Review*, 20, 163–178.

Faust, L. J., Long, S. T., Perišin, K. and Simonis, J. L. (2019). Uncovering challenges to sustainability of AZA Animal Programs by evaluating the outcomes of breeding and transfer recommendations with PMCTrack. *Zoo Biology*, 38, 24–35.

Fazio, E., Medica, P., Bruschetta, G. and Ferlazzo, A. (2014). Do handling and transport stress influence adrenocortical response in the tortoises (*Testudo hermanni*)? *International Scholarly Research Notes Veterinary Science*, 2014. http://dx.doi.org/10.1155/2014/798273.

Fekete, J. M., Norcross, J. L. and Newman, J. D. (2000). Artificial turf foraging boards as environmental enrichment for pair-housed female squirrel monkeys. *Journal of the American Association for Laboratory Animal Science*, 39(2), 22–26.

Fell, L. R. and Shutt, D. A. (1986). Adrenocortical response of calves to transport stress as measured by salivary cortisol. *Canadian Journal of Animal Science*, 66, 637–641.

Fernandez, E. J., Kinley, R. C. and Timberlake, W. (2019). Training penguins to interact with enrichment devices for lasting effects. *Zoo Biology*, 38, 481–489.

Fernandez, L. T., Bashaw, M. J., Sartor, R. L. Bouwens, N. R. and Maki, T. S. (2008). Tongue twisters: feeding enrichment to reduce oral stereotypy in giraffe. *Zoo Biology*, 27, 200–212.

Fernández-Bellon, D. and Kane, A. (2020). Natural history films raise species awareness: a big data approach. *Conservation Letters*, 13, e12678.

Fernández-Bellon, H., Rodon, J., Fernández-Bastit, L., et al. (2021). Monitoring natural SARS-CoV-2 infection in lions (*Panthera leo*) at the Barcelona Zoo: viral dynamics and host responses. *Viruses*, 13(9), 1683.

Fernando, P., Janaka, H. K., Ekanayaka, S. K. K., Nishantha, H. G. and Pastorini, J. (2009). A simple method for assessing elephant body condition. *Gajah*, 31, 29–31.

Ferrie, G. M., Becker, K. K., Wheaton, C. J., Fontenot, D. and Bettinger, T. (2011). Chemical and surgical interventions to alleviate intraspecific aggression in male collared lemurs (*Eulemur collaris*). *Journal of Zoo and Wildlife Medicine*, 42, 214–221.

FFI (2020). *Annual Report and Accounts 2020*. Fauna & Flora International, Cambridge.

Fidgett, A. L. and Webster, M. (2011). Managing zoo diet information: what do we need from the next generation software? *Zooquaria Nutrition News, the European Association of Zoos and Aquaria (EAZA)*, 5, 12–13.

Finch, K., Williams, L. and Holmes, L. (2020). Using longitudinal data to evaluate the behavioural impact of a switch to carcass feeding on an Asiatic lion (*Panthera leo persica*). *Journal of Zoo and Aquarium Research*, 8, 283–287.

Fine, L., Barnes, C., Niedbalski, A. and Deem, S. L. (2022). Staff perceptions of COVID-19 impacts on wildlife conservation at a zoological institution. *Zoo Biology*. https://doi.org/10.1002/zoo.21669.

Finlay, T., James, L. R. and Maple, T. L. (1988). People's perceptions of animals: the influence of zoo environment. *Environment and Behavior*, 20, 508–528.

Fischer-Tenhagen, C., Wetterholm, L., Tenhagen, B.-A. and Heuwieser, W. (2011). Training dogs on a scent platform for oestrus detection in cows. *Applied Animal Behaviour Science*, 131, 63–70.

Fisken, F. A. (ed.) (2007). *International Zoo Yearbook. Vol. 41. Animal Health and Conservation*. Zoological Society of London, London.

Fitch, H. M. and Fagan, D. A. (1982). Focal palatine erosion associated with dental malocclusion in captive cheetahs. *Zoo Biology*, 1, 295–310.

Flies, A. S., Mansfield, L. S., Grant, C. K., Weldele, M. L. and Holekamp, K. E. (2015). Markedly elevated antibody responses in wild versus captive spotted hyenas show that environmental and ecological factors are important modulators of immunity. *PLoS ONE*, 10, e0137679.

Florens, F. B. V., Daby, D. and Jones, C. (1998). Impact of weeding and rat control on the diversity and abundance of land snails in the Mauritian upland forest. *Journal of Conchology, Special Publication*, 2, 87–88.

Föllmi, J., Steiger, A., Walzer, C., et al. (2007). A scoring system to evaluate physical condition and quality of life in geriatric zoo mammals. *Animal Welfare*, 16, 309–318.

Fooden, J. and Izor, R. J. (1983). Growth curves, dental emergence norms, and supplementary morphological observations in known-age captive orangutans. *American Journal of Primatology*, 5, 285–301.

Fooks, A. R., Brookes, S. M., Johnson, N., McElhinney, L. M. and Hutson, A. M. (2003). European bat lyssaviruses: an emerging zoonosis. *Epidemiology & Infection*, 131, 1029–1039.

Forrester, G. S., Quaresmini, C., Leavens, D. A., Spiezio, C. and Vallortigara, G. (2012). Target animacy influences chimpanzee handedness. *Animal Cognition*, 15, 1121–1127.

Forthman, D. L. and Bakeman, R. (1992). Environmental and social influences on enclosure use and activity patterns of captive sloth bears (*Ursus ursinus*). *Zoo Biology*, 11, 405–415.

Fowler, G. S., Wingfield, J. C. and Boersma, P. D. (1995). Hormonal and reproductive effects of low levels of petroleum fouling in Magellanic penguins (*Spheniscus magellanicus*). *Auk*, 112, 382–389.

Francis, D., Esson, M. and Moss, A. (2007). The use of visitor tracking to evaluate 'Spirit of the Jaguar' at Chester Zoo. *International Zoo Educators Journal*, 43, 20–24.

Frankham, R., Ballou, J. D. and Brisoe, D. A. (2002). *Introduction to Conservation Genetics*. Cambridge University Press, Cambridge.

Franks, V. R., Andrews, C. E., Ewen, J. G., et al. (2019). Changes in social groups across reintroductions and effects on post-release survival. *Animal Conservation*, 23, 443–454.

Friedman, C. R., Torigian, C., Shillam, P. J., et al. (1998). An outbreak of salmonellosis among children attending a reptile exhibit at a zoo. *The Journal of Pediatrics*, 132, 802–807.

Frynta, D., Lišková, S., Bültmann, S. and Burda, H. (2010) Being attractive brings advantages: the case of parrot species in captivity. *PLoS ONE*, 5(9), e12568.

Frynta, D., Šimková, O, Lišková, S. and Landová, E. (2013). Mammalian collection on Noah's Ark: the effects of beauty, brain and body size. *PLoS ONE*, 8(5), e63110.

Fukui, D., Nagano, M., Nakamura, R., et al. (2013). The effects of frequent electroejaculation on the semen characteristics of a captive Siberian tiger (*Panthera tigris altaica*). *Journal of Reproduction and Development*, 59, 491–495.

Furrer, S. C. and Corredor, G. (2008). Conservation of threatened amphibians in Valle del Cauca, Colombia: a cooperative project between Cali Zoological Foundation, Colombia, and Zoo Zürich, Switzerland. *International Zoo Yearbook*, 42, 158–164.

Galama, W., King, C. and Brouwer., K. (2002). *EAZA Hornbill Management and Husbandry Guidelines* (1st ed.). EAZA Hornbill TAG, National Foundation for Research in Zoological Gardens, Amsterdam Zoo, Amsterdam.

Gans, C. and Mix, H. (1974). A sequential insect dispenser for behavioral experiments. *BioScience*, 24, 88–89.

Garcia, G., Cunningham, A. A., Horton, D. L., et al. (2007). Mountain chickens *Leptodactylus fallax* and sympatric amphibians appear to be disease free on Montserrat. *Oryx*, 41, 398–401.

Garcia, V. C. and de Almeida-Santos, S. M. (2022). Reproductive cycles of neotropical boid snakes evaluated by ultrasound. *Zoo Biology*, 41, 74–83.

Garcia-Pelegrin, E., Clark, F. and Miller, R. (2022). Increasing animal cognition research in zoos. *Zoo Biology*. https://doi.org/10.1101/2021.11.24.469897.

Garner, J. P. (2005). Stereotypies and other abnormal repetitive behaviors: potential impact on validity, reliability, and replicability of scientific outcomes. *ILAR Journal*, 46, 106–117.

Garner, J. P., Mason, G. J. and Smith, R. (2003). Stereotypic route-tracing in experimentally caged songbirds correlates with general behavioural disinhibition. *Animal Behaviour*, 66, 711–727.

Gartland, K., McDonald, M., Braccini Slade, S., White, F. and Sanz, C. (2018). Behavioral changes following alterations in the composition of a captive bachelor group of Western lowland gorillas (*Gorilla gorilla gorilla*). *Zoo Biology*, 37, 391–398.

Gartland, K. N., Carrigan, J. and White, F. J. (2021). Preliminary relationship between over-night separation and wounding in bachelor groups of Western lowland gorillas (*Gorilla gorilla gorilla*). *Applied Animal Behaviour Science*, 241, 105388.

Gartner, M. C. and Powell, D. (2012). Personality assessment in snow leopards (*Uncia uncia*). *Zoo Biology*, 31, 151–165.

Gartner, M. C. and Weiss, A. (2013). Scottish wildcat (*Felis silvestris grampia*) personality and subjective well-being: implications for captive management. *Applied Animal Behaviour Science*, 147, 261–267.

Gatti, R. C., Velichevskaya, A., Gottesman, B. and Davis, K. (2021). Grey wolf may show signs of self-awareness with the sniff test of self-recognition. *Ethology, Ecology & Evolution*, 33, 444–467.

Gentz, E. J. (1990). Employment of veterinarians by zoos in North America: a survey. *Journal of Zoo and Wildlife Medicine*, 21, 24–26.

Gerits, I., Wydooghe, E., Peere, S., et al. (2022). Semen collection, evaluation, and cryopre-servation in the bonobo (*Pan paniscus*). *BMC Zoology*, 7, 1–9.

Gerstell, R. (1936). The elk in Pennsylvania: its extermination and reintroduction. *Pennsylvania Game News*, 7, 6–7.

Gilbert, T., Gardner, R., Kraaijeveld, A. R. and Riordan, P. (2017). Contribution of zoos and aquariums to reintroductions: historical reintroduction efforts in the context of changing conservation perspectives. *International Zoo Yearbook*, 51, 15–31.

Gili, C., Meijer, G. and Lacave, G. (eds.) (2018). *EAZA and EAAM Best Practice Guidelines for Otariidae and Phocidae (Pinnipeds)* (1st ed.). Acquario di Genova, Italy.

Giljov, A., Karenina, K., Ingram, J. and Malashichev, Y. (2015). Parallel emergence of true handedness in the evolution of marsupials and placentals. *Current Biology*, 25, 1878–1884.

Gill, V. (2022). Vaccine trial for killer elephant virus begins. BBC News, 3 February. www.bbc.co.uk/news/science-environment-60222464.

Gilloux, I., Gurnell, J. and Shepherdson, D. (1992). An enrichment device for great apes. *Animal Welfare*, 1, 279–289.

Gimmel, A., Öfner, S. and Liesegang, A. (2021). Body condition scoring (BCS) in corn snakes (*Pantherophis guttatus*) and comparison to pre-existing body condition index (BCI) for snakes. *Journal of Animal Physiology and Animal Nutrition*, 105, 24–28.

Gingrich-Philbrook, C. (2016). On the execution of the young giraffe, Marius, by the Copenhagen Zoo: Conquergood's 'Lethal Theatre' and posthumanism. *Text and Performance Quarterly*, 36, 200–211.

Ginsburg, H. J. and Miller, S. M. (1982). Sex differences in children's risk-taking behavior. *Child Development*, 53, 426–428.

Gippoliti, S. and Kitchener, A. C. (2007). The Italian zoological gardens and their role in mammal systematic studies, conservation biology and museum collections. *Hystrix, the Italian Journal of Mammalogy*, 18, 173–184.

Glaeser, S. S., Klinck, H., Mellinger, D. K., et al. (2009). A vocal repertoire of Asian elephants (*Elephas maximus*) and comparison of call classification methods. *The Journal of the Acoustical Society of America*, 125, 2709.

Glatston, A. R. (1998). The control of zoo populations with special reference to primates. *Animal Welfare*, 7, 269–281.

Glatston, A. R. and Roberts, M. (1988). The current status and future prospects of the red panda (*Ailurus fulgens*) studbook population. *Zoo Biology*, 7, 47–59.

Glatt, S. E., Francl, K. E. and Scheels, J. L. (2008). A survey of dental problems and treatment in zoo animals. *International Zoo Yearbook*, 41, 206–213.

Goff, C., Howell, S. M., Fritz, J. and Nankivell, B. (1994). Space use and proximity of captive chimpanzee (*Pan troglodytes*) mother/offspring pairs. *Zoo Biology*, 13, 61–68.

Gollakner, R. and Capua, I. (2020). Is COVID-19 the first pandemic that evolves into a panzootic? *Veterinaria Italiana*, 56, 11–12.

Gómez, M. C., Earle Pope, C., Giraldo, A., et al. (2004). Birth of African wildcat cloned kittens born from domestic cats. *Cloning and Stem Cells*, 6, 247–258.

Goodenough, A. E., McDonald, K., Moody, K. and Wheeler, C. (2019). Are 'visitor effects' overestimated? Behaviour in captive lemurs is mainly driven by co-variation with time and weather. *Journal of Zoo and Aquarium Research*, 7, 59–66.

Goodman, G., Girling, S., Pizzi, R., et al. (2012). Establishment of a health surveillance program for reintroduction of the Eurasian beaver (*Castor fiber*) into Scotland. *Journal of Wildlife Diseases*, 48, 971–978.

Goodman, G., Meredith, A., Girling, S. et al. (2017). Outcomes of a 'One Health' monitoring approach to a five-year beaver (*Castor fiber*) reintroduction trial in Scotland. *EcoHealth*, 14, 139–143.

Gore, M., Hutchins, M. and Ray, J. (2006). A review of injuries caused by elephants in captivity: an examination of predominant factors. *International Zoo Yearbook*, 40, 51–62.

Gorges, M. A., Martinez, K. M., Labriola, N. F., et al. (2022). Effects of tricaine methanesulfonate in a managed collection of moon jellyfish (*Aurelia aurita*). *Journal of Zoo and Wildlife Medicine*, 53, 100–107.

Gori, S., Agrillo, C., Dadda, M. and Bisazza, A. (2014). Do fish perceive illusory motion? *Scientific Reports*, 4, 6443.

Gosling, S. D. (2001). From mice to men: what can we learn about personality from animal research? *Psychological Bulletin*, 127, 45–86.

Gould, E. and Bres, M. (1986). Regurgitation and reingestion in captive gorillas: description and intervention. *Zoo Biology*, 5, 241–250.

Gozalo, A. and Montoya, E. (1991). Mortality causes of the moustached tamarin (*Saguinus oedipus*) in captivity. *Journal of Medical Primatology*, 21, 35–38.

Graham, M. S. and Dow, P. R. (1990). Dental care for a captive killer whale, *Orcinus orca*. *Zoo Biology*, 9, 325–330.

Grand, A. P., Leighty, K. A., Cory, L. J., et al. (2013). The neighbor effect in bachelor and breeding groups of Western lowland gorillas (*Gorilla gorilla gorilla*). *International Journal of Comparative Psychology*, 26, 26–36.

Grandin, T. (2000). Habituating antelope and bison to cooperate with veterinary procedures. *Journal of Applied Animal Welfare Science*, 3, 253–261.

The Graphic (1872). Transferring the hairy rhinoceros from her travelling den to her cage. *The Graphic*, 2 March 1872, p. 208.

The Graphic (1882a). Advertisement. *The Graphic*. 25 February 1882, p.179.

The Graphic (1882b). Jumbo, the big African elephant at the zoological gardens. *The Graphic*. 25 February 1882, pp. 179–180.

Gray, S. M., Faust, L. J., Kuykendall, N. A., et al. (2021). Reasons for unfulfilled breeding and transfer recommendations in zoos and aquariums *Zoo Biology*, 41, 143–156.

Greco, B. J., Meehan, C. L., Heinsus, J. L. and Mench, J. A. (2017). Why pace? The influence of social, housing, management, life history, and demographic characteristics on locomotor stereotypy in zoo elephants. *Applied Animal Behaviour Science*, 194, 104–111.

Greenberg, J. A., DiMenna, M. A., Hanelt, B. and Hofkin, B. V. (2012). Analysis of post-blood meal flight distances in mosquitoes utilizing zoo animal blood meals. *Journal of Vector Ecology*, 37, 83–89.

Greenwood, A. G., Cusdin, P. A. and Radford, M. J. (2001). *Effectiveness Study of the Dangerous Wild Animals Act 1976*. DEFRA, London.

Gregory, N. G. and Robins, J. K. (1998). A body condition scoring system for layer hens. *New Zealand Journal of Agricultural Research*, 41, 555–559.

Greig, J. E., Carnie, J. A., Tallis, G. F., et al. (2004). An outbreak of Legionnaires' disease at the Melbourne Aquarium, April 2000: investigation and case–control studies. *Medical Journal Australia*, 180, 566–572.

Griffin, A. S., Blumstein, D. T. and Evans, C. S. (2000). Training captive-bred or translocated animals to avoid predators. *Conservation Biology*, 14, 1317–1326.

Griffioen, J. A., Flower, J. E., Nelson, P. J., et al. (2022). Baseline hematologic and biochemical values and correlations to environmental parameters in managed Japanese spider crabs (*Macrocheira kaempferi*). *Journal of Zoo and Wildlife Medicine*, 53, 173–186.

Griffith, B., Scott, M., Carpenter, J. W. and Reed, C. (1989). Translocation as a species conservation tool: status and strategy. *Science*, 245, 477–480.

Griffith, M. P., Clase, T., Toribio, P., et al. (2020). Can a botanic garden metacollection better conserve wild plant diversity? A case study comparing pooled collections with an ideal sampling model. *International Journal of Plant Sciences*, 181(5). https://doi.org/10.1086/707729.

Griffiths, C. J., Jones, C. G., Hansen, D. M., et al. (2010). The use of extant non-indigenous tortoises as a restoration tool to replace extinct ecosystem engineers. *Restoration Ecology*, 18, 1–7.

Griner, L. A. (1983). *Pathology of Zoo Animals: a Review of Necropsies Conducted Over a Fourteen-Year Period at the San Diego Zoo and San Diego Wildlife Park*. Zoological Society of San Diego, San Diego.

Gruen, L. (ed.) (2014). *The Ethics of Captivity*. Oxford University Press, Oxford.

The Guardian (2021). Mammoth journey ahead as elephants leave Kent zoo for the Kenyan savannah. *The Guardian*. www.theguardian.com/environment/2021/jul/05/elephants-leave-kent-zoo-for-the-kenyan-savannah-aoe (accessed 7 July 2021).

Gunasekera, C. A. (2018). The ethics of killing 'surplus' zoo animals. *Journal of Animal Ethics*, 8, 93–102.

Gurusamy, V., Tribe, A., Toukhsati, S. and Phillips, C. J. C. (2015). Public attitudes in India and Australia toward elephants in zoos. *Anthrozoös*, 28, 87–100.

Gusset, M. and Dick, G. (2010). 'Building a future for wildlife'? Evaluating the contribution of the world zoo and aquarium community to in situ conservation. *International Zoo Yearbook*, 44, 183–191.

Gusset, M. and Dick, G. (2011). The global reach of zoos and aquariums in visitor numbers and conservation expenditure. *Zoo Biology*, 30, 566–569.

Gutierrez, S., Canington, S. L., Eller, A. R., Herrelko, E. S. and Sholts, S. B. (2021). The intertwined history of non-human primate health and human medicine at the Smithsonian's National Zoo and Conservation Biology Institute. Notes and records. *The Royal Society Journal of the History of Science*. https://doi.org/10.1098/rsnr.2021.0009

Gutnick, T., Weissenbacher, A. and Kuba, M. J. (2020). The underestimated giants: operant conditioning, visual discrimination and long-term memory in giant tortoises. *Animal Cognition*, 23, 159–167.

Haaretz (2013). Berlin Zoo comes to terms with Nazi Past, seeks out former Jewish shareholders. *Haaretz*, 3 December.

Hacker, C. E. and Miller, L. J. (2016). Zoo visitor perceptions, attitudes, and conservation intent after viewing African elephants at the San Diego Zoo Safari Park. *Zoo Biology*, 35, 355–361.

Hadfield, C. A. and Clayton, L. A. (2011). Fish quarantine: current practices in public zoos and aquaria. *Journal of Zoo and Wildlife Medicine*, 42, 641–650.

Hall, B. A., McGill, D. M., Sherwen, S. L. and Doyle, R. E. (2021).Cognitive enrichment in practice: a survey of factors affecting its implementation in zoos globally. *Animals*, 11, 1721. https://doi.org/10.3390/ani11061721.

Hall, M. (2017). Exploring the cultural dimensions of environmental victimization. *Palgrave Communications*, 3, 17076. https://doi.org/10.1057/palcomms.2017.76.

Hambrecht, S. and Reichler, S. (2013). Group dynamics of young Asian elephant bulls (*Elephas maximus* Linnaeus, 1758) in Heidelberg zoo: integration of a newcomer in an established herd. *Der Zoologische Garten*, 82, 267–292.

Hamerton, A. E. (1941). Report on the deaths occurring in the Society's gardens during the years 1939–1940. *Proceedings of the Zoological Society of London*, B111, 151–185.

Hanzlíková, V., Pluháček, J. and Čulik, L. (2014). Association between taxonomic relatedness and interspecific mortality in captive ungulates. *Applied Animal Behaviour Science*, 153, 62–67.

Harcourt, A. H. and Stewart, K. J. (1978). Coprophagy by wild mountain gorilla. *African Journal of Ecology*, 16, 223–225.

Hardie, S. M. (1997). Exhibiting mixed-species groups of sympatric tamarins *Saguinus* spp. at Belfast Zoo. *International Zoo Yearbook*, 35, 261–266.

Harding, G., Griffiths, R. A. and Pavajeau, L. (2016). Developments in amphibian captive breeding and reintroduction programs. *Conservation Biology*, 30, 340–349.

Hardy, D. F. (1996). Current research activities in zoos. In: Kleiman, D. G., Allen, M. E., Thompson, K. V. and Lumpkins, S. (eds.), *Wild Mammals in Captivity. Principles and Techniques*. University of Chicago Press, Chicago, pp. 531–536.

Hare, V. J., Rich, B. and Worley, K. E. (2007). Enrichment gone wrong! In *Proceedings of the Eighth International Conference on Environmental Enrichment*, Vienna, pp. 35–45.

Harley, E. H., Knight, M. H., Lardner, C., Wooding, B. and Gregor, M. (2009). The Quagga project: progress over 20 years of selective breeding. *African Journal of Wildlife Research*, 39, 155–163.

Harley, J. J., Power, A. and Stack, J. D. (2019). Investigation of the efficacy of the GnRH agonist deslorelin in mitigating intraspecific aggression in captive male Amur leopards (*Panthera pardus orientalis*). *Zoo Biology*, 38, 214–219.

Haron, A. W., Ming, Y. and Zainuddin, Z. Z. (2000). Evaluation of semen collected by electroejaculation from captive lesser Malay chevrotain (*Tragulus javanicus*). *Journal of Zoo and Wildlife Medicine*, 31, 164–167.

Harris, M., Sherwin, C. and Harris, S. (2008). *The Welfare, Housing and Husbandry of Elephants in UK Zoos. Final Report, November 10th 2008*. University of Bristol, Bristol.

Harrison, R. A. and Whiten, A. (2018). Chimpanzees (*Pan troglodytes*) display limited behavioural flexibility when faced with a changing foraging task requiring tool use. *PeerJ*, 6, e4366.

Harrison, R. M. and Nystrom, P. (2010). Handedness in captive gorillas (*Gorilla gorilla*). *Primates*, 51, 251–261.

Hartley, M., Wood, A. and Yon, L. (2019). Facilitating the social behaviour of bull elephants in zoos. *International Zoo Yearbook*, 53, 62–77.

Hartup, B. K., Olsen, G. H. and Czekala, N. M. (2005). Fecal corticoid monitoring in whooping cranes (*Grus americana*) undergoing reintroduction. *Zoo Biology*, 24, 15–28.

Hashmi, A. and Sullivan, M. (2020). The visitor effect in zoo-housed apes: the variable effect on behaviour of visitor number and noise. *Journal of Zoo and Aquarium Research*, 8, 268–282.

Hassan, K. H. (2014). Occupational and animals safety in zoos: a legal narrative. *American Journal of Animal and Veterinary Sciences*, 9, 1–5.

Hatley, J. (1986). Zoos: resource or recreation? *Journal of Biological Education*, 20, 231–234.

Hattingh, J. (1977). Blood sugar as an indicator of stress in the freshwater fish *Labeo capensis* (Smith). *Journal of Fish Biology*, 10, 191–195.

Hawkins, S. A., Brady, D. B., Mayhew, M., et al. (2020). Community perspectives on the reintroduction of Eurasia lynx (*Lynx lynx*) to the UK. *Restoration Ecology*, 28, 1408–1418.

Hays, D. W. and Stinson, D. W. (2019). *Draft Periodic Status Review for the Oregon Silverspot in Washington*. Washington Department of Fish and Wildlife, Olympia.

Haywood, M. (2014). Human places for large non-humans: from London's Imperial Elephant Stables to Copenhagen's Postmodern Glasshouse. In: Pauknerova, K., Stella, M., Gibas, P., et al. (eds.), *Non-Humans in Social Science: Ontologies, Theories and Case Studies*. Pavel Mervart, Červenỳ Kostelec, pp. 201–218.

Heape, W. (1898). On menstruation and ovulation in monkeys and in the human female. *British Medical Journal*, 2(1982), 1868–1869.

Heaver, J. and Waters, M. (2019). A retrospective study of mortality in Eurasian lynx (*Lynx lynx*) in UK zoos. *Zoo Biology*, 38, 200–208.

Hebb, D. O. (1949). Temperament in chimpanzees: I. Method of analysis. *Journal of Comparative and Physiological Psychology*, 42, 192–206.

Hebert, P. L. and Bard, K. (2000). Orangutan use of vertical space in an innovative habitat. *Zoo Biology*, 19, 239–251.

Hedeen, S. E. (1982). Utilization of space by captive groups of lowland gorillas (*Gorilla g. gorilla*). *Ohio Journal of Science*, 82, 27–30.

Hedeen, S. E. (1983). The use of space by lowland gorillas (*Gorilla g. gorilla*) in an outdoor enclosure. *Ohio Journal of Science*, 83, 183–185.

Hediger, H. (1950). *Wild Animals in Captivity*. Butterworth Scientific Publications, London.

Hediger, H. (1969). *Psychology and Behaviour of Animals in Zoos and Circuses*. Dover Publications, New York.

Heidegger, E. M., von Houwald, F., Steck, B. and Clauss, M. (2016). Body condition scoring system for greater one-horned rhino (*Rhinoceros unicornis*): development and application, *Zoo Biology*, 35, 432–443.

Heinz, J., Anderson, K. and Wolf, K. (2022). Retrospective mortality review of tufted puffins (*Fratercula cirrhata*) at a single institution (1982–2017). *Journal of Zoo and Wildlife Medicine*, 53, 11–18.

Hejna, P., Zátopková, L. and Šafr, M. (2011). A fatal elephant attack. *Journal of Forensic Sciences*, 57. https://doi.org/10.1111/j.1556-4029.2011.01967.

Hellmuth, H., Augustine, L., Watkins, B. and Hope, K. (2012). Using operant conditioning and desensitization to facilitate veterinary care with captive reptiles. *Veterinary Clinics: Exotic Animal Practice*, 15, 425–443.

Hemsworth, P. H. and Coleman, G. J. (1998). *Human–Livestock Interactions: The Stockperson and the Productivity and Welfare of Intensively Farmed Animals*. CABI, London.

Henry, B. A., Power, M. L., Maslanka, M. T., Rencken, C. A. and Nollman, J. A. (2022). Challenges of devising a milk recipe in a hand-reared hippopotamus (*Hippopotamus amphibius*). *Zoo Biology*. https://doi.org/10.1002/zoo.21680.

Henry Doorly Zoo (2008). www.omahazoo.com/exhibits/jungle.htm (accessed 18 July 2008).

Herbert, H. (2014). How the Indianapolis Zoo preserved their visitor experience with dynamic pricing. https://npengage.com/nonprofit-management/how-the-indianapolis-zoo-preserved-the-visitor-experience-with-dynamic-pricing/ (accessed 15 August 2017).

Hermes, R., Behr, B., Hildebrandt, T. B., et al. (2009a). Sperm sex-sorting in the Asian elephant (*Elephas maximus*). *Animal Reproduction Science*, 112, 390–396.

Hermes, R., Göritz, F., Portas, T. J., et al. (2009b). Ovarian superstimulation, transrectal ultrasound-guided oocyte recovery, and IVF in rhinoceros. *Theriogenology*, 72, 959–968.

Hermes, R., Lecu, A., Potier, R., et al. (2022). Cryopreservation of giraffe epidydimal spermatozoa using different extenders and cryoprotectants. *Animals*, 12(7), 857.

Herrelko, E. S., Vick, S. J. and Buchanan-Smith, H. M. (2012). Cognitive research in zoo-housed chimpanzees: influence of personality and impact on welfare. *American Journal of Primatology*, 74, 828–840.

Herrelko, E. S., Buchanan-Smith, H. M. and Vick, S. J. (2015). Perception of available space during chimpanzee introductions: number of accessible areas is more important than enclosure size. *Zoo Biology*, 34, 397–405.

Herrick, J. R. (2019). Assisted reproductive technologies for endangered species conservation: developing sophisticated protocols with limited access to animals with unique reproductive mechanisms. *Biology of Reproduction*, 100, 1158–1170.

Herrick, J. R., Campbell, M. K. and Swanson, W. F. (2002). Electroejaculation and semen analysis in the La Plata three-banded armadillo (*Tolypeutes matacus*). *Zoo Biology*, 21, 481–487.

Hewer, M. J. (2020). Determining the effect of extreme weather events on human participation in recreation and tourism: a case study of Toronto Zoo. *Atmosphere*, 11, 99.

Hewer, M. J. and Gough, W. A. (2016a). Weather sensitivity for zoo visitation in Toronto, Canada: a quantitative analysis of historical data. *International Journal of Biometeorology*, 60, 1645–1660.

Hewer, M. J. and Gough, W. A. (2016b). Assessing the impact of projected climate change on zoo visitation in Toronto (Canada). *Journal of Geography and Geology*, 8, 30–48.

Hewer, M. J. and Gough, W. A. (2016c). The effect of seasonal climatic anomalies on zoo visitation in Toronto (Canada) and the implications for projected climate change. *Atmosphere*, 7, 71.

Heywood, V. H. (ed.) (1995). *Global Biodiversity Assessment*. United Nations Environment Programme and Cambridge University Press, Cambridge.

Hiem, A. B. and Holt, E. A. (2022). Staring at signs: biology undergraduates pay attention to signs more othen than animals at the zoo. *Curator: The Museum Journal.* https://doi.org/10.1111/cura.12480.

Hildebrandt, T. B., Göritz, F., Pratt, N. C., et al. (2000a). Ultrasonography of the urogenital tract in elephants (*Loxodonta africana* and *Elephas maximus*): an important tool for assessing female reproductive function. *Zoo Biology*, 19, 321–332.

Hildebrandt, T. B., Hermes, R., Pratt, N. C., et al. (2000b). Ultrasonography of the urogenital tract in elephants (*Loxodonta africana* and *Elephas maximus*): an important tool for assessing male reproductive function. *Zoo Biology*, 19, 333–345.

Hill, D. J., Langley, R. L. and Morrow, W. M. (1998). Occupational injuries and illnesses reported by zoo veterinarians in the United States. *Journal of Zoo and Wildlife Medicine*, 29, 371–385.

Hinkson, K. M. and Poo, S. (2020). Inbreeding depression in sperm quality in a critically endangered amphibian. *Zoo Biology*, 39, 197–204.

Hioki, A. and Inaba, R. (2021). Occupational fatalities due to mammal-related accidents in Japan, 2000–2019. *Wilderness & Environmental Medicine*, 32, 19–26.

Hirshi, K. and Screven, C. (1990). Effects of questions on visitor reading behavior. *ILVS Review: A Journal of Visitor Behavior*, 1, 50–61.

Hirskyj-Douglas, I., Gray, S. and Piitulainen, R. (2021). ZooDesign: methods for understanding and facilitating children's education at zoos. In: *2021 ACM Interaction Design and Children Conference (IDC 2021)*. https://doi.org/10.1145/3459990.3460697.

Hitchens, P., Hultgren, J., Frössling, J., Emanuelson, U. and Keeling, L. (2017). Circus and zoo animal welfare in Sweden: an epidemiological analysis of data from regulatory inspections by the official competent authorities. *Animal Welfare*, 26, 373–382.

HMSO (1965). *Report of the Technical Committee to Enquire into the Welfare of Animals kept under Intensive Livestock Husbandry Systems.* HMSO, London.

Hoage, R. J. and Deiss, W. A. (eds.) (1996). *New Worlds, New Animals: From Menagerie to Zoological Park in the Nineteenth Century.* Johns Hopkins University Press, Baltimore.

Hoehfurtner, T., Wilkinson, A., Nagabaskaaran, G. and Burman, O. H. P. (2021). Does the provision of environmental enrichment affect the behaviour and welfare of captive snakes? *Applied Animal Behaviour Science*, 239, 105324. https://doi.org/10.1016/j.applanim.2021.105324.

Hoff, M. P. and Maple, T. L. (1982). Sex and age differences in the avoidance of reptile exhibits by zoo visitors. *Zoo Biology*, 1, 263–269.

Hogan, E. S., Houpt, K. A. and Sweeney, K. (1988). The effect of enclosure size on social interactions and daily activity patterns of the captive Asiatic wild horse (*Equus przewalskii*). *Applied Animal Behaviour Science*, 21, 147–168

Höhn, M., Kronschnabl, M. and Gansloβer, U. (2000). Similarities and differences in activities and agonistic behaviour of male Eastern grey kangaroos (*Macropus giganteus*) in captivity and the wild. *Zoo Biology*, 19, 529–539.

Hollamby, S., Murphy, D. and Schiller, C. A. (2000). An epizootic of amoebiasis in a mixed species collection of juvenile tortoises. *Journal of Herpetological Medicine and Surgery*, 10, 9–15.

Hollén, L. I. and Manser, M. B. (2007). Persistence of alarm-call behaviour in the absence of predators: a comparison between wild and captive-born meerkats (*Suricata suricatta*). *Ethology*, 113, 1038–1047.

Hollister, N. (1917). Some effects of environment and habit on captive lions. *Proceedings of the United States National Museum*, 53, 177–193.

Holmes, B. (1998). Day of the sparrow. *New Scientist*, 2140, 32.

Holtorf, C. (2013). The zoo as a realm of memory. *Anthropological Journal of European Cultures*, 22, 98–114.

Holtorf, C. and Ortman, O. (2008). Endangerment and conservation ethos in natural and cultural heritage: the case of zoos and archaeological sites. *International Journal of Heritage Studies*, 14, 74–90.

Holzer, D., Scott, D. and Bixler, R. D. (1998). Socialization influences on adult zoo visitation. *Journal of Applied Recreation Research*, 23, 43–62.

Honess, P. E., Gimpel, J. L., Wolfensohn, S. E. and Mason, G. J. (2005). Alopecia scoring: the quantitative assessment of hair loss in captive macaques. *Alternatives to Laboratory Animals*, 33, 193–206.

Hopper, L. M. (2017). Cognitive research in zoos. *Current Opinion in Behavioral Sciences*, 16, 100–110.

Horvath, K., Angeletti, D., Nascetti, G. and Carere, C. (2013). Invertebrate welfare: an overlooked issue. *Annali dell'Istituto superiore di sanità*, 49, 9–17.

Hosey, G. R. and Skyner, L. J. (2007). Self-injurious behaviour in zoo primates. *International Journal of Primatology*, 28, 1431–1437.

Hosey, G., Hill, S. P. and Lherbier, M. L. (2012). Can zoo records help answer behavioural research questions? The case of the left-handed lemurs (*Lemur catta*). *Zoo Biology*, 31, 189–196.

Hosey, G., Melfi, V., Formella, I., et al. (2016). Is wounding aggression in zoo-housed chimpanzees and ring-tailed lemurs related to zoo visitor numbers? *Zoo Biology*, 35, 205–209.

Hosey, G., Birke, L., Shaw, W. S. and Melfi, V. (2018). Measuring the strength of human–animal bonds in zoos. *Anthrozoös*, 31, 273–281.

Hosey, G., Melfi, V. and Ward, S. J. (2020a). Problematic animals in the zoo: the issue of charismatic megafauna. In: Angelici, F. and Rossi, L. (eds.), *Problematic Wildlife II*. Springer, Cham, pp. 485–508.

Hosey. G., Ward, S. J., Ferguson, A., Jenkins, H. and Hill, S. P. (2020b). Zoo-housed mammals do not avoid giving birth on weekends. *Zoo Biology*, 40, 3–8.

Houser, A., Gusset, M., Bragg, C. J., Boast, L. K. and Somers, M. J. (2011). Pre-release hunting training and post-release monitoring are key components in the rehabilitation of orphaned large felids. *South African Journal of Wildlife Research*, 41, 11–20.

Houser, D. S., Martin, S., Crocker, D. E. and Finneran, J. J. (2020). Endocrine response to simulated U.S. Navy mid-frequency sonar exposures in the bottlenose dolphin (*Tursiops truncatus*). *Journal of the Acoustical Society of America*, 147, 1681–1687.

Howard, J. G., Bush, M., de Voss, V. and Wildt, D. E. (1989). Electroejaculation, semen characteristics and serum testosterone concentration of free ranging African elephants (*Loxodonta africana*). *Journal of Reproduction and Fertility*, 72, 187–195.

Howletts (2022). African elephant. www.aspinallfoundation.org/howletts/animals/African-elephant (accessed 17 December 2022).

Hoy, J. M., Murray, P. J. and Tribe, A. (2010). Thirty years later: enrichment practices for captive mammals. *Zoo Biology*, 29, 303–316.

Huang, B., Tian, X., Maheshwari, A., et al. (2022). The destiny of living animals imported into Chinese zoos. *Diversity*, 15, 335. https://doi.org/1-.3390/d14050335.

Hugo, C., Seier, J., Mdhluli, C., et al. (2003). Fluoxetine decreases stereotypic behaviour in primates. *Progress in Neuro-Psychopharmacology and Biological Psychiatry*, 27, 639–643.

Hunter, S. C., Gusset, M., Miller, L. J. and Somers, M. J. (2014). Space use as an indicator of enclosure appropriateness in African wild dogs (*Lycaon pictus*). *Journal of Applied Animal Welfare Science*, 17, 98–110.

Hussain, Z., Ali, Z., Nemat, A., et al. (2015). Enclosure size of animals of Lahore Zoological Garden in comparison of international norms. *Journal of Animal and Plant Sciences*, 25 (3 Suppl. 2), 500–508.

Hutchins, M. (2006). Variation in nature; its implications for zoo elephant management. *Zoo Biology*, 25, 161–171.

Hutchins, M. (2007). The animal rights–conservation debate: can zoos and aquariums play a role? In: Zimmerman, A., Hatchwell, M., Dickie, L. and West, C. (eds.), *Zoos in the 21st Century*. Cambridge University Press, Cambridge, pp. 92–109.

Hutchins, M. and Ballentine, J. (2001). Fueling the conservation engine: fund-raising and public relations. In: Conway, W. G., Hutchins, M., Souza, M., Kapetanakos, Y. and Paul, E. (eds.), *AZA Field Conservation Resource Guide*. Wildlife Conservation Society and Zoo Atlanta, Georgia, pp. 268–271.

Hutchins, M. and Keele, M. (2006). Elephant importation from range countries: ethical considerations for accredited zoos. *Zoo Biology*, 25, 219–233.

Hutchins, M. and Souza, M. (2001). AZA's conservation endowment fund: zoos and aquariums supporting conservation action. In: Conway, W. G., Hutchins, M., Souza, M., Kapetanakos, Y. and Paul, E. (eds.), *AZA Field Conservation Resource Guide*. Wildlife Conservation Society and Zoo Atlanta, Georgia, pp. 291–302.

Hutchinson, J. R., Schwerda, D., Famini, D. J., et al. (2006). The locomotor kinematics of Asian and African elephants: changes with speed and size. *Journal of Experimental Zoology*, 209, 3812–3827.

Ichino, J., Isoda, K., Hanai, A. and Ueda, T. (2013). Effects of the display angle in museums on user's cognition, behavior, and subjective responses. *Proceedings of the SIGCHI Conference on Human Factors in Computing Systems*. https://doi.org/10.1145/2470654.2481413

IDA (2022). Honoring animals purposely killed by zoos on World Zoothanasia Day. www.idausa.org/campaign/elephants/latest-news/honoring-animals-purposely-killed-by-zoos-on-world-zoothanasia-day/ (accessed 5 April 2022).

Idaho Fish and Game Commission. (1997). Fish and Game Commission policy paper: grizzly bear recovery. Idaho Fish and Game Commission, Boise.

ILAR (1998). *The Psychological Well-Being of Nonhuman Primates*. Committee on Well-Being of Nonhuman Primates, Institute for Laboratory Animal Research, Commission on Life Sciences, National Research Council, National Academy Press, Washington, DC.

The Illustrated London News (1847a). Wombwell's Menagerie at Windsor Castle. *The Illustrated London News*. 6 November, pp. 297–298.

The Illustrated London News (1847b). Death of the zoological society's elephant. *The Illustrated London News*, 19 June, pp. 1–2.

The Illustrated London News (1874). The Berlin Zoological Gardens. *The Illustrated London News*, 17 October, p. 378.

The Illustrated London News (1876). The New Lion House. *The Illustrated London News*, 1 April, p. 325.

The Illustrated London News (1882). Attempt to remove Jumbo, the great elephant, from the zoological gardens. 25 February, p. 200.

The Illustrated Sporting and Dramatic News (1875). The Zoological Society. *The Illustrated Sporting and Dramatic News*, 22 May, p. 190.

The Independent (2014). The killing of Marius the giraffe opens an important debate about genetics, animal rights and zoo inbreeding. www.independent.co.uk/news/world/europe/the-killing-of-marius-the-giraffe-opens-an-important-debate-about-genetics-animal-rights-and-zoo-9120219.html (accessed 14 August 2017).

Inglett, B. J., French, J. A., Simmons, L. G. and Vires, K. W. (1989). Dynamics of intrafamily aggression and social reintegration in lion tamarins. *Zoo Biology*, 8, 67–78.

Ings, R., Waran, N. K. and Young, R. J. (1997). Attitude of zoo visitors to the idea of feeding live prey to zoo animals. *Zoo Biology*, 16, 343–347.

ISIS (2006a). ISIS abstracts: *Loxodonta africana*. https://app.isis.org/abstracts/abs.asp (accessed 27 October 2006).

ISIS (2006b). ISIS abstracts: *Elephas maximus*. https://app.isis.org/abstracts/abs.asp (accessed 27 October 2006).

Ito, T. (2014). *London Zoo and the Victorians*. Boydell & Brewer, Rochester.

Itoh, K., Ide, K., Kojima, Y. and Terada, M. (2010). Hibernation exhibit for Japanese black bear *Ursus thibetanus japonicus* at Ueno Zoological Gardens. *International Zoo Yearbook*, 44, 55–64.

Itoh, M. (2010). *Japanese Wartime Zoo Policy: The Silent Victims of World War II*. Palgrave Macmillan, New York.

IUCN (1995). *IUCN/SSC Guidelines for Re-Introductions*. http://iucn.org/themes/ssc/pubs/policy/reinte.htm.

IUCN/SSC (2013). *Guidelines for Reintroductions and Other Conservation Translocations. Version 1.0.* IUCN Species Survival Commission. Gland.

IUCN/SSC (2014). *IUCN Species Survival Commission Guidelines on the Use of Ex Situ Management for Species Conservation. Version 2.0.* IUCN Species Survival Commission, Gland.

IUDZG and the Captive Breeding Specialist Group of IUCN/SSC (1993). *The World Zoo Conservation Strategy: the role of the zoos and aquaria of the world in global conservation*. www.waza.org/conservation/wczs.php (accessed 21 October 2003).

Jachowski, D. S. and Pizzaras, C. (2005). Introducing an innovative semi-captive environment for the Philippine tarsier (*Tarsius syrichta*). *Zoo Biology*, 24, 101–109.

Jain, N., Santymire, R. M. and Wark, J. D. (2021). Evaluating physiological and behavioural responses to social changes and construction in two zoo-housed female giraffes. *Journal of Zoo and Aquarium Research*, 9, 228–238.

Jakob-Hoff, R., Harley, D., Magrath, M., Lancaster, M. L. and Kuchling, G. (2015). Advances in the contribution of zoos to reintroduction programs. In Armstrong, D., Hayward, M., Moro, D. and Seddon, P. (eds.), *Advances in Reintroduction Biology in Australian and New Zealand Fauna*. CSIRO Publishing, Melbourne, pp. 201–215.

Jakob-Hoff, R., Kingan, M., Fenemore, C., et al. (2019). Potential impact of construction noise on selected animal species. *Animals*, 9, 504. https://doi.org/10.3390/ani9080504.

James, C., Nicholls, A., Freeman, M., Hunt, K. and Brereton, J. E. (2021). Should zoo foods be chopped: macaws for consideration. *Journal of Zoo and Aquarium Research*, 9, 200–207.

Jang, Y. H., Lee, S. J., Lim, J. G., et al. (2008). The rate of *Salmonella* spp. infection in zoo animals at Seoul Grand Park, Korea. *Journal of Veterinary Science*, 9, 177–181.

Jansson, M., Amundin, M. and Laikre, L. (2015). Genetic contribution from a zoo population can increase genetic variation in the highly inbred wild Swedish wolf population. *Conservation Genetics*, 16, 1501–1505.

Januszczak, I. S., Bryant, Z., Tapley, B., et al. (2016). Is behavioural enrichment always a success? Comparing food presentation strategies in an insectivorous lizard (*Plica plica*). *Applied Animal Behaviour Science*, 183, 95–103.

Jarvis, C. (1967). The value of zoos for science and conservation. *Oryx*, 9, 127–136.

Jarvis, C. and Morris, D. (eds.) (1960). *International Zoo Yearbook, Vol. 2, Elephants, Hippopotamuses and Rhinoceroses in Captivity*. Zoological Society of London, London.

Javed, R. M. and Khan, B. N. (2005). Saving river dolphin: report of an in situ conservation project of Lahore Zoo carried out in association with Sindh Wildlife Department, Pakistan. *Zoos' Print*, 6, 9–11.

Jennings, H. (1996). Focus groups with zoo visitors who are blind or have low vision: how can we deliver our message to those who cannot see signs? *Visitor Studies*, 9, 171–175.

Jensen, E. (2014). Evaluating children's conservation biology learning at the zoo. *Conservation Biology*, 28, 1004–1011.

Jensvold, M. L. A. (2008). Chimpanzee (*Pan troglodytes*) responses to caregiver use of chimpanzee behaviours. *Zoo Biology*, 27, 345–359.

Jepson, P. D. and Deaville, R. (2008). *Investigation of the Common Dolphin Mass Stranding Event in Cornwall, 9th June 2008*. UK Cetacean Strandings Investigation Programme, Zoological Society of London, London.

Jewell, P. A. (1976). Should domestic animals be kept in zoos? *International Zoo Yearbook*, 16, 249–251.

Jewgenow, K., Wiedemann, C., Bertelsen, M. F. and Ringleb, J. (2011). Cryopreservation of mammalian ovaries and oocytes. *International Zoo Yearbook*, 45, 124–132.

Jodidio, R. L. (2020). The Animal Welfare Act is lacking: how to update the federal statute to improve zoo animal welfare. *Golden Gate University Environmental Law Journal*, 12, 53.

Johnson, W. (1994). *The Rose-Tinted Menagerie*. Heretic Books Ltd, Farnham.

Johnston, R. J. (1998a). Estimating demand for wildlife viewing in zoological parks: an exhibit-specific, time allocation approach. *Human Dimensions of Wildlife*, 3, 16–33.

Johnston, R. J. (1998b). Exogenous factors and visitor behaviour: a regression analysis of viewing time. *Environment and Behavior*, 31, 322–347.

Jones, C. G. (1984). The captive management and biology of the Mauritius kestrel: *Falco punctatus*. *International Zoo Yearbook*, 23, 76–82.

Jones, M. L. (1962). Mammals in captivity: primate longevity. *Laboratory Primate Newsletter*, 1, 3–13.

Jones, R. C. (2013). Science, sentience, and animal welfare. *Biology & Philosophy*, 28, 1–30.

Jones, R. W. (1997). The sight of creatures strange to our clime: London Zoo and the consumption of the exotic. *Journal of Victorian Culture*, 2, 1–26.

Jose, D., Bradfield, K., Power, V. and Lambert, C. (2011). Predator awareness training at Perth Zoo: a review. *Thylacinus*, 35(3), 2–7.

Jule, K. R., Leaver, L. A. and Lea, S. E. (2008). The effects of captive experience on reintroduction survival in carnivores: a review and analysis. *Biological Conservation*, 141, 355–363.

Jule, K. R., Lea, S. E. G. and Leaver, L. A. (2009). Using a behaviour discovery curve to predict optimal observation time. *Behaviour*, 146, 1531–1542.

Kagan, R. L. (2013). Challenges of zoo animal welfare: the path from good care to great welfare – keynote. *Journal of Applied Animal Welfare Science*, 16, 381.

Kamel, A. A. and Abdel-Latef, G. K. (2021). Prevalence of intestinal parasites with molecular detection and identification of *Giardia duodenalis* in fecal samples of mammals, birds and zookeepers at Beni-Seuf Zoo, Egypt. *Journal of Parasitic Diseases*, 45, 695–705.

Kane, L. P., O'Connor, M. R. and Papich, M. G. (2022). Pharmacokinetics of a single dose of intramuscular and oral meloxicam in yellow stingrays (*Urobatis jamaicensis*). *Journal of Zoo and Wildlife Medicine*, 53, 153–158.

Kappelhof, J. and Weerman, J. (2020). The development of the red panda *Ailurus fulgens* EEP: from a failing captive population to a stable population that provides effective support to in situ conservation. *International Zoo Yearbook*, 54, 102–112.

Karanikola, P., Panagopoulos, T., Tampakis, S. and Tampakis, A. (2020). Visitor preferences and satisfaction in Attica Zoological Park, Greece. *Heliyon*, 6(9), e04935.

Karstad, L. and Sileo, L. (1971). Causes of death in captive wild waterfowl in the Kortright Waterfowl Park, 1967–1970. *Journal of Wildlife Diseases*, 7, 236–241.

Katz, B. (2019). A German circus uses stunning holograms instead of live animal performers. www.smithsonianmag.com/smart-news/german-circus-uses-stunning-holograms-instead-live-animal-performers-180972376/ (accessed 10 September 2019).

Kaufman, A. B., Bashaw, M. J. and Maple, T. L. (eds.) (2019). *Scientific Foundations of Zoos and Aquariums: Their Role in Conservation and Research.* Cambridge University Press, Cambridge.

Kawata, K. (2008). Zoo animal feeding: a natural history viewpoint. Zootierfütterung: ein naturgeschichtlicher Standpunkt. *Der Zoologische Garten*, 78, 17–42.

Keiper, R. R. (1969). Causal factors of stereotypies in caged birds. *Animal Behaviour*, 17, 114–119.

Keller, M. (2017). Feeding live invertebrate prey in zoos and aquaria: are there welfare concerns? *Zoo Biology*, 36, 316–322.

Kellert, S. R. and Dunlap, J. (1989). *Informal Learning at the Zoo: A Study of Attitude and Knowledge Impacts.* Zoological Society of Philadelphia, Philadelphia.

Kelling, N. and Kelling, A. (2014). Zooar: zoo based augmented reality signage. *Proceedings of the Human Factors and Ergonomics Society Annual Meeting*, 58, 1099–1103.

Kelly, K. R. and Ocular, G. (2021). Family smartphone practices and parent–child conversations during informal science learning at an aquarium. *Journal of Technology in Behavioral Science*, 6, 114–123.

Kelly, L. A. D., Luebke, J. F., Clayton, S., et al. (2014). Climate change attitudes of zoo and aquarium visitors: implications for climate literacy education. *Journal of Geoscience Education*, 62, 502–510.

Kendall, S. K. (2003). The American Livestock Breeds Conservancy and the Rare Breeds Survival Trust. *Journal of Agricultural & Food Information*, 5, 3–10.

Kerr, K. C. (2021). Zoo animals as 'proxy species' for threatened sister taxa: defining a novel form of species surrogacy. *Zoo Biology*, 40, 65–75.

Key, B. (2016). Why fish do not feel pain. *Animal Sentience*, 3(1), 2016003.

Kiers, A., Klarenbeek, A., Mendelts, B., Van Soolingen, D. and Koëter, G (2008). Transmission of *Mycobacterium pinnipedii* to humans in a zoo with marine mammals. *The International Journal of Tuberculosis and Lung Disease*, 12, 1469–1473.

Kim, J.-Y., Oh, S. H., Kim, Y. B., et al. (2010). Capture myopathy in a red-necked wallaby (*Macropus rufogriseus*). *Journal of Veterinary Clinics*, 27, 198–201.

Kim, S. H. (2008). Animal escape artists. *Risk Management*, 55, 30.

Kim, Y. I., Kim, S. G., Kim, S. M., et al. (2020). Infection and rapid transmission of SARS-CoV-2 in ferrets. *Cell Host & Microbe*, 27, 704–709.

Kinder, J. M. (2013). Zoo animals and modern war: captive casualties, patriotic citizens, and good soldiers. In: Hediger, R. (ed.). *Animals in War: Studies of Europe and North America.* Brill, Leiden, pp. 45–75.

King, C. E. (1994). Black stork *Ciconia nigra*: management in Europe. *International Zoo Yearbook*, 33, 49–54.

King, N. E. and Mellen, J. D. (1994). The effects of early experience on adult copulatory behavior in zoo-born chimpanzees (*Pan troglodytes*). *Zoo Biology*, 13, 51–59.

King, S. L. and Janik, V. M. (2013). Bottlenose dolphins can use learned vocal labels to address each other. *Proceedings of the National Academy of Sciences*, 110, 13216–13221.

King, T., Boyen, E. and Muilerman, S. (2003). Variation in reliability of measuring behaviours of reintroduced orphan gorillas. *International Zoo News*, 50, 288–297.

Kinville, C. (1968). Oklahoma City Zoo's education programme. *International Zoo Yearbook*, 8, 171–172.

Kirchshofer, R. (1968). Frankfurt Zoo's education programme. *International Zoo Yearbook*, 8, 169–171.

Kisling, V. N. (1993). Libraries and archives in the historical and professional development of American zoological parks. *Libraries & Culture*, 28, 247–265.

Kisling, V. N. (ed.) (2000). *Zoo and Aquarium History: Ancient Animal Collections to Zoological Gardens*. CRC Press, Boca Raton.

Kitchener, A. C. (2004). The problem of old bears in zoos. *International Zoo News*, 51, 282–293.

Kitchener, A. C. (2020). Small carnivorans, museums and zoos. *International Zoo Yearbook*, 54, 43–52.

Kitchener, A. C. and Conroy, J. W. H. (1997). The history of the Eurasian Beaver *Castor fiber* in Scotland. *Mammal Review*, 27, 95–108.

Kleespies, M. W., Gübert, J., Popp, A., et al. (2020). Connecting high school students with nature: how different guided tours in the zoo influence the success of extracurricular educational programs. *Frontiers in Psychology*, 11, 1804. https://doi.org/10.3389/fpsyg .2020.0180.

Kleespies, M. W., Montes, N. Á., Bambach, A. M., et al. (2021). Identifying factors influencing attitudes towards species conservation: a transnational study in the context of zoos. *Environmental Education Research*, 27, 1421–1439.

Kleiman, D. G. (1983). Ethology and reproduction of captive giant pandas (*Ailuropoda melanoleuca*). *Zeitschrift für Tierpsychologie*, 62, 1–46.

Kleiman, D. G. (1989). Reintroduction of captive mammals for conservation. *BioScience*, 39, 152–161.

Kleiman, D. G. (1992). Behavior research in zoos: past, present and future. *Zoo Biology*, 11, 301–312.

Kleiman, D. G. and Mallinson, J. J. (1998). Recovery and management committees for lion tamarins: partnerships in conservation planning and implementation. *Conservation Biology*, 12, 27–38.

Knežević, M., Žučko, I. and Ljuština, M. (2016). Who is visiting the Zagreb Zoo: visitors' characteristics and motivation. *Sociologija i prostor: časopis za istraživanje prostornoga i sociokulturnog razvoja*, 54, 169–184.

Knight, M. (1967). *How to keep an elephant*. Wolfe Publishing Ltd, London.

Knowles, T. G., Warriss, P. D. and Vogel, K. (2014). Stress physiology of animals during transport. In Grandin, T. (ed.), *Livestock Handling and Transport* (4th ed.). CABI. Wallingford, pp. 399–420.

Kohler, I. V., Preston, S. H. and Lackey, L. B. (2006). Comparative mortality levels among selected species of captive animals. *Demographic Research*, 15, 413–434.

Koldewey, H., Christie, S., Curnick, D., et al. (2020). A response to Welden et al. (2020): The contributions of EAZA zoos and aquaria to peer-reviewed scientific research. *Journal of Zoo and Aquarium Research*, 8, 133–138.

Kolmstetter, C., Munson, L. and Ramsay, E. C. (2000). Degenerative spinal disease in large felids. *Journal of Zoo and Wildlife Medicine*, 31, 15–19.

Koot, S., Kapteijn, C. M., Huiskes, R. H. and Kranendonk, G. (2016). A note on the social compatibility of an all-male group of Hamadryas baboons (*Papio hamadryas*). *Journal of Zoo and Aquarium Research*, 4, 7–13.

Kowalska, Z. (1969). A note on bear hybrids *Thalarctos maritimus* × *Ursus arctos* at Łódź Zoo. *International Zoo Yearbook*, 9, 89–89.

Kragness, B. J., Graham, J. E., Bedenice, D., Restifo, M. M. and Boudrieau, R. J. (2016). Surgical correction of a cervical spinal fracture in a Bennett's wallaby (*Macropus rufogriseus*). *Journal of Zoo and Wildlife Medicine*, 47, 379–382.

Krajewska, M., Załuski, M., Zabost, A., et al. (2015a). Tuberculosis in antelopes in a zoo in Poland: problem of public health. *Polish Journal of Microbiology*, 4, 405–407.

Krajewska, M., Czujkowska, A., Weiner, M., Lipiec, M. and Szulowski, K. (2015b). Avian tuberculosis in a captive cassowary (*Casuarius casuarius*). *Bulletin of the Veterinary Institute Pulawy*, 59, 483–487.

Krajewska-Wedzina, M., Augustynowicz-Kopec, E., Weiner, M. and Szulowski, K. (2018). Treatment for active tuberculosis in giraffe (*Giraffa camelopardalis*) in a zoo and potential consequences for public health: case report. *Annals of Agricultural and Environmental Medicine*, 25, 593–595.

Kreger, M. D., Estevez, I., Hatfield, J. S. and Gee, G. F. (2004). Effects of rearing treatment on the behavior of captive whooping cranes (*Grus americana*). *Applied Animal Behaviour Science*, 89, 243–261.

Krief, S., Jamart, A. and Hladik, C.-M. (2004). On the possible adaptive value of coprophagy in free-ranging chimpanzees. *Primates*, 45, 141–145.

Kroshko, J., Clubb, R., Harper, L., et al. (2016). Stereotypic route tracing in captive Carnivora is predicted by species-typical home range sizes and hunting styles. *Animal Behaviour*, 117, 197–209.

Kuhar, C. W. (2008). Group differences in captive gorillas' reaction to large crowds. *Applied Animal Behaviour Science*, 110, 377–385.

Kuhar, C. W., Bettinger, T. L., Sironen, A. L., Shaw, J. H. and Lasley, B. L. (2003). Factors affecting reproduction in zoo-housed Geoffroy's tamarins (*Saguinus geoffroyi*). *Zoo Biology*, 22, 545–559.

Kuhar, C. W., Miller, L. J., Lehnardt, J., et al. (2010). A system for monitoring and improving animal visibility and its implications for zoological parks. *Zoo Biology*, 29, 68–79.

Kuhar, C. W., Fuller, G. A. and Dennis, P. M. (2013). A survey of diabetes prevalence in zoo-housed primates. *Zoo Biology*, 32, 63–69.

Kumar, A., Rai, U., Roka, B., Jha, A. K. and Reddy, P. A. (2016). Genetic assessment of captive red panda (*Ailurus fulgens*) population. *SpringerPlus*, 5(1), 1750.

Kumar, V., Pruthvishree, B., Pande, T., et al. (2020). SARS-CoV-2 (COVID-19): zoonotic origin and susceptibility of domestic and wild animals. *Journal of Pure and Applied Microbiology*, 14 (Suppl.1), 741–747.

Kurt, F. (1995). Asian elephants *(Elephas maximus)* in captivity and the role of captive propagation for maintenance of the species. In: Spooner, N. G. and Whitear, J. A. (eds.), *Proceedings of the Eighth UK Elephant Workshop*. North of England Zoological Society, Chester Zoo, Chester, pp. 69–96.

Kvapil, P., Nemec, A., Zadravec, M. and Račnik, J. (2018). Oral and dental examination findings in a family of zoo suricates (*Suricata suricatta*). *Journal of Veterinary Dentistry*, 35, 114–120.

KWS (2022). Conservation fees 1st January, 2022 to 30th June, 2022. Kenya Wildlife Service. www.kws.go.ke.

Laméris, D. W., Staes, N., Salas, M., et al. (2021). The influence of sex, rearing history, and personality on abnormal behaviour in zoo-housed bonobos (*Pan paniscus*). *Applied Animal Behaviour Science*, 234, 105178.

Lamglait, B. (2018). Retrospective study of mortality in captive Struthioniformes in a French zoo. *Journal of Zoo and Wildlife Medicine*, 49, 967–976.

Lamglait, B. (2020). A retrospective review of causes of mortality in captive springboks (*Antidorcas marsupialis*) at the Réserve Africaine de Sigean, France, from 1990 to 2015. *Journal of Zoo and Aquarium Research*, 8. https://doi.org/10.19227/jzar.v8i3.344.

Lamglait, B., Moresco, A., Couture, É. L., Ferrell, S. T. and Lair, S. (2022). Vaginal foreign bodies in six nonhuman primates with underlying pathological conditions. *Zoo Biology*. https://doi.org/10.1002/zoo.21689.

Landová, E., Poláková, P., Rádlová, S., et al. (2018). Beauty ranking of mammalian species kept in the Prague Zoo: does beauty of animals increase the respondents' willingness to protect them? *The Science of Nature*, 105(11), 1–14.

Langman, V. A., Roberts, T. J., Black, J., et al. (1995). Moving cheaply: energetics of walking in the African elephant. *Journal of Experimental Biology*, 198, 629–632.

Langman, V. A., Rowe, M., Forthman, D., et al. (1996). Thermal assessment of zoological exhibits I: sea lion enclosure at the Audubon Zoo. *Zoo Biology*, 15, 403–411.

Langman, V. A., Rowe, M., Forthman, D., et al. (2003). Quantifying shade using a standard environment. *Zoo Biology*, 22, 253–260.

Langman, V. A., Rowe, M. F., Roberts, T. J., Langman, N. V. and Taylor, C. R. (2012). Minimum cost of transport in Asian elephants: do we really need a bigger elephant? *Journal of Experimental Biology*, 215, 1509–1514.

Lankard, J. R. (ed.) (2001). *AZA Annual Report on Conservation and Science 1999–2000. Volume III: Member Institution Publications*. American Zoo and Aquarium Association, Silver Spring.

Lanza, R. P., Cibelli, J. B., Diaz, F., et al. (2000). Cloning of an endangered species (*Bos gaurus*) using interspecies nuclear transfer. *Cloning*, 2, 79–90.

Latham, N. R. and Mason, G. J. (2008). Maternal deprivation and the development of stereotypic behaviour. *Applied Animal Behaviour Science*, 110, 84–108.

Lauderdale, L. K., Shorter, K. A., Zhang, D., et al. (2021). Bottlenose dolphin habitat and management factors related to activity and distance traveled in zoos and aquariums. *PLoS ONE*, 18, e0250687. http://doi.org/10.1371/Journal.pone.0250687.

Law, G. and Kitchener, A. C. (2020). Twenty years of the tiger feeding pole: review and recommendations. *International Zoo Yearbook*, 54, 174–190.

Law, S., Prankel, S., Schwitzer, C. and Dutton, J. (2021). Inter-species interactions involving *Lemur catta* housed in mixed-species exhibits in UK zoos. *Journal of Zoo and Aquarium Research*, 9, 247–258.

Learmonth, M. J. (2020). Human–animal interactions in zoos: what can compassionate conservation, conservation welfare and duty of care tell us about the ethics of interacting, and avoiding unintended consequences. *Animals*, 10, 2037. https://doi.org/10.3390/ani10112037.

Learmonth, M. J., Sherwen, S. and Hemsworth, P. H. (2018). The effects of zoo visitors on quokka (*Setonix brachyurus*) avoidance behaviour in a walk-through exhibit. *Zoo Biology*, 37, 223–228.

Leeds, A., Boyer, D., Ross, S. R. and Lukas, K. E. (2015). The effects of group type and young silverbacks on wounding rates in Western lowland gorilla (*Gorilla gorilla gorilla*) groups in North American zoos. *Zoo Biology*, 34, 296–304.

Lees, C. M. and Wilcken, J. (2009). Sustaining the Ark: the challenges faced by zoos in maintaining viable populations. *International Zoo Yearbook*, 43, 6–18.

Lefeuvre, M., Gouat, P., Mulot, B., Cornette, R. and Pouydebat, E. (2022). Analogous laterality in trunk movements in captive African elephants: a pilot study. *Laterality*, 27, 101–126.

Lefevre, C. E., Wilson, V. A., Morton, F. B., et al. (2014). Facial width-to-height ratio relates to alpha status and assertive personality in capuchin monkeys. *PLoS ONE*, 9(4), e93369.

Leggett, K. (2004). Coprophagy and unusual thermoregulatory behaviour in desert-dwelling elephants of north-western Namibia. *Pachyderm*, 36, 113–115.

Lehman, S. M., Ratsimbazafy, J., Rajaonson, A. and Day, S. (2006). Decline of *Propithecus diadema edwardsi* and *Varecia variegata variegata* (Primates: Lemuridae) in south-east Madagascar. *Oryx*, 40, 108–111.

Lehnhardt, J. (1991). Elephant handling: a problem of risk management and resource allocation. *AAZPA Regional Conference Proceedings*. American Association of Zoological Parks and Aquariums, Wheeling, pp. 569–575.

Lehocká, K., Hanusová, J. and Kadlečík, O. (2018). Genetic diversity of Barbary lion based on genealogic analysis. *Acta Fytotechnica et Zootechnica*, 21, 113–118.

Leighty, K. A., Soltis, J. and Savage, A. (2010). GPS assessment of the use of exhibit space and resources by African elephants (*Loxodonta africana*). *Zoo Biology*, 29, 210–220.

Leinwand, J. G., Moyse, J. A., Hopper, L. M., Leahy, M. and Ross, S. R. (2021). The use of biofloors in great ape zoo exhibits. *Journal of Zoo and Aquarium Research*, 9, 41–48.

Leonardi, R., Buchanan-Smith, H. M., Dufour, V., MacDonald, C. and Whiten, A. (2010). Living together: behavior and welfare in single and mixed species groups of capuchin (*Cebus apella*) and squirrel monkeys (*Saimiri sciureus*). *American Journal of Primatology: Official Journal of the American Society of Primatologists*, 72, 33–47.

Leong, K. M., Terrell, S. P. and Savage, A. (2004). Causes of mortality in captive cotton-top tamarins (*Saguinus oedipus*). *Zoo Biology*, 23, 127–137.

Lernould, J. M., Kierulff, M. C. M. and Canale, G. (2012). Yellow-breasted capuchin *Cebus xanthosternos*: support by zoos for its conservation – a success story. *International Zoo Yearbook*, 46, 71–79.

Less, E. H., Lukas, K. E., Kuhar, C. W. and Stoinski, T. S. (2010). Behavioral response of captive Western lowland gorillas (*Gorilla gorilla gorilla*) to the death of silverbacks in multi-male groups. *Zoo Biology*, 29, 16–29.

Lethmate, J. (1979). Instrumental learning of zoo orang-utans. *Journal of Human Evolution*, 8, 741–744.

Leuck, B. E. (1977). Differential use of space by eight species of birds in a free-flight zoological park aviary. *Applied Animal Ethology*, 3, 105–126.

Levá, M. and Pluháček, J. (2020). Does social facilitation affect suckling behaviour in zebras? *Behavioural Processes*, 185, 104347.

Levin, A. (2015). Zoo animals as specimens, zoo animals as friends: the life and death of Marius the giraffe. *Environmental Philosophy*. https://doi.org/10.5840/envirophil201552622.

Lewbart, G. A. (ed.) (2022). *Invertebrate Medicine*. Wiley-Blackwell, Chichester.

Lewis, K., Descovich, K. and Jones, M. (2017). Enclosure utilisation and activity budgets of disabled Malayan sun bears (*Helarctos malayanus*). *Behavioural Processes*, 145, 65–72.

Li, B. V. and Pimm, S. L. (2016). China's endemic vertebrates sheltering under the protective umbrella of the giant panda. *Conservation Biology*, 30, 329–339.

Li, C., Jiang, Z., Tang, S. and Zeng, Y. (2007). Influence of enclosure size and animal density on fecal cortisol concentration and aggression in Père David's deer stags. *General and Comparative Endocrinology*, 151, 202–209.

Li, M. F., Swaisgood, R. R., Owen, M. A., et al. (2022). Consequences of nescient mating: artificial insemination increases cub rejection in the giant panda (*Ailuropoda melanoleuca*). *Applied Animal Behaviour Science*, 247, 105565. https://doi.org/10.1016/j.applanim.2022.105565.

Li, S., McShea, W. J., Wang, D., et al. (2020). Retreat of large carnivores across the giant panda distribution range. *Nature Ecology & Evolution*, 4, 1327–1331.

Lima, G. L., Barros, F. F., Costa, L. L., et al. (2009). Determination of semen characteristics and sperm cell ultrastructure of captive coatis (*Nasua nasua*) collected by electroejaculation. *Animal Reproduction Science*, 115, 225–230.

Lind, A. L., Lai, Y. Y., Mostovoy, Y., et al. (2019). Genome of the Komodo dragon reveals adaptations in the cardiovascular and chemosensory systems of monitor lizards. *Nature Ecology & Evolution*, 3, 1241–1252.

Line, S. W., Morgan, K. N. and Markowitz, H. (1991). Simple toys do not alter the behavior of aged rhesus monkeys. *Zoo Biology*, 10, 473–484.

Linseele, V., Van Neer, W., Friedman, R. (2009). Special animals from a special place? The fauna from HK29A at predynastic Hierakonpolis. *Journal of the American Research Center in Egypt*, 45, 105–136.

Liptovszky, M., Dobbs, P., Moittie, S., et al. (2021). Assessing the educational value of a zoo placement for veterinary students: a report on student feedback and perceptions. *Journal of Veterinary Medical Education*, 49, 236–240.

Litchfield, P. (2005). Leaders and matriarchs: a new look at elephant social hierarchies. *International Zoo News*, 52, 338–339.

Litwak, J. (1996). Visitors learn more from labels that ask questions. *Current Trends in Audience Research*, 10, 40–50.

Live Blackpool. (2022). History of Blackpool Tower. www.liveblackpool.info/about/history/history-of-blackpool-tower (accessed 11 June 2022).

Llewellyn, T. and Rose, P. E. (2021). Education is entertainment? Zoo science communication on YouTube. *Journal of Zoological and Botanical Gardens*, 2, 250–264.

Lloyd, M., Walsh, N. D. and Johnson, B. (2021). Investigating visitor activity on a safari drive. *Journal of Zoological and Botanical Gardens*, 2, 576–585.

Lochmiller, R. L. and Grant, W. E. (1982). Intraspecific aggression results in death of a collared peccary. *Zoo Biology*, 1, 161–162.

Loeb, J. (2020). Keeping dangerous pets. *Veterinary Record*, 186, 333.

Loeb, J. and Leeming, S. (2020). Dangerous snake laws need constricting. *Veterinary Record*, 186, 336–337.

Loeding, E., Thomas, J., Bernier, D. and Santymire, R. (2011). Using faecal hormone and behavioral analysis to evaluate the introduction of two sable antelope at Lincoln Park Zoo. *Journal of Applied Animal Welfare Science*, 14, 220–246.

Long, J. L. (1981). *Introduced Birds of the World: The Worldwide History, Distribution and Influence of Birds Introduced to New Environments*. Food and Agriculture Organisation, Rome.

Loomis, R. (1987). *Museum Visitor Evaluation: New Tool for Management*. American Association for State and Local History, Nashville.

Louv, R. (2005). *Last Child in the Woods: Nature Deficit Disorder*. Algonquin Books, Chapel Hill.

Lowenstine, L. J. and Montali, R. J. (2006). Historical perspective and future directions in training of veterinary pathologists with an emphasis on zoo and wildlife species. *Journal of Veterinary Medical Education*, 33, 338–345.

Lücker, H. (2003). Haltung von schwach-elektrischen Fischen in Schauaquarien eine "spannungsvolle" Sache. *Zoologische Garten*, 73, 284–295.

Lugosi, Z. and Lee, P. (2021). A case study exploring the use of virtual reality in the zoo context. *Animal Behavior and Cognition*, 8, 576–588.

Luis, P. L. J., Milanes, S., Ramón, D., et al. (2009). Evaluation of semen obtained by electroejaculation in nonhuman primates Anubis baboon (*Papio anubis*) held at the national zoo Cuba. *REDVET*, 10(10), 100916.

Lukas, K. E. and Ross, S. R. (2005). Zoo visitor knowledge and attitudes towards gorillas and chimpanzees. *The Journal of Environmental Education*, 36, 33–48.

Lukas, K. E., Hoff, M. P. and Maple, T. L. (2003). Gorilla behaviour in response to systematic alternation between enclosures. *Applied Animal Behaviour Science*, 81, 367–386.

Lynggaard, C., Bertelsen, M. F., Jensen, C. V., et al. (2022). Airborne environmental DNA for terrestrial vertebrate community monitoring. *Current Biology*, 32. https://doi.org/10.1016/j.cub2021.12.014.

Macdonald, C. and Whiten, A. (2011). The 'Living Links to Human Evolution' research centre in Edinburgh Zoo: a new endeavour in collaboration. *International Zoo Yearbook*, 45, 7–17.

Macdonald, D. W., Tattersall, F. H., Rushton, S., et al. (2000). Reintroducing the beaver (*Castor fiber*) to Scotland: a protocol for identifying and assessing suitable release sites. *Animal Conservation*, 3, 125–133.

Maceda-Veiga, A., Domínguez-Domínguez, O., Escribano-Alacid, J. and Lyons, J. (2016). The aquarium hobby: can sinners become saints in freshwater fish conservation? *Fish and Fisheries*, 17, 860–874.

Mäekivi, N. and Maran, T. (2016). Semiotic dimensions of human attitudes towards other animals: a case of zoological gardens. *Sign Systems Studies*, 44, 209–230.

Mahat, T. J. and Koirala, M. (2006). Economic valuation of the Central Zoo of Nepal. *9th Biennial Conference of the International Society for Ecological Economics*, pp. 15–18.

Malamud, R. (1998). *Reading Zoos: Representations of Animals and Captivity*. New York University Press, New York.

Malamud, R., Broglio, R., Marino, L., Lilienfeld, S. O. and Nobis, N. (2010). Do zoos and aquariums promote attitude change in visitors? A critical evaluation of the American zoo and aquarium study. *Society & Animals*, 18, 126–138.

Malecki, I. A., Rybnik, P. K. and Martin, G. B. (2008). Artificial insemination technology for ratites: a review. *Australian Journal of Experimental Agriculture*, 48, 1284–1292.

Mallapur, A., Qureshi, Q. and Chellam, R. (2002). Enclosure design and space utilization by Indian leopards (*Panthera pardus*) in four zoos in Southern India. *Journal of Applied Animal Welfare Science*, 5, 111–124.

Mallapur, A., Waran, N. and Sinha, A. (2005). Use of enclosure space by captive lion-tailed macaques (*Macaca silenus*) housed in Indian Zoos. *Journal of Applied Animal Welfare Science*, 8, 175–186.

Mallory, H. S., Howard, A. F. and Weiss, M. R. (2016). Timing of environmental enrichment affects memory in the house cricket, *Acheta domesticus*. *PLoS ONE*, 11(4), e0152245.

Maloney, M. A., Meiers, S. T., White, J. and Romano, M. A. (2006). Effects of three food enrichment items on the behavior of the black lemurs (*Eulemur macaco macaco*) and ringtail lemurs(*Lemur catta*) at the Henson Robinson Zoo, Springfield, Illinois. *Journal of Applied Animal Welfare Science*, 9, 111–127.

Mancera, K. F., Murray, P. J., Gao, Y. N., Lisle, A. and Phillips, C. J. C. (2014). The effects of simulated transport on the behaviour of eastern blue-tongued lizards (*Tiliqua scincoides*). *Animal Welfare*, 23, 239–249.

Maple, T. L. and Perdue, B. M. (2013). *Zoo Animal Welfare*. Springer, Berlin.

Mar, K. U., Maung, M., Thein, M., et al. (1995). Electroejaculation and semen characteristics in Myanmar timber elephants. In: Daniel, J. C. and Datye, H. (eds.), *A Week with Elephants: Proceedings of the International Seminar on the Conservation of Asian elephant*. Bombay Natural History Society, Oxford University Press, Oxford, pp. 473–482.

Marcellini, D. L. and Jenssen, T. A. (1988). Visitor behavior in the National Zoo's reptile house. *Zoo Biology*, 7, 329–338.

Marešová, J. and Frynta, D. (2008). Noak's Ark is full of common species attractive to humans: the case of boid snakes in zoos. *Ecological Economics*, 64, 554–558.

Margodt, K. (2000). *The Welfare Ark: Suggestions for a Renewed Policy in Zoos*. VUB Press, Brussels.

Marinkovich, M., Wallace, C., Morris, P. J., Rideout, B. and Pye, G. W. (2016). Lessons from a retrospective analysis of a 5-yr period of preshipment testing at San Diego Zoo: a risk-based approach to preshipment testing may benefit animal welfare. *Journal of Zoo and Wildlife Medicine*, 47, 297–300.

Marker, L., Honig, M., Pfeiffer, L., Kuypers, M. and Gervais, K. (2021). Captive rearing of orphaned African wild dogs (*Lycaon pictus*) in Namibia: a case study. *Zoo Biology*. https://doi.org/10.1002/zoo.21662.

Markowitz, H. (1982). *Behavioral Enrichment at the Zoo*. Van Nostrand Reinhold, New York.

Markowitz, H. (2011). *Enriching Animal Lives*. Mauka Press, USA.

Markwell, K., Weiler, B., Skibins, J. C. and Saunders, R. (2019). Sympathy for the devil? Uncovering inhibitors and enablers of emotional engagement between zoo visitors and the Tasmanian devil, *Sarcophilus harrisi*. *Visitor Studies*, 22, 84–103.

Maroldo, G. K. (1982). Zoo animal protection in the event of thermonuclear catastrophes. *Zoo Biology*, 1, 363–369.

Marolf, B., McElligott, A. G. and Müller, A. E. (2007). Female social dominance in two eulemur species with different social organizations. *Zoo Biology*, 26, 201–214.

Marranzino, A. (2013). The use of positive reinforcement in training zebra sharks (*Stegostoma fasciatum*). *Journal of Applied Animal Welfare Science*, 16, 239–253.

Marrow, J. C., Woc-Colburn, M., Hayek, L. A. C., Marker, L. and Murray, S. (2015). Comparison of two α2-adrenergic agonists on urine contamination of semen collected by electroejaculation in captive and semi–free-ranging cheetah (*Acinonyx jubatus*). *Journal of Zoo and Wildlife Medicine*, 46, 417–420.

Marshall, C. E. (1984). Considerations for the cryopreservation of semen. *Zoo Biology*, 3, 343–356.

Marti-Colombas, M., Sánchez-Calabuig, M. J., Castaño, C., et al. (2022). Optimization of semen cryopreservation in black-footed (*Spheniscus demersus*) and gentoo (*Pygoscelis papua*) penguins using dimethylacetamide and dimethylsulphoxide. *Animal Reproduction Science*, 237, 106933.

Martin, P. and Bateson, P. (2007). *Measuring Behaviour: An Introductory Guide*. Cambridge University Press, Cambridge.

Martinez, S. G., Spiro, S., Guthrie, A., et al. (2022). Cholelithiasis in captive mountain chicken frogs (*Leptodactylus fallax*): diagnostic imaging and histopathological features. *Journal of Zoo and Wildlife Medicine*, 53, 19–30.

Martínez-de la Puente, J., Soriguer, R., Senar, J. C., et al. (2020). Mosquitoes in an urban zoo: identification of blood meals, flight distances of engorged females, and avian malaria infections. *Frontiers in Veterinary Science*, 7, 460. http://doi.org/10.3389/fvets.2020.00460.

Martínez-Torres, M., Álvarez-Rodríguez, C., Luis, J. and Sánchez-Rivera, U. Á. (2019). Electroejaculation and semen evaluation of the viviparous lizard *Sceloporus torquatus* (Squamata: Phrynosomatidae). *Zoo Biology*, 38, 393–396.

Mary, S., Benbow, P. and Hallman, B. C. (2008). Reading the zoo map: cultural heritage insights from popular cartography. *International Journal of Heritage Studies*, 14, 30–42.

Maslak, R., Sergiel, A. and Hill, S. P. (2013). Some aspects of locomotory stereotypies in spectacled bears (*Tremarctos ornatus*) and changes in behavior after relocation and dental treatment. *Journal of Veterinary Behavior*, 8, 335–341.

Mason, G. J. and Veasey, J. S. (2010a). How should the psychological well-being of zoo elephants be objectively investigated? *Zoo Biology*, 29, 237–255.

Mason, G. J. and Veasey, J. S. (2010b). What do population-level welfare indices suggest about the well-being of zoo elephants? *Zoo Biology*, 29, 256–273.

Mason, P. (2007). Roles of the modern zoo: conflicting or complementary? *Tourism Review International*, 11, 251–263.

Masters, N. J., Burns, F. M. and Lewis, J. C. (2007). Peri-anaesthetic and anaesthetic-related mortality risks in great apes (Hominidae) in zoological collections in the UK and Ireland. *Veterinary Anaesthesia and Analgesia*, 34, 431–442.

Masui, M., Hiramatsu, H., Nose, N., et al. (1989). Successful artificial insemination in the giant panda (*Ailuropoda melanoleuca*) at Ueno Zoo. *Zoo Biology*, 8, 17–26.

Mather, F. (1878). Feeding of fishes in confinement. *Transactions of the American Fisheries Society*, 7, 67–72.

Mather, F. (1879). The management of public aquaria, with a plan for reducing their running expenses. *Transactions of the American Fisheries Society*, 8, 46–51.

Mather, J. A. and Anderson, R. C. (2007). Ethics and invertebrates: a cephalopod perspective. *Diseases of Aquatic Organisms*, 75, 119–129.

Matrai, E., Kwok, S. T., Boos, M. and Pogány, Á. (2022). Testing use of the first multi-partner cognitive enrichment devices by a group of male bottlenose dolphins. *Animal Cognition*. https://doi.org/10.1007/s10071-022-01605-9.

Mattison, S. (2012). Training birds and small mammals for medical behaviors. *Veterinary Clinics: Exotic Animal Practice*, 15, 487–499.

Maynard, L., Jacobson, S. K., Monroe, M. C. and Savage, A. (2020). Mission impossible or mission accomplished: do zoos' organizational missions influence conservation practices? *Zoo Biology*, 39, 304–314.

McAloose, D., Laverack, M., Wang, L., et al. (2020). From people to *Panthera*: natural SARS-CoV-2 infection in tigers and lions at the Bronx Zoo. *Host-Microbe Biology*, 11(5), e02220-20.

McCann, C. M. and Rothman, J. M. (1999). Changes in nearest-neighbor associations in a captive group of Western lowland gorillas after the introduction of five hand-reared infants. *Zoo Biology*, 18, 261–278.

McCrimmon, H. R. (1950). The reintroduction of Atlantic salmon into tributary streams of Lake Ontario. *Transactions of the American Fisheries Society*, 78, 128–132.

McEntire, M. S. and Sanchez, C. R. (2017). Multimodal drug therapy and physical rehabilitation in the successful treatment of capture myopathy in a lesser flamingo (*Phoeniconaias minor*). *Journal of Avian Medicine and Surgery*, 31, 232–238.

McGreevy, P. D., Cripps, P. J., French, N. P., Green, L. E. and Nicol, C. J. (1995). Management factors associated with stereotypic and redirected behaviour in the Thoroughbred horse. *Equine Veterinary Journal*, 27, 86–91.

McKay, G. M. (1973). Behavior and ecology of the Asian elephant in Southeastern Ceylon. *Smithsonian Contributions to Zoology*, 125, 1–113.

McKenna, K. (2019). Failed million pound bid for Scottish panda 'has been a disgrace'. *The Guardian*. www.theguardian.com/world/2019/dec/15/end-in-sight-for-edinburgh-panda-breeding-programme-that-shames-scotland (accessed 9 July 2021).

McLean, K. M., Schook, M. W. and Pye, G. W. (2021). Comparison between standard zoo quarantine practices and risk-based management of animal transfers: a retrospective analysis of avian acquisition morbidity and mortality (2013–2018). *Journal of Zoo and Wildlife Medicine*, 51, 1017–1020.

McPhee, M. E. (2002). Intact carcasses as enrichment for large felids: effects on on- and off-exhibit behaviors. *Zoo Biology*, 21, 37–47.

McPhee, M. E. (2004). Generations in captivity increases behavioral variance: considerations for captive breeding and reintroduction programs. *Biological Conservation*, 115, 71–77.

McRee, A. E., Tully, T. N., Nevarez, J. G., et al. (2018). Effect of routine handling and transportation on blood leukocyte concentrations and plasma corticosterone in captive Hispaniolan Amazon parrots (*Amazona ventralis*). *Journal of Zoo and Wildlife Medicine*, 49, 396–403.

Medina, C. (2010). Age, rearing history and relatedness as determinants of gorilla dominance behaviors in bachelor groups at Sedgwick County Zoo. Unpublished thesis, University of Colorado Boulder.

Meehan, C. L., Mench, J. A., Carlstead, K. and Hogan, J. N. (2016). Determining connections between the daily lives of zoo elephants and their welfare: an epidemiological approach. *PLoS ONE*, 11(7), e0158124.

Melfi, V. A. (2009). There are big gaps in our knowledge, and thus approach, to zoo animal welfare: a case for evidence-based zoo animal management. *Zoo Biology*, 28, 574–588.

Melfi, V. (2013). Is training zoo animals enriching? *Applied Animal Behaviour Science*, 147, 299–305.

Melfi, V., Skyner, L., Birke, L., et al. (2021). Furred and feathered friends: how attached are zookeepers to the animals in their care? *Zoo Biology*, 41. http://doi.org/10.1002/zoo.21656.

Mellen, J. D. (1997). *Minimum Husbandry Guidelines for Mammals; Small Felids*. American Association of Zoos and Aquariums, Silver Spring.

Meller, C. L., Croney, C. C. and Sheperdson, D. (2007). Effects of rubberized flooring on Asian elephant behaviour in captivity. *Zoo Biology*, 26, 51–61.

Mellish, S., Sanders, B., Litchfield, C. A. and Pearson, E. L. (2017*)*. An investigation of the impact of Melbourne Zoo's 'Seal-the-Loop' donate call-to-action on visitor satisfaction and behavior. *Zoo Biology*, 36, 237—242.

Mellish, S., Ryan, J. C., Pearson, E. L. and Tuckey, M. R. (2019a). Research methods and reporting practices in zoo and aquarium conservation-education evaluation. *Conservation Biology*, 33, 40–52.

Mellish, S., Pearson, E. L., McLeod, E. M., Tuckey, M. R. and Ryan, J. C. (2019b). What goes up must come down: an evaluation of a zoo conservation-education program for balloon litter on visitor understanding, attitudes, and behaviour. *Journal of Sustainable Tourism*, 27, 1393–1415.

Mellor, D. J. and Reid, C. S. W. (1994). Concepts of animal well-being and predicting the impact of procedures on experimental animals. In: Baker, R. M., Jenkin, G. and Mellor, D. J. (eds.), *Improving the Well-being of Animals in the Research Environment*. Australian and New Zealand Council for the Care of Animals in Research and Teaching, Glen Osmond, pp. 3–18.

Mellor, D. J., Patterson-Kane, E. and Stafford, K. J. (2009). *The Sciences of Animal Welfare*. Wiley-Blackwell, Oxford.

Mellor, E., Brilot, B. and Collins, S. (2018). Abnormal repetitive behaviour in captive birds: a Tinbergian review. *Applied Animal Behaviour Science*, 198, 109–120.

Melville, D. F., Crichton, E. G., Paterson-Wimberley, T. and Johnston, S. D. (2008). Collection of semen by manual stimulation and ejaculate characteristics of the black flying-fox (*Pteropus alecto*). *Zoo Biology*, 27, 159–164.

Mendl, M., Paul, E. S. and Chittka, L. (2011). Animal behaviour: emotion in invertebrates. *Current Biology*, 21, R463–R465.

Micheletta, J. and Waller, B. M. (2012). Friendship affects gaze following in a tolerant species of macaque, *Macaca nigra*. *Animal Behaviour*, 83, 459–467.

Mihailovic, Z., Savic, S., Damjanjuk, I., Stanojevic, A. and Milosevic, M. (2011). A case of fatal Himalayan black bear attack in the zoo. *Journal of Forensic Sciences*, 56, 806–809.

Mikota, S. K. (2006). Preventative health care and physical examination. In: Fowler, E. F. and Mikota, S. K. (eds.), *Biology, Medicine, and Surgery of Elephants*. Blackwell Publishing, Oxford, pp. 67–73.

Mikota, S. K., Larsen, R. S. and Montali, R. J. (2000). Tuberculosis in elephants in North America. *Zoo Biology*, 19, 393–403.

Miller, A. and Kuhar, C. W. (2008). Long-term monitoring of social behavior in a grouping of six female tigers (*Panthera tigris*). *Zoo Biology*, 27, 89–99.

Miller, L. J., Lauderdale, L. K., Mellon, J. D., Walsh, M. T. and Granger, D. A. (2021). Relationships between animal management and habitat characteristics with two potential indicators of welfare for bottlenose dolphins under professional care. *PLoS ONE*, 16(8), e0252861.

Miller, L. J., Luebke, J. F. and Matiasek, J. (2018). Viewing African and Asian elephants at accredited zoological institutions: conservation intent and perceptions of animal welfare. *Zoo Biology*, 37, 466–477.

Miller, M., Michael, A., van Helden, P. and Buss, P. (2017). Tuberculosis in rhinoceros: an underrecognized threat? *Transboundary and Emerging Diseases*, 64, 1071–1078.

Mitchell, G., Obradovich, S., Sumner, D., et al. (1990). Cage location effects on visitor attendance at three Sacramento Zoo mangabey enclosures. *Zoo Biology*, 9, 55–63.

Mitchell, P. C. (1929). *Centenary History of the Zoological Society of London*. Zoological Society of London, London.

Mkono, M. and Holder, A. (2019). The future of animals in tourism recreation: social media as spaces of collective moral reflexivity. *Tourism Management Perspectives*, 29, 1–8.

Moberg, G. P. (2000). Biological response to stress: implications for animal welfare. In: Moberg, G. P. and Mench, J. A. (eds.), *The Biology of Animal Stress*. CABI, Wallingford, pp. 1–21.

Mollineau, W. M., Adogwa, A. O. and Garcia, G. W. (2010). Improving the efficiency of the preliminary electroejaculation technique developed for semen collection from the agouti (*Dasyprocta leporina*). *Journal of Zoo and Wildlife Medicine*, 41, 633–637.

Montali, R. J., Bush, M., Hess, J., et al. (1995). *Ex situ* diseases and their control for reintroduction of the endangered lion tamarin species (*Leontopithecus* spp.). *Erkrankungen der Zootiere*, 37, 93–98.

Montali, R. J., Richman, L. K. and Hildebrandt, T. B. (1998). Highly fatal disease of Asian elephants in North America and Europe is attributed to a newly recognised endotheliotropic herpesvirus. *Elephant Journal*, 1(3–4), 3.

Mooney, A., Conde, D. A., Healy, K. and Buckley, Y. M. (2020). A system wide approach to managing zoo collections for visitor attendance and in-situ conservation. *Nature Communications*, 11(1). http://doi.org/10.1038/s41467-020-14303-2.

Moore, F. L. and Miller, L. J. (1984). Stress-induced inhibition of sexual behavior: corticosterone inhibits courtship behaviors of a male amphibian (*Taricha granulosa*). *Hormones and Behavior*, 18, 400–410.

Moore, I. T., Lemaster, M. P. and Mason, R. T. (2000). Behavioural and hormonal responses to capture stress in the male red-sided garter snake, *Thamnophis sirtalis parietalis*. *Animal Behaviour*, 59, 529–534.

Morfeld, K. A. and Brown, J. L. (2014). Ovarian acyclicity in zoo African elephants (*Loxodonta africana*) is associated with high body condition scores and elevated serum insulin and leptin. *Reproduction, Fertility and Development*, 28, 640–647.

Morfeld, K. A., Lehnhardt, J., Alligood, C., Bolling, J. and Brown, J. L. (2014). Development of a body condition scoring index for female African elephants validated by ultrasound measurements of subcutaneous fat. *PLoS ONE*, 9(4), e93802.

Morgan, J. M. and Gramann, J. H. (1989). Predicting effectiveness of wildlife education programs: a study of students' attitudes and knowledge towards snakes. *Wildlife Society Bulletin*, 17, 501–509.

Morgan-Davies, A. M. (1980). Translocating crocodiles. *Oryx*, 15, 371–373.

Morris, C. and Reed, M. (2007). From burgers to biodiversity? The McDonaldization of on-farm nature conservation in the UK. *Agriculture and Human Values*, 24, 207–218.

Morris, D. (1960). Automatic seal feeding apparatus at London Zoo. *International Zoo Yearbook*, 2, 70.

Morton, F. B., Lee, P. C., Buchanan-Smith, H. M., et al. (2013a). Personality structure in brown capuchin monkeys (*Sapajus apella*): comparisons with chimpanzees (*Pan troglodytes*), orangutans (*Pongo* spp.), and rhesus macaques (*Macaca mulatta*). *Journal of Comparative Psychology*, 127, 282–298.

Morton, F. B., Lee, P. C. and Buchanan-Smith, H. M. (2013b). Taking personality selection bias seriously in animal cognition research: a case study in capuchin monkeys (*Sapajus apella*). *Animal Cognition*, 16, 677–684.

Mosaferi, S., Niasari-Naslaji, A., Abarghani, A., Gharahdaghi, A. A. and Gerami, A. (2005). Biophysical and biochemical characteristics of Bactrian camel semen collected by artificial vagina. *Theriogenology*, 63, 92–101.

Moscardo, G., Woods, B. and Saltzer, R. (2004). The role of interpretation in wildlife tourism. In: Higginbottom, E. (ed.). *Wildlife Tourism: Impacts, Management and Planning*. Common Ground Publishing, University of Illinois, Urbana, pp. 231–251.

Moss, A. and Esson, M. (2010). Visitor interest in zoo animals and the implications for collection planning and zoo education programmes. *Zoo Biology*, 29, 715–731.

Moss, A. and Esson, M. (2013). The educational claims of zoos: where do we go from here? *Zoo Biology*, 32, 13–18.

Moss, A., Francis, D. and Esson, M. (2008). The relationship between viewing area size and visitor behaviour in an immersive Asian elephant exhibit. *Visitor Studies*, 11, 26–40.

Moss, A., Jensen, E. and Gusset, M. (2015). Evaluating the contribution of zoos and aquariums to Aichi Biodiversity Target 1. *Conservation Biology*, 29, 537–544.

Moss, A., Littlehales, C., Moon, A., Smith, C. and Sainsbury, C. (2017). Measuring the impact of an in-school zoo education programme. *Journal of Zoo and Aquarium Research*, 5, 33–37.

Moss, C. (1988). *Elephant Memories: Thirteen Years in the Life of an Elephant Family*. William Collins Sons & Co. Ltd, Glasgow.

Mullan, B. and Marvin, G. (1999). *Zoo Culture*. University of Illinois Press, Urbana.

Mullar, S. L., Bissell, S. L., Cunningham, K. M. and Strasser, R. (2021). How do you behave at the zoo? A look at visitor perceptions of other visitors' behaviour at the zoo. *Animal Behavior and Cognition*, 8, 619–631.

Mumby. H. S., Courtiol, A., Mar, K. U. and Lummaa, V. (2013). Climatic variation and age-specific survival in Asian elephants from Myanmar. *Ecology*, 94, 1131–1141.

Murray, S., Tell, L. A. and Bush, M. (1997). Zinc toxicosis in a Celebes ape (*Macaca nigra*) following ingestion of pennies. *Journal of Zoo and Wildlife Medicine*, 28, 101–104.

Myers, Jr, O. E., Saunders, C. D. and Birjulin, A. A. (2004). Emotional dimensions of watching zoo animals: an experience sampling study building on insights from psychology. *Curator: The Museum Journal*, 47, 299–321.

Naish, D. (2017). An ode to Britains toy animals. *Scientific American*, 23 August. https://blogs.scientificamerican.com/tetrapod-zoology/an-ode-to-britains-toy-animals/ (accessed 11 June 2022).

Napier, J. E., Loskutoff, N. M., Simmons, L. G. and Armstrong, D. L. (2011). Comparison of carfentanil-xylazine and thiafentanil-medetomidine in electroejaculation of captive gaur (*Bos gaurus*). *Journal of Zoo and Wildlife Medicine*, 42, 430–436.

Narayan, E. J., Webster, K., Nicolson, V., Mucci, A. and Hero, J. M. (2013). Non-invasive evaluation of physiological stress in an iconic Australian marsupial: the koala (*Phascolarctos cinereus*). *General and Comparative Endocrinology*, 187, 39–47.

Nath, B. G. and Chakraborty, A. (2013). Traumatic injury and stress related death of non-human primates in Assam State Zoo. *Zoo's Print*, 8, 28–29.

National Bison Association (2021). *Bison by the Numbers. Data and Statistics*. National Bison Association, Westminster, CO. www.bisoncentral.com (accessed 16 December 2021).

National Research Council (2008). *Recognition and Alleviation of Distress in Laboratory Animals*. National Academies Press, Washington, DC.

Nature (2017). News and views. 50 & 100 years ago. From Nature 23 December 1967. *Nature* 552, 341.

Naylor, W. and Parsons, E. C. M. (2019). An international online survey on public attitudes towards the keeping of whales and dolphins in captivity. *Tourism in Marine Environments*, 14, 133–142.

Nechay, G. (1996) Editorial. *Naturopa*, 82, 3.

Nekaris, K. A. I., Campera, M., Nijman, V., et al. (2020). Slow lorises use venom as a weapon in intraspecific competition. *Current Biology*, 30, R1252–R1953.

Nekolný, L. and Fialová, D. (2018). Zoo tourism: what actually is a zoo? *Czech Journal of Tourism*, 7, 153–166.

Nekolný, L. and Fialová, D. (2021). Attendance and perceived constraints to attendance at zoological gardens during the spring 2020 COVID-19 re-opening: the Czechia case. *Journal of Zoological and Botanical Gardens*, 2, 234–249.

New York Times (1906). Bushman shares a cage with Bronx Park apes. *New York Times*, 9 September.

NEZS (1970). *Report of the Council and Statement of Accounts 1970*. The North of England Zoological Society, Chester.

NEZS (1980). *Annual Report for 1980*. The North of England Zoological Society, Chester.

NFCA (2022). The tower. National Fairground and Circus Archive, University of Sheffield. www.sheffield.ac.uk/nfca/projects/towerhistory (accessed 11 June 2022).

Nicholls, R. (1992). *The Belle Vue Story*. Neil Richardson, Manchester.

Nieuwenhuijsen, K. and de Waal, F. B. M. (1982). Effects of spatial crowding on social behavior in a chimpanzee colony. *Zoo Biology*, 1, 5–28.

Nobles, G. (2017). *John James Audubon: The Nature of the American Woodsman*. University of Pennsylvania Press. Philadelphia.

Norris, M. (2014). World War zoo gardens: wartime zoos, the challenging future and the use of zoo history in visitor engagement. *International Zoo Educators Association Journal*, 50, 42–47.

North Carolina Zoo (2008). Elephant exhibit. www.nczoo.com/give/corporate/elephant_exhibit (accessed 18 July 2008).

Novotny, J. F. and Beeman, J. W. (1990). Use of a fish health condition profile in assessing the health and condition of juvenile chinook salmon. *The Progressive Fish-Culturist*, 52, 162–170.

Nowak, R. M. (1999). *Walker's Mammals of the World* (6th ed.). Johns Hopkins University Press, Baltimore.

NSW Department of Primary Industries (1986). *Policy on the Management of Solitary Elephants in New South Wales. Exhibited Animals Protection Act, 1986*. A publication of the Director General, NSW Department of Primary Industries (pursuant to Clause 8(1) of the Exhibited Animals Protection Regulation, 2005).

Nyhus, P. J., Tilson, R. L. and Tomlinson, J. L. (2003). Dangerous animals in captivity: ex situ tiger conflict and implications for private ownership of exotic animals. *Zoo Biology*, 22, 573–586.

O'Brien, J. K. and Robeck, T. R. (2006). Development of sperm sexing and associated assisted reproductive technology for sex preselection of captive bottlenose dolphins (*Tursiops truncatus*). *Reproduction, Fertility and Development*, 18, 319–329.

O'Brien, J. K., Hollinshead, F. K., Evans, K. M., Evans, G. and Maxwell, W. M. C. (2004). Flow cytometric sorting of frozen–thawed spermatozoa in sheep and non-human primates. *Reproduction, Fertility and Development*, 15, 367–375.

O'Brien, J. K., Stojanov, T., Crichton, E. G., et al. (2005). Flow cytometric sorting of fresh and frozen-thawed spermatozoa in the Western lowland gorilla (*Gorilla gorilla gorilla*). *American Journal of Primatology*, 66, 297–315.

O'Brien, J. K., Steinman, K. J., Schitt, T. and Robeck, T. R. (2008). Semen collection, characterisation and artificial insemination in the beluga (*Delphinapterus leucas*) using liquid-stored spermatozoa. *Reproduction, Fertility and Development*, 20, 770–783.

O'Grady, E., Doyle, M., Fitzgerald, C. W. R., Mortell, A. and Murray, D. (2014). Animal attack: an unusual case of multiple trauma in childhood. *Irish Medical Journal*, 107, 328–329.

O'Hara, K., Kindberg, T., Glancy, M., et al. (2007). Collecting and sharing location-based content on mobile phones in a zoo visitor experience. *Computer Supported Cooperative Work (CSCW)*, 16, 11–44.

O'Malley, M. O., Woods, J. M., Bryant, J. and Miller. L. J. (2021). How is Western lowland gorilla (*Gorilla gorilla gorilla*) behavior and physiology impacted by 360° visitor viewing access? *Animal Behavior and Cognition*, 8, 468–480.

O'Malley, R. C. and McGrew, W. C. (2006). Hand preferences in captive orangutans (*Pongo pygmaeus*). *Primates*, 47, 279–283.

O'Regan, H. J. and Kitchener, A. C. (2005). The effects of captivity on the morphology of captive, domesticated and feral mammals. *Mammal Review*, 31, 60–65.

O'Regan, H., Turner, A. and Sabin, R. (2006). Medieval big cat remains from the Royal Menagerie at the Tower of London. *International Journal of Osteoarchaeology*, 16, 385–394.

Oakleaf, B., Luce, B. and Thome, E. T. (1992). Black-footed ferret reintroduction in Wyoming: project description and 1992 protocol. Wyoming Game and Fish Department, Laramie, Wyoming (unpublished).

Ödberg, F. O. (1978). Abnormal behavior: stereotypies. In: *Proceedings of the First World Congress of Ethology Applied to Zootechnics*, Industrias Grafices Espana, Madrid, pp. 475–480.

Odum, R. A. and Reinert, H. K. (2015). The Aruba Island rattlesnake *Crotalus unicolor* Species Survival Plan: a case history in ex situ and in situ conservation. *International Zoo Yearbook*, 49, 104–112.

Okuyama, J., Shimizu, T., Abe, O., Yoseda, K. and Arai, N. (2010). Wild versus head-started hawksbill turtles *Eretmochelys imbricata*: post-release behaviour and feeding adaptations. *Endangered Species Research*, 10, 181–190.

Olson, D. and Wiese, R. J. (2000). State of the North American African elephant population and projections for the future. *Zoo Biology*, 19, 311–320.

Orban, D. A., Siegford, J. M. and Snider, R. J. (2016). Effects of guest feeding programs on captive giraffe behaviour. *Zoo Biology*, 35, 157–166.

Orban, D. A., Soltis, J., Perkins, L. and Mellen, J. D. (2017). Sound at the zoo: using animal monitoring, sound measurement, and noise reduction in zoo animal management. *Zoo Biology*, 36, 231–236.

Owen, A., Wilkinson, R. and Sözer, R. (2014). In situ conservation breeding and the role of zoological institutions and private breeders in the recovery of highly endangered Indonesian passerine birds. *International Zoo Yearbook*, 48, 199–211.

Owen, M. A., Swaisgood, R. R., Czekala, N. M., Steinman, K. and Lindburg, D. G. (2004). Monitoring stress in captive giant pandas (*Ailuropoda melanoleuca*): behavioral and hormonal responses to ambient noise. *Zoo Biology*, 23, 147–164.

Owen, M. A., Hall, S., Bryant, L. and Swaisgood, R. R. (2013). The influence of ambient noise on maternal behaviour in a Bornean sun bear (*Helarctos malayanus euryspilus*). *Zoo Biology*, 33, 49–53.

Owen, R. (1834). On the anatomy of the cheetah, *Felis jubata*, Schreb. *The Transactions of the Zoological Society of London*, 1, 129–136.

Owen, R. (1839a). Osteological contributions to the natural history of the orang utans (Simia, Erxleben). *The Transactions of the Zoological Society of London*, 2, 165–172.

Owen, R. (1839b). Notes on the anatomy of the Nubian giraffe. *The Transactions of the Zoological Society of London*, 2, 217–243.

Özkan, B., Koenhemsi, L., Sönmez, K. and Arun, S. S. (2018). Capture myopathy accompanied with severe enteritis in a female lion. *Journal of Istanbul Veterinary Sciences*, 2, 1–7.

Palmer, C., Morrin, H. and Sandøe, P. (2019). Defensible zoos and aquariums. In: Fischer, B. (ed.). *The Routledge Handbook of Animal Ethics*. Routledge, Abingdon, pp. 394–406.

Palomino, J. M., Mastromonaco, G. F., Cervantes, M. P., et al. (2020). Effect of season and superstimulatory treatment on in vivo and in vitro embryo production in wood bison (*Bison bison athabascae*). *Reproduction in Domestic Animals*, 55, 54–63.

Panagiotopoulou, O., Pataky, T. C., Hill, Z. and Hutchinson, J. R. (2012). Statistical parametric mapping of the regional distribution and ontogenetic scaling of foot pressures during walking in Asian elephants (*Elephas maximus*). *Journal of Experimental Biology*, 215, 1584–1593.

Paradise Wildlife Park (2022). Paradise Wildlife Park history. www.pwpark.com/about/history (accessed 27 April 2022).

Parker, S. T. (1994). Incipient mirror self-recognition in zoo gorillas and chimpanzees. In: Parker, S. T., Mitchell, R. W. and Boccia, M. L. (eds.), *Self-Awareness in Animals and Humans: Developmental Perspectives*. Cambridge University Press, Cambridge, pp. 301–307.

Parsons, R., Aldous-Mycock, C. and Perrin, M. R. (2007). A genetic index for stripe-pattern reduction in the zebra: the quagga project. *African Journal of Wildlife Research*, 37, 105–116.

Pastorino, Q. G., Viau, A., Curone, G., et al. (2017). Role of personality in behavioural responses to new environments in captive Asiatic lions (*Panthera leo persica*). *Veterinary Medicine International*.. https://doi.org/10.1155/2017/6585380.

Patrick, P. G. (2006). Mental models students hold of zoos. Doctoral dissertation, University of North Carolina at Greensboro.

Patrick, P. G. and Caplow, S. (2018). Identifying the foci of mission statements of the zoo and aquarium community. *Museum Management and Curatorship*, 33, 120–135.

Patrick, P. G., Matthews, C. E., Ayers, D. F. and Tunnicliffe, S. D. (2007). Conservation and education: prominent themes in zoo mission statements. *Journal of Conservation Education*, 38, 53–60.

Patton, M. L., White, A. M., Swaisgood, R. R., et al. (2001). Aggression control in a bachelor herd of fringe-eared oryx (*Oryx gazella callotis*), with melengestrol acetate: behavioral and endocrine observations. *Zoo Biology,* 20, 375–388.

Paukner, A., Anderson, J. R. and Fujita, K. (2004). Reactions of capuchin monkeys (*Cebus apella*) to multiple mirrors. *Behavioural Processes*, 66, 1–6

Pavlov, I. P. (1906). The scientific investigation of the psychical faculties or processes in the higher animals. *Science*, 24, 613–619.

Pearce-Kelly, P., Clarke, D., Robertson, M. and Andrews, C. (1991). The display, culture and conservation of invertebrates at London Zoo. *International Zoo Yearbook*, 30, 21–30.

Pearce-Kelly, P., Morgan, R., Honan, P., et al. (2007). The conservation value of insect breeding programmes: rationale, evaluation tools and example programme case studies. In: Stewart, A. J. A., New, T. R. and Lewis, O. T. (eds.), *Insect Conservation Biology: Proceedings of the Royal Entomological Society's 23rd Symposium*, The Royal Entomological Society, London, pp. 57–73.

Pellet, S. and Bushell, M. (2015). Emergency care and first aid of invertebrates. *Companion Animal*, 20, 182–186.

Penfold, L. M., Munson, L., Plotka, E. and Citino, S. B. (2007). Effect of progestins on serum hormones, semen production, and agonistic behavior in the gerenuk (*Litocranius walleri walleri*). *Zoo Biology*, 26, 245–257.

Pennsylvania Game Commission (2013). *Elk in Pennsylvania. Past, Present & Future*. Pennsylvania Game Commission, Harrisburg.

The Penny Illustrated Paper and Illustrated Times (1874). The gallant rescue of keepers from the infuriated rhinoceros at the Zoological Gardens (illustration). *The Penny Illustrated Paper and Illustrated Times*, 5 December, p. 353 (front page).

Pereira, A. S., Kavanagh, E., Hobaiter, C., Slocombe, K. E. and Lameira, A. R. (2020). Chimpanzee lip-smacks confirm primate continuity for speech-rhythm evolution. *Biology Letters*, 16(5), 20200232.

Pérez, S., Encinoso, M., Corbera, J. A., et al. (2021). Cranial structure of *Varanus komodoensis* as revealed by computed-tomographic imaging. *Animals*, 11, 1078. https://doi.org/10.3390/ani11041078.

Pérez-Granados, C. and Schuchmann, K.-L. (2020). Vocalisations of the greater Rhea (*Rhea americana*): an allegedly silent ratite. *Bioacoustics*, 30, 564–574.

Perkins, D. R. and Debbage, K. G. (2016). Weather and tourism: thermal comfort and zoological park visitor attendance. *Atmosphere*, 7, 44.

Perrin, K. L., Nielsen, S. S., Martinussen, T. and Bertelsen, M. F. (2021). Quantification and risk factor analysis of elephant endotheliotropic herpesvirus-haemorrhagic disease fatalities in Asian elephants *Elephas maximus* in Europe (1985–2017). *Journal of Zoo and Aquarium Research*, 9(1). http://doi.org/10.19227/jzar.v9i1.553.

Perry, C. J. and Baciadonna, L. (2017). Studying emotion in invertebrates: what has been done, what can be measured and what they can provide. *Journal of Experimental Biology*, 220, 3856–3868.

Persson, T., Sauciuc, G.-A. and Madsen, E. A. (2018). Spontaneous cross-species imitation in interactions between chimpanzees and zoo visitors. *Primates*, 59, 19–29.

Perzanowski, K. and Olech, W. (2013). Restoration of wisent population within the Carpathian eco-region. In: Soorae, P. S. (ed.), *Global Re-introduction Perspectives: 2013 – Further Case-Studies from Around the Globe*. IUCN and Environment Agency, Gland and Abu Dhabi, pp. 190–193.

PETA (2021). Dangerous animal incidents. www.peta.org/issues/animals-in-entertainment/dangerous-animal-incidents (accessed 8 June 2022).

Petersen, H. H., Stenbak, R., Blaabjerg, C., et al. (2022). Development of a quantitative immunoassay for serum haptoglobin as a putative disease marker in the Southern white rhinoceros (*Ceratotherium simum simum*). *Journal of Zoo and Wildlife Medicine*, 53, 141–152.

Pfistermüller, R., Walzer, C. and Licka, T. (2011). First evidence of the use of kinetic gait analysis as a clinical tool for early detection of chronic foot disease in an Indian rhino (*Rhinoceros unicornis*). Rotterdam Zoo and International Elephant Foundation. www.rhinoresourcecenter.com/pdf_files/134/1344675214.pdf.

Philipps, M. R., Johannsen, B. F., Andersen, T. D., Levinsen, H. and Foss, K. K. (2019). Feasible ways to personal meaning mapping in out-of-school contexts? In *European Conference on e-Learning*, ACPIL, pp. 476–485.

Pickering, S., Creighton, E. and Stevens-Wood, B. (1992). Flock size and breeding success in flamingos. *Zoo Biology*, 11, 229–234.

Piitulainen, R. and Hirskyj-Douglas, I. (2020). Music for monkeys: building methods to design with white-faced sakis for animal-driven audio enrichment devices. *Animals*, 10, 1768.

Plotnik, J. M., de Waal, F. B. M. and Reiss, D. (2006). Self-recognition in an Asian elephant. *Proceedings of the National Academy of Sciences*, 103(45), 17053–17057.

Pohlin, F., Hooijberg, E. H. and Meyer, L. C. R. (2021). Challenges to animal welfare during transportation of wild mammals (1990–2020). *Journal of Zoo and Wildlife Medicine*, 52, 1–13.

Poindexter, S. A., Reinhardt, K. D., Nijman, V. and Nekaris, K. A. I. (2018). Slow lorises (*Nycticebus* spp.) display evidence of handedness in the wild and in captivity. *Laterality: Asymmetries of Body, Brain and Cognition*, 23, 705–721.

Polgár, Z., Wood, L. and Haskell, M. J. (2017). Individual differences in zoo-housed squirrel monkeys' (*Saimiri sciureus*) reactions to visitors, research participation, and personality ratings. *American Journal of Primatology*, 79(5), e22639.

Pollott, G. E. and Kilkenny, J. B. (1976). A note on the use of condition scoring in commercial sheep flocks. *Animal Science*, 23, 261–264.

Popp, J. W. (1984). Interspecific aggression in mixed ungulate species exhibits. *Zoo Biology*, 3, 211–219.

Poulsen, E. M. B., Honeyman, V., Valentine, P. A. and Teskey, G. C. (1996). Use of fluoxetine for the treatment of stereotypical pacing behavior in a captive polar bear. *Journal of the American Veterinary Association*, 209, 1470–1474.

Powell, D. M. (2019). Collection planning for the next 100 years: what will we commit to save in zoos and aquariums? *Zoo Biology*, 38, 139–148.

Powell, D. M. and Ardaiolo, M. (2016). Survey of U.S. zoo and aquarium animal care staff attitudes regarding humane euthanasia for population management. *Zoo Biology*, 35, 187–200.

Powell, D. M. and Svoke, J. T. (2008). Novel environmental enrichment may provide a tool for rapid assessment of animal personality: a case study with giant pandas (*Ailuropoda melanoleuca*). *Journal of Applied Animal Welfare Science*, 11, 301–318.

Powell, D. M., Speeg, B., Li, S., Blumer, E. and McShea, W. (2013). An ethogram and activity budget of captive Sichuan takin (*Budorcas taxicolor tibetana*) with comparisons to other Bovidae. *Mammalia*, 77, 391–401.

Powell, D. M., Lan, J. and Eng, C. (2018). Survey of US-based veterinarians' attitudes on population management euthanasia. *Zoo Biology*, 37, 478–487.

Powell, D. M., Dorsey, C. L. and Faust, L. J. (2019). Advancing the science behind animal program sustainability: an overview of the special issue. *Zoo Biology*, 38, 5–11.

Prado-Oviedo, N. A., Bonaparte-Saller, M. K., Malloy, E. J., et al. (2016). Evaluation of demographics and social life events of Asian (*Elephas maximus*) and African elephants (*Loxodonta africana*) in North American zoos. *PLoS ONE*, 11(7), e0154750.

Prates, H. M. and Bicca-Marques, J. C. (2005). Coprophagy in captive brown capuchin monkeys (*Cebus appella*). *Neotropical Primates*, 13, 18–21.

Princée, P. G. and Glatston, A. R. (2016). Influence of climate on the survivorship of neonatal red pandas in captivity. *Zoo Biology*, 35, 104–110.

Proctor, H. S., Carder, G. and Cornish, A. R. (2013). Searching for animal sentience: a systematic review of the scientific literature. *Animals*, 3(3), 882–906.

Prokop, P. and Tunnicliffe, S. D. (2010). Effects of having pets at home on children's attitudes toward popular and unpopular animals. *Anthrozoös*, 23, 21–35.

Prokop, P., Prokop, M. and Tunnicliffe, S. D. (2008). Effects of keeping animals as pets on children's concepts of vertebrates and invertebrates. *International Journal of Science Education*, 30, 431–449.

Proskuryakova, A. A., Kulemzina, A. I., Perelman, P. L., et al. (2017). X chromosome evolution in Cetartiodactyla. *Genes*, 8, 216. https://doi.org/10.3390/genes8090216.

Prosser, N. S., Gardner, P. C., Smith, J. A., et al. (2016). Body condition scoring of Bornean banteng in logged forests. *BMC Zoology*, 1. http://doi.org/10.1186/s40850-016-0007-5.

Pruetz, J. D. and Bloomsmith, M. A. (1992). Comparing two manipulable objects as enrichment for captive chimpanzees. *Animal Welfare*, 1, 127–137.

Puan, C. L. and Zakaria, M. (2007). Perception of visitors towards the role of zoos: a Malaysian perspective. *International Zoo Yearbook*, 41, 226–232.

Pullen, P. K. (2005). Preliminary comparisons of male/male interactions within bachelor and breeding groups of Western lowland gorillas (*Gorilla gorilla gorilla*). *Applied Animal Behaviour Science* 90, 143–153.

Pullen, P. K. (2009). Male–male social interactions in breeder and bachelor groups of gorillas (*Gorilla gorilla*): an indication of behavioural flexibility. PhD thesis, University of Exeter.

Putnam, A. S., Ferrie, G. M. and Ivy, J. A. (2022). Ex situ breeding programs benefit from science-based co-operative management. *Zoo Biology*. https://doi.org/10.1002/zoo.21700.

Pye, G. W., Adkesson, M. J., Guthrie, A., Clayton, L. A. and Janssen, D. L. (2018). Risk analysis: changing the quarantine paradigm? *Journal of Zoo and Wildlife Medicine*, 49, 513–519.

Pyott, B. E. and Schulte-Hostedde, A. I. (2020). Peer-reviewed scientific contributions from Canadian zoos and aquariums. *FACETS*, 5, 381–392.

Quadros, S., Goulart, V. D. L., Passos, L., Vecci, M. A. M. and Young, R. J. (2014). Zoo visitor effect on mammal behaviour: does noise matter? *Applied Animal Behaviour Science*, 156, 78–84.

Rachels, J. (1986). *The End of Life*. Oxford University Press, New York.

Radford, H. V. (1906). Bringing back the beaver: its successful reintroduction to the Adirondack region. *Four-Track-News*, April.

Raines, J. A. and Fried, J. J. (2016). Use of deslorelin acetate implants to control aggression in a multi-male group of rock hyrax (*Procavia capensis*). *Zoo Biology*, 35, 201–204.

Ralls, K. and Ballou, J. D. (1992). Managing genetic diversity in captive breeding and reintroduction programs. In *Transactions of the Fifty-Seventh North American Wildlife and Natural Resources Conference*, pp. 263–282.

Ramakrishnan, B., Ilakkia, M. and Ramkumar, K. (2014). Use of shade trees for Asian elephant (*Elephas maximus*) in the Mudumalai Tiger Reserve, Nilgiris, Tamilnadu, Southern India. *Scientific Transactions in Environment and Technovation*, 7, 144–150.

Rammelmeyer, A. and Porterfield, K. (1978). How do you zoo? Beauvoir first-graders use National Zoo as extension of their classroom. *Roundtable Reports*. www.jstor.org/stable/40479408.

Ramsey, M. A. and Stirling. I. (1988). Reproductive biology and ecology of female polar bears (*Ursus maritimus*). *Journal of Zoology*, 214, 601–633.

Randerson, J. (2003). Wide-roaming carnivores suffer most in zoos. *New Scientist*, 1 October. www.newscientist.com/article/dn4221-wideroaming-carnivores-suffer-most-in-zoos.html (accessed 27 May 2008).

Rawski, M. and Józefiak, D. (2014). Body condition scoring and obesity in captive African sideneck turtles (Pelomedusidae). *Annals of Animal Science*, 14, 573–584.

Reade, L. S. and Waran, N. K. (1996). The modern zoo: how do people perceive zoo animals? *Applied Animal Behaviour Science*, 47, 109–118.

Reading, R. P., Miller, B. and Shepherdson, D. (2013). The value of enrichment to reintroduction success. *Zoo Biology*, 32, 332–341.

Redford, K. H. (1985). Feeding and food preference in captive and wild giant anteaters (*Myrmecophaga tridactyla*). *Journal of Zoology*, 205, 559–572.

Redford, K. H. and Dorea, J. G. (1984), The nutritional value of invertebrates with emphasis on ants and termites as food for mammals. *Journal of Zoology*, 203, 385–395.

Redmond, J. and Lamperez, A. (2004). Leading limb preference during brachiation in the gibbon family member, *Hylobates syndactylus* (siamangs): a study of the effects of singing on lateralisation. *Laterality: Asymmetries of Body, Brain and Cognition*, 9, 381–396.

Rees, P. A. (1977). Some aspects of the feeding ecology of the African elephant (*Loxodonta africana africana*, Blumenbach 1797) in captivity. Unpublished BSc dissertation, University of Liverpool, Liverpool.

Rees, P. A. (1982). Gross assimilation efficiency and food passage time in the African elephant. *African Journal of Ecology*, 20, 193–198.

Rees, P. A. (1983). Synchronization of defaecation in the African elephant (*Loxodonta africana*). *Journal of Zoology*, 201, 581–585.

Rees, P. A. (2000). Are elephant enrichment studies missing the point? *International Zoo News*, 47, 369–371.

Rees, P. A. (2001). Is there a legal obligation to reintroduce animal species into their former habitats? *Oryx*, 35, 216–223.

Rees, P. A. (2002). Asian elephants (*Elephas maximus*) dust bathe in response to an increase in environmental temperature. *Journal of Thermal Biology*, 27, 353–358.

Rees, P. A. (2003). Asian elephants in zoos face global extinction: should zoos accept the inevitable? *Oryx*, 37, 20–22.

Rees, P. A. (2004a). Low environmental temperature causes an increase in stereotypic behaviour in captive Asian elephants (*Elephas maximus*). *Journal of Thermal Biology*, 29, 37–43.

Rees, P. A. (2004b). Some preliminary evidence of the social facilitation of mounting behavior in the Asian elephant (*Elephas maximus*). *Journal of Applied Animal Welfare Science*, 7, 49–58.

Rees, P. A. (2004c). Unreported appeasement behaviours in the Asian elephant (*Elephas maximus*). *Journal of the Bombay Natural History Society*, 101, 71–74.

Rees, P. A. (2005a). The EC Zoos Directive: a lost opportunity to implement the Convention on Biological Diversity. *Journal of International Wildlife Law and Policy*, 8, 51–62.

Rees, P. A. (2005b). Will the EC Zoos Directive increase the conservation value of zoo research? *Oryx*, 39, 128–131.

Rees, P. A. (2009a). Activity budgets and the relationship between feeding and stereotypic behaviours in Asian elephants (*Elephas maximus*) in a zoo. *Zoo Biology*, 28, 79–97.

Rees, P. A. (2009b). The sizes of elephant groups in zoos: implications for elephant welfare. *Journal of Applied Animal Welfare Science*, 12, 44–60.

Rees, P. A. (2011). *An Introduction to Zoo Biology and Management*. Wiley-Blackwell, Chichester.

Rees, P. A. (2013). *Dictionary of Zoo Biology and Animal Management*. Wiley-Blackwell, Chichester.

Rees, P. A. (2015). *Studying Captive Animals: A Workbook of Methods in Behaviour, Welfare and Ecology*. Wiley-Blackwell, Chichester.

Rees, P. A. (2017). *The Laws Protecting Animals and Ecosystems*. Wiley-Blackwell, Chichester.

Rees, P. A. (2021). *Elephants under Human Care: The Behaviour, Ecology, and Welfare of Elephants in Captivity*. Academic Press, San Diego.

Reeve, C. L., Spitzmüller, C., Rogelberg, S. G., et al. (2004). Employee reactions and adjustments to euthanasia-related work: identifying turning-point events through retrospective narratives. *Journal of Applied Animal Welfare Science*, 7, 1–25.

Regaiolli, B., Rizzo, A., Ottolini, G., et al. (2019). Motion illusions as environmental enrichment for zoo animals: a preliminary investigation on lions (*Panthera leo*). *Frontiers in Psychology*, 10, 2220.

Regaiolli, B., Bolcato, S., Ottolini, G., et al. (2021). Preliminary investigation of foot preference for a string-pulling task in zoo macaws. *Applied Animal Behaviour Science*, 238, 105307.

Reid, G. M. (2007). Science in zoos and aquariums. *Science in Parliament*, 64, 6–7.

Reinhardt, V. and Roberts, A. (1997). Effective feeding enrichment for non-human primates: a brief review. *Animal Welfare*, 6, 265–272.

Reinhardt, V., Houser, W. D., Cowley, E. and Champoux, M. (1987). Preliminary comments on environmental enrichment with branches for individually caged rhesus monkeys. *Laboratory Primate Newsletter*, 26, 1–3.

Reiss, D. and Marino, L. (2001). Mirror self-recognition in the bottlenose dolphin: a case of cognitive convergence. *Proceedings of the National Academy of Sciences*, 98, 5937–5942.

Remis, M. J. (2000). Initial studies on the contributions of body size and gastrointestinal passage rates to dietary flexibility among gorillas. *American Journal of Physical Anthropology*, 112, 171–180.

Remport, L., Sós-Koroknai, V., Hoitsy, M. and Sós, E. (2022). Mandibular fractures in giraffes (*Giraffa camelopardalis*) in European zoos. *Journal of Zoo and Wildlife Medicine*, 53, 448–454.

Reuter, H. O. and Adcock, K. (1998). Standardised body condition scoring system for black rhinoceros (*Diceros bicornis*). *Pachyderm*, 26, 116–121.

Rhoads, D. L. and Goldsworthy, R. J. (1979). The effects of zoo environments on public attitudes toward endangered wildlife. *International Journal of Environmental Studies*, 13, 283–287.

Richardson, D. M. (2000). Euthanasia: a nettle we need to grasp. *Ratel*, 27, 80–88.

Richmond, D. J., Sinding, M.-H. S., Thomas, M. and Gilbert, P. (2016). The potential and pitfalls of de-extinction. *Zoologica Scripta*, 45, 22–36.

Rickman, L. K., Montali, R. J., Garber, R. L., et al. (1999). Novel endotheliotropic herpesviruses fatal for Asian and African elephants. *Science*, 283, 1171.

Riggio, G., Pirrone, F., Lunghini, E., Gazzano, A. and Mariti, C. (2020). Zookeepers' perception of zoo canid welfare and its effect on job satisfaction, worldwide. *Animals*, 10(5), 916.

Ritvo, S. E. and MacDonald, S. E. (2016). Music as enrichment for Sumatran orangutans (*Pongo abelii*). *Journal of Zoo and Aquarium Research*, 4, 156–163.

Ritzer, G. (2008). *The McDonaldization of Society 5*. Pine Forge Press, Thousand Oaks.

Robbins, C. T., Tollefson, T. N., Rode, K. D., Erlenbach, J. A. and Ardente, A. J. (2022). New insights into dietary management of polar bears (*Ursus maritimus*) and brown bears (*U. arctos*). *Zoo Biology*, 41, 166–175.

Robbins, L. and Margulis, S. W. (2014). The effects of auditory enrichment on gorillas. *Zoo Biology*, 33, 197–203.

Robbins, L. and Margulis, S. W. (2016). Music for the birds: effects of auditory enrichment on captive bird species. *Zoo Biology*, 35, 29–34.

Robbins, M. M., Gray, M., Fawcett, K. A., et al. (2011). Extreme conservation leads to recovery of the Virunga mountain gorillas. *PLoS ONE*, 6(6), e19788.

Robbins, R. L. and Sheridan, J. (2021). Effect of enclosure expansion on the activity budgets of Eastern black-and-white colobus monkeys, *Colobus guereza*. *Zoo Biology*, 40, 115–123.

Robert, A., Colas, B., Guigon, I., et al. (2015). Defining reintroduction success using IUCN criteria for threatened species: a demographic assessment. *Animal Conservation*, 18, 397–406.

Robert, S. (1986). Ontogeny of mirror behavior in two species of great apes. *American Journal of Primatology*, 10, 109–117.

Roberts, M. and Cunningham, B. (1986). Space and substrate use in captive Western tarsiers (*Tarsius bancanus*). *International Journal of Primatology*, 7, 113–130.

Robinson, L. M., Altschul, D. M., Wallace, E. K., et al. (2017). Chimpanzees with positive welfare are happier, extraverted, and emotionally stable. *Applied Animal Behaviour Science*, 191, 90–97.

Roe, K., McConney, A. and Mansfield, C. F. (2014). The role of zoos in modern society: a comparison of zoos' reported priorities and what visitors believe they should be. *Anthrozoös*, 27, 529–541.

Rohr, L. (1989). A survey of American zoo and aquarium archives. *Science & Technology Libraries*, 9, 75–84.

Roocroft, A. (2007). Protected contact training of elephants in Europe. *International Zoo News*, 54, 6–26.

Rose, P. E. and Croft, D. P. (2015). The potential of social network analysis as a tool for the management of zoo animals. *Animal Welfare*, 24, 123–138.

Rose, P. and Robert, R. (2013). Evaluating the activity patterns and enclosure usage of a little-studied zoo species, the sitatunga (*Tragelaphus spekii*). *Journal of Zoo and Aquarium Research*, 1, 14–19.

Rose, P. E., Brereton, J. E. and Croft, D. P. (2018). Measuring welfare in captive flamingos: activity patterns and exhibit usage in zoo-housed birds. *Applied Animal Behaviour Science*, 205, 115–125.

Rose, P. E., Brereton, J. E., Rowden, L. J., de Figueiredo, R. L. and Riley, L. M. (2019). What's new from the zoo? An analysis of ten years of zoo-themed research output. *Palgrave Communications*, 5, 128. https://doi.org/10.1057/s41599-019-0345-3.

Rosenzweig, M. R., Bennett, E. L., Hebert, M. and Morimoto, H. (1978). Social grouping cannot account for cerebral effects of enriched environments. *Brain Research*, 153, 563–576.

Rosier, R. L. and Langkilde, T. (2011). Does environmental enrichment really matter? A case study using the Eastern fence lizard, *Sceloporus undulatus*. *Applied Animal Behaviour Science*, 131, 71–76.

Ross, S. R. and Gillespie, K. L. (2008). Influences on visitor behavior at a modern immersive zoo exhibit. *Zoo Biology*, 28, 462–472.

Ross, S. R. and Lukas, K. E. (2006). Use of space in a non-naturalistic environment by chimpanzees and lowland gorillas. *Applied Animal Behaviour Science*, 96, 143–152.

Ross, S. R, Schapiro, S. J., Hau, J. and Lukas, K. E. (2009). Space use as an indicator of enclosure appropriateness: a novel measure of captive animal welfare. *Applied Animal Behaviour Science*, 121, 42–50.

Roth, A. M. and Cords, M. (2020). Zoo visitors affect sleep, displacement activities, and affiliative and aggressive behaviours in captive ebony langurs (*Trachypithecus auratus*). *Acta Ethologica*, 23, 61–68.

Roth, T. L., Stoops, M. A., Atkinson, M. W., et al. (2005). Semen collection in rhinoceroses (*Rhinoceros unicornis, Diceros bicornis, Ceratotherium simum*) by electroejaculation with a uniquely designed probe. *Journal of Zoo and Wildlife Medicine*, 36, 617–627.

Rothwell, E. S., Bercovitch, F. B., Andrews, J. R. M. and Anderson, M. J. (2011). Estimating daily walking distance of captive African elephants using an accelerometer. *Zoo Biology*, 30, 579–591.

Rowell, T. A. A. D., Magrath, M. J. L. and Magrath, R. D. (2020). Predator-awareness training in terrestrial vertebrates: progress, problems and possibilities. *Biological Conservation*, 252. https://doi.org/10.1016/j.biocon.2020.108740.

Rowlands, A. N., Capel, T., Rowden, L. and Dow, S. (2021). Burrowing in captive juvenile Desertas wolf spiders *Hogen ingens*. *Journal of Zoo and Aquarium Research*, 9, 66–72.

RSPB (2021). *Annual Report 2020–2021*. Royal Society for the Protection of Birds, Sandy, Bedfordshire.

Ruby, S. and Buchanan-Smith, H. M. (2015). The effects of individual cubicle research on the social interactions and individual behavior of brown capuchin monkeys (*Sapajus apella*). *American Journal of Primatology*, 77, 1097–1108.

Rushen, J. (1993). The 'coping' hypothesis of stereotypic behaviour. *Animal Behaviour*, 45, 613–615.

Rushen, J., de Passillé, A. M. and Munksgaard, L. (1999). Fear of people by cows and effects on milk yield, behavior and heart rate at milking. *Journal of Dairy Science*, 82, 720–727.

Russow, L. M. (1994). Why do species matter? In: Westphal, D. and Westphal, F. (eds.), *Planet in Peril: Essays in Environmental Ethics*. Holt, Reinhart and Winston, Orlando, pp. 149–170.

Rutowski, P. (1990). Theater techniques in an aquarium or a natural history museum. *Journal of Museum Education*, 15, 5–7.

Ryan, C. and Saward, J. (2004). The zoo as ecotourism attraction: visitor reactions, perceptions and management implications. The case of Hamilton Zoo, New Zealand. *Journal of Sustainable Tourism*, 12, 245–266.

Ryan, S., Thompson, S. D., Roth, A. M. and Gold, K. C. (2002). Effects of hand-rearing on the reproductive success of Western lowland gorillas in North America. *Zoo Biology* 21, 389–401.

Sach, F., Fitzpatrick, M., Masters, N. and Field, D. (2019). Financial planning required to keep elephants in zoos in the United Kingdom in accordance with the Secretary of State's Standards of Modern Zoo Practice for the next 30 years. *International Zoo Yearbook*. https://doi.org/10.1111/izy.12213.

Sadhukhan, C., Root-Gutteridge, H. and Habib, B. (2021). Identifying unknown wolves by their distinctive howls: its potential as a non-invasive survey method. *Scientific Reports*, 11, 7309.

Safadi, R. and Ram, J. (2020). Creating an effective electric eel display for the Belle Isle Aquarium. https://papers.ssrn.com/sol3/papers.cfm?abstract_id=3729126.

Saha, J. (2016). Murder at London Zoo: late colonial sympathy in interwar Britain. *The American Historical Review,* 121, 1468–1491.

Saint Jalme, M., Lecoq, R., Seigneurin, F., Blesbois, E. and Plouzeau, E. (2003). Cryopreservation of semen from endangered pheasants: the first step towards a cryobank for endangered avian species. *Theriogenology*, 59(3–4), 875–888.

Saito, Y. (1998). Appreciating nature on its own terms. *Environmental Ethics*, 20, 135–149.

Sakagami, T. and Ohta, M. (2010). The effect of visiting zoos on human health and quality of life. *Animal Science Journal*, 81, 129–134.

Samways, M. J. (2019). *Insect Conservation: A Global Synthesis*. CABI, Oxford.

Sanderson, E. W. (2006). How many animals do we want to save? The many ways of setting population target levels for conservation. *Bioscience*, 56, 911–922.

Sandstrom, P. A., Phan, K. O., Switzer, W. M., et al. (2000). Simian foamy virus infection among zoo keepers. *The Lancet*, 355(9203), 551–552.

Santymire, R., Branvold-Faber, H. and Marinari, P. E. (2014). Recovery of the black-footed ferret. In: Fox, J. G. and Marini, R. P. (eds.), *Biology and Diseases of the Ferret* (3rd ed.). Wiley & Sons, Hoboken, pp. 219–231.

Sayer, E. C., Whitham, J. C. and Margulis, S. W. (2007). Who needs a forelimb anyway? Locomotor, postural and manipulative behaviour in a one-armed gibbon. *Zoo Biology*, 26, 215–222.

Schaefer, S. A. and Steklis, H. D. (2014). Personality and subjective well-being in captive male Western lowland gorillas living in bachelor groups. *American Journal of Primatology*, 76, 879–889.

Schaffer, N. E., Beehler, B., Jeyendran, R. S. and Balke, B. (1990). Methods of semen collection in an ambulatory greater one-horned rhinoceros (*Rhinoceros unicornis*). *Zoo Biology*, 9, 211–221.

Schalinski, S., Hollmann, T. and Tsokos, M. (2008). Fatal attacks of zoo animals on humans: a case report. *Archiv fur Kriminologie*, 221, 113–119.

Schaller, G. (1972). *The Serengeti Lion*. University of Chicago Press, Chicago.

Schlaich, J. and Schober, H. (1997). Glass roof for the Hippo House at the Berlin Zoo. *Structural Engineering International*, 7, 252–254.

Schmidt, H. and Kappelhof, J. (2019). Review of the management of the Asian elephant *Elephas maximus* EEP: current challenges and future solutions. *International Zoo Yearbook*, 53, 31–44.

Schmitt, D. (2008). View from the big top:. why elephants belong in North American circuses. In: Wemmer, C. and Christen C. A. (eds.), *Elephants and Ethics: Towards a Morality of Coexistence*. Johns Hopkins University Press, Baltimore, pp. 227–234.

Schmitt, D. L. (1998). Report of a successful artificial insemination in an Asian elephant. In *Third International Elephant Research Symposium*, Springfield.

Schmitt, E. C. (1988). Effects of conservation legislation on the professional development of zoos. *International Zoo Yearbook*, 27, 3–9.

Schmitt, T. L., St. Aubin, D. J., Schaefer, A. M. and Dunn, J. L. (2010). Baseline, diurnal variations, and stress-induced changes of stress hormones in three captive beluga whales, *Delphinapterus leucas*. *Marine Mammal Science*, 26, 635–647.

Schneider, M., Nogge, G. and Kolter, L. (2014). Implementing unpredictability in feeding enrichment for Malayan sun bears (*Helarctos malayanus*). *Zoo Biology*, 33, 54–62.

Schneider, S. and Dierkes, P. W. (2021). Localize Animal Sound Events Reliably (LASER): a new software for sound localization in zoos. *Journal of Zoological and Botanical Gardens*, 2, 146–163.

Schneider, W. H. (2008). The ethnographic exhibitions of the Jardin Zoologique d'Acclimation. In: Blanchard, P., Bancel, N., Boëtsch, G., et al. (eds.), *Human Zoos: Science and Spectacle in the Age of Colonial Empires*. Liverpool University Press, Liverpool, pp. 142–150.

Schneiderová, I. and Vodička, R. (2021). Bioacoustics as a tool to monitor the estrus cycle in a female slow loris (*Nycticebus* sp.). *Zoo Biology*, 40, 575–583.

Schneiderová, I., Zouhar, J., Štefanská, L., et al. (2016).Vocal activity of lesser galagos (*Galago* spp.) at zoos. *Zoo Biology*, 35, 147–156.

Schneiders, A., Sonksen, J. and Hodges, J. K. (2004). Penile vibratory stimulation in the marmoset monkey: a practical alternative to electro-ejaculation, yielding ejaculates of enhanced quality. *Journal of Medical Primatology*, 33, 98–104.

Schomberg, G. (1970). The value of zoological collections in conservation. *Journal of Small Animal Practice*, 11, 55–62.

Schreier, A. L., Readyhough, T. S., Moresco, A., Davis, M. and Joseph, S. (2021). Social dynamics of a newly integrated bachelor group of Asian elephants (*Elephas maximus*): welfare implications. *Journal of Applied Animal Welfare Science*. http://doi.org/10.1080/10888705.2021.1908141.

Schultz, J. G. W. and Joordens, S. (2014). The effect of visitor motivation on the success of environmental education at the Toronto Zoo. *Environmental Education Research*, 20, 753–775.

Schwartz, G. T., Reid, D. J., Dean, M. C. and Zihlman, A. L. (2006). A faithful record of stressful life events recorded in the dental developmental record of a juvenile gorilla. *International Journal of Primatology*, 27, 1201–1219.

Schwarzenberger, F., Möstl, E., Palme, R. and Bamberg, E. (1996). Faecal steroid analysis for non-invasive monitoring of reproductive status in farm, wild and zoo animals. *Animal Reproduction Science*, 42, 515–526.

Schwitzer, C. and Kaumanns, W. (2001). Body weights of ruffed lemurs (*Varecia variegata*) in European zoos with reference to the problem of obesity. *Zoo Biology*, 20, 261–269.

Scoleri, V. P., Johnson, C. N., Vertigan, P. and Jones, M. E. (2020). Conservation trade-offs: island introduction of a threatened predator suppresses invasive mesopredators but eliminates a seabird colony. *Biological Conservation*, 248, 108635.

Scottish Beaver Trial (2007). Trial reintroduction of the European beaver to Knapdale, Mid-Argyll. Scottish Beaver Trial. Local consultation report: 1 October–30 November 2007.

Scotto-Lomassese, S., Strambi, C., Strambi, A., et al. (2000). Influence of environmental stimulation on neurogenesis in the adult insect brain. *Journal of Neurobiology*, 45, 162–171.

Seeber, P. A., Morrison, T., Ortega, A., et al. (2020). Immune differences in captive and free-ranging zebras (*Equus quagga* and *E. zebra*). *Mammalian Biology*, 100, 155–164.

Seidensticker, J. and Doherty, J. G. (1996). Integrating animal behavior and exhibit design. In: Kleiman, D. G., Allen, M. E., Thompson, K. V. and Lumpkins, S. (eds.), *Wild Mammals in Captivity: Principles and Techniques*. University of Chicago Press, Chicago, pp. 180–190.

Seidlitz, A., Bryant, K. A., Armstrong, N. J., Calver, M. and Wayne, A. F. (2021). Optimising camera trap height and model increases detection and individual identification rates for a small mammal, the numbat (*Myrmecobius fasciatus*). *Australian Mammalogy*, 43, 226–234.

Semple, S. (2002). Analysis of research projects conducted in Federation collections to 2000. *Federation Research Newsletter*, 3(1), 3.

Senn, H. V., Ghazali, M., Kaden, J., et al. (2019). Distinguishing the victim from the threat: SNP-based methods reveal the extent of introgressive hybridization between wildcats and domestic cats in Scotland and inform future in situ and ex situ management options for species restoration. *Evolutionary Applications*, 12, 399–414.

Shannon, D., Kitchener. A. C. and Macdonald, A. A. (1995). The preputial glands of the coati, *Nasau nasau. Journal of Zoology*, 236, 319–357.

Shapiro, B. (2017). Pathways to de-extinction: how close can we get to resurrection of an extinct species? *Functional Ecology*, 31, 996–1002.

Shapland, A. (2004). Where have all the monkeys gone? The changing nature of the Monkey Temple at Bristol Zoo. *Anthrozoös*, 17, 194–209.

Shapland, A. and Van Reybrouck, D. (2008). Competing natural and historical heritage: the Penguin Pool at London Zoo. *International Journal of Heritage Studies*, 14, 10–29.

Shchukina, M., Korobkov, A., Byrkova, P. and Pavlova, N. (2019). Image of St. Petersburg public oceanarium in the minds of its visitors. *Scientific Notes Journal of St. Petersburg State Institute of Psychology and Social Work*, 31, 32–41.

Sheldon, J. D., Adkesson, M. J., Allender, M. C., et al. (2020). Objective gait analysis in Humboldt penguins (*Spheniscus humboldti*) using a pressure-sensitive walkway. *Journal of Zoo and Wildlife Medicine*, 50, 910–916.

Shepherdson, D. J. (1998). Introduction: tracing the path of environmental enrichment in zoos. In: Shepherdson, D. J., Mellen, J. D. and Hutchins, M. (eds.), *Second Nature: Environmental Enrichment for Captive Animals*. Smithsonian Institution Press, Washington, DC, pp. 1–12.

Shepherdson, D. J., Carlstead, K., Mellen, J. D. and Seidensticker, J. (1993). The influence of food presentation on the behavior of small cats in confined environments. *Zoo Biology*, 12, 203–216.

Shepherdson, D. J., Mellen, J. D. and Hutchins, M. (1998). *Second Nature: Environmental Enrichment for Captive Animals*. Smithsonian Institution Press, Washington.

Sherwen, S. L., Magrath, M. J. L., Butler, K. L., Phillips, C. J. C. and Hemsworth, P. H. (2014). A multi-enclosure study investigating the behavioural response of meerkats to zoo visitors. *Applied Animal Behaviour Science*, 156, 70–77.

Sherwen, S. L., Magrath, M. J. L., Butler, K. L. and Hemsworth, P. H. (2015). Little penguins, *Eudyptula minor*, show increased avoidance, aggression and vigilance in response to zoo visitors. *Applied Animal Behaviour Science*, 168, 71–76.

Sherwood, K. P., Rallis, S. F. and Stone, J. (1989). Effects of live vs. preserved specimens on student learning. *Zoo Biology*, 8, 99–104.

Shettel-Neuber, J. and O'Reilly, J. (1981). *Now Where? A Study of Visitor Orientation and Circulation at the Arizona-Sonora Desert Museum*. Technical Report No. 87-25. Psychology Institute, Jacksonville State University, Jacksonville.

Shier, D. (2016). Manipulating behavior to ensure reintroduction success. In: Berger-Tal, O. and Saltz, D. (eds.), *Conservation Behavior: Applying Behavioral Ecology to Wildlife Conservation Management*. Cambridge University Press, Cambridge, pp. 275–304.

Shier, D. M. and Swaisgood, R. R. (2012). Fitness costs of neighborhood disruption in translocations of a solitary mammal. *Conservation Biology*, 26, 116–123.

Shopland, S., Barbon, A. R., Cotton, S., Whitford, H. and Barrows, M. (2020). Retrospective review of mortality in captive pink pigeons (*Nesoenas mayeri*) housed in European collections: 1977–2018. *Journal of Zoo and Wildlife Medicine*, 51, 159–169.

Shora, J., Myhill, M. and Brereton, J. E. (2018). Should zoo foods be coati chopped. *Journal of Zoo and Aquarium Research*, 6, 22–25.

Shoshani, J. and Eisenberg, J. F. (1982). Elephas maximus. *Mammalian Species*, 182, 1–8.

Shyne, A. (2006). Meta-analytic review of the effects of enrichment on stereotypic behavior in zoo mammals. *Zoo Biology*, 25, 317–337.

Sickler, J. and Fraser, J. (2009). Enjoyment in zoos. *Leisure Studies*, 28, 313–331.

Siew, K. M., Ramachandran, S., Siow, M. L., Shuib, A. and Kunasekaran, P. (2018). Visitors' level of awareness on safety instructions at Giant Panda Conservation Centre (GPCC), Zoo Negara, Malaysia. *International Journal of Business and Society*, 19, 103–116.

Silinski, S., Walzer, C., Schwarzenberger, F., Slotta-Bachmayr, L. and Stolla, R. (2002). Pharmacological methods of enhancing penile erection for ex-copula semen collection in standing white rhinoceros (*Ceratotherium simium simum*). In *European Association of Zoo and Wildlife Veterinarians (EAZWV) 4th Scientific Meeting Joint With the Annual Meeting of the European Wildlife Diseases Association (EXDA)*, Heidelberg, pp. 391–394.

Silva, G. F., Gomes, J. E., Cunha, R., et al. (2022). Fatal congenital and traumatic cervical spine injuries in a newborn plains zebra (*Equus quagga*). *Open Veterinary Journal*, 12, 75–79.

Sim, R. R., Stringer, E., Donovan, D., et al. (2017). Use of composite materials as a component of tusk fracture management in an Asian elephant (*Elephas maximus*) and an African elephant (*Loxodonta africana*). *Journal of Zoo and Wildlife Medicine*, 48, 891–896.

Singer, P. (1975). *Animal Liberation: A New Ethics of Our Treatment of Animals*. Jonathan Cape Ltd, London.

Singer, P. (1995). *Animal Liberation* (2nd ed.). Pimlico, London.

Sitompul, A. F., Griffin, C. R., Rayl, N. D. and Fuller, T. K. (2013). Spatial and temporal habitat use of an Asian elephant in Sumatra. *Animals*, 3, 670–679.

Skibiel, A. L., Trevino, H. S. and Naugher, K. (2007). Comparison of several types of enrichment for captive felids. *Zoo Biology*, 26, 371–381.

Smart, T., Counsell, G. and Quinnell, R. J. (2021). The impact of immersive exhibit design on visitor behaviour and learning at Chester Zoo, UK. *Journal of Zoo and Aquarium Research*, 9, 139–149.

Smith, A., Lindburg, D. G. and Vehrencamp, S. (1989). Effect of food preparation on feeding behavior of lion-tailed macaques. *Zoo Biology*, 8, 57–65.

Smith, B. and Hutchins, M. (2000). The value of captive breeding programmes to field conservation: elephants as an example. *Pachyderm*, 28, 101–109.

Smith, B., Hutchins, M., Allard, R. and Warmolts, D. (2002). Regional collection planning for speciose taxonomic groups. *Zoo Biology*, 21, 313–320.

Smith, K. M., Murray, S. and Sanchez, C. (2005). Successful treatment of suspected exertional myopathy in a rhea (*Rhea americana*). *Journal of Zoo and Wildlife Medicine*, 36, 316–320.

Smith, L. and Broad, S. (2008). Do zoo visitors attend to conservation messages? A case study of an elephant exhibit. *Tourism Review International*, 11, 225–235.

Smith, L., Broad, S. and Weiler, B. (2008). A closer examination of the impact of zoo visits on visitor behaviour. *Journal of Sustainable Tourism*, 16, 544–562.

Smith, L., Weiler, B. Smith, A. and van Dijk, P. (2012). Applying visitor preference criteria to choose pro-wildlife behaviours to ask of zoo visitors. *Curator: The Museum Journal*, 55, 453–466.

Smith, R. J. and Barley, R. (eds.) (2019). *Living Collections Strategy 2019*. Royal Botanic Gardens, Kew.

Smits, K., Hoogewijs, M., Woelders, H., Daels, P. and Van Soom, A. (2012). Breeding or assisted reproduction? Relevance of the horse model applied to the conservation of endangered equids. *Reproduction in Domestic Animals*, 47, 239–248.

Sommerfeld, R., Bauert, M., Hillman, E. and Stauffacher, M. (2006). Feeding enrichment by self-operated food boxes for white-fronted lemurs (*Eulemur fulvus albifrons*) in the Masoala exhibit of the Zurich Zoo. *Zoo Biology*, 25, 145–154.

Sonnweber, R., Ravignani, A. and Fitch, W. (2015). Non-adjacent visual dependency learning in chimpanzees. *Animal Cognition*, 18, 733–745.

South, A., Rushton, S. and Macdonald, D. (2000). Simulating the proposed reintroduction of the European beaver (*Castor fiber*) to Scotland. *Biological Conservation*, 93, 103–116.

Sowińska, N. (2021). The domestic cat as a research model in the assisted reproduction procedures of wild felids. *Postepy Biochemii*, 67, 362–369.

Spalton, A. (1993). A brief history of the reintroduction of the Arabian oryx *Oryx leucoryx* into Oman 1980–1992. *International Zoo Yearbook*, 32, 81–90.

Spalton, J. A., Lawrence, M. W. and Brend, S. A. (1999). Arabian oryx reintroduction in Oman: successes and setbacks. *Oryx*, 33, 168–175.

Spooner, S. L., Jense, E. A., Tracey, L. and Marshall, A. R. (2021). Evaluating the effectiveness of live animal shows at delivering information to zoo audiences. *International Journal of Science Education, Part B*, 11, 1–16.

Sripiboon, S., Tanhaew, P., Lungka, G. and Thitaram, C. (2013). The occurrence of elephant endotheliotropic herpesvirus in captive Asian elephants (*Elephas maximus*): first case of EEHV4 in Asia. *Journal of Zoo and Wildlife Medicine*, 44, 100–104.

Stanley-Price, M. R. (1989). *Animal Reintroductions: the Arabian Oryx in Oman*. Cambridge University Press, Cambridge.

Stevens, E. F. and Pickett, C. (1994). Managing the social environments of flamingos for reproductive success. *Zoo Biology*, 13, 501–507.

Stevens, P. E., Hill, H. M. and Bruck, J. N. (2021). Cetacean acoustic welfare in wild and managed-care settings: gaps and opportunities. *Animals*, 11, 3312. http://doi.org/10.3390/ani11113312.

STL Zoo (2014). Saint Louis Zoo. www.stlzoo.org/about/economicimpact/ (accessed 15 August 2017).

Stoinski, T. S. and Beck, B. B. (2004). Changes in locomotor and foraging skills in captive-born, reintroduced golden lion tamarins (*Leontopithecus roselia roselia*). *American Journal of Primatology*, 62, 1–13.

Stoinski, T. S., Lukas, K. E. and Maple, T. L. (1998). A survey of research in North American zoos and aquariums. *Zoo Biology*, 17, 167–180.

Stoinski, T. S., Daniel, E. and Maple, T. L. (2000). A preliminary study of the behavioral effects of feeding enrichment on African elephants. *Zoo Biology*, 19, 485–493.

Stoinski, T. S., Hoff, M. P. and Maple, T. L. (2001). Habitat use and structural preferences of captive Western lowland gorillas (*Gorilla gorilla gorilla*): effects of environmental and social variables. *International Journal of Primatology*, 22, 431–447.

Stoinski, T. S., Allen, M. T., Bloomsmith, M. A., Forthman, D. L. and Maple, T. L. (2002). Educating zoo visitors about complex environmental issues: should we do it and how? *Curator: The Museum Journal*, 45, 129–143.

Stoinski, T. S., Lukas, K. E., Kuhar, C. W. and Maple, T. L. (2004). Factors influencing the formation and maintenance of all-male gorilla groups in captivity. *Zoo Biology*, 23, 189–203.

Stoinski, T. S., Lukas, K. E. and Kuhar, C. W. (2013). Effects of age and group type on social behaviour of Western gorillas (*Gorilla gorilla gorilla*) in North American Zoos. *Applied Animal Behaviour Science*, 147, 316–323.

Suddendorf, T. and Collier-Baker, E. (2009). The evolution of primate visual self-recognition: evidence of absence in lesser apes. *Proceedings of the Royal Society B: Biological Sciences*, 276(1662), 1671–1677.

Sulser, C. E., Steck, B. L. and Baur, B. (2008). Effects of construction noise on behaviour of and exhibit use by snow leopards *Uncia uncia* at Basel Zoo. *International Zoo Yearbook*, 42, 199–205.

Summers, L., Clingerman, K. J. and Yang, X. (2012). Validation of a body condition scoring system in rhesus macaques (*Macaca mulatta*): assessment of body composition by using dual-energy X-ray absorptiometry. *Journal of the American Association for Laboratory Animal Science*, 51, 88–93.

Suter, I. C., Maurer, G. P. and Baxter, G. (2014). Population viability of captive Asian elephants in the Lao PDR. *Endangered Species Research*, 24, 1–7.

Suzuki, M., Hirako, K., Saito, S., et al. (2008). Usage of high-performance mattresses for transport of Indo-Pacific bottlenose dolphin. *Zoo Biology*, 27, 331–340.

Swaisgood, R. R., Dickman, D. and White, A. M. (2006). A captive population in crisis: testing hypotheses for reproductive failure in captive-born southern white rhinoceros females. *Biological Conservation*, 129, 468–476.

Swaisgood, R. R., Wang, D. and Wei, F. (2018). Panda downlisted but not out of the woods. *Conservation Letters*, 11(1), e12355.

Swanagan, J. S. (2000). Factors influencing zoo visitors' conservation attitudes and behavior. *Journal of Environmental Education*, 31, 26–31.

Swanson, W. F., Stoops, M. A., Magarey, G. M. and Herrick, J. R. (2007). Sperm cryopreservation in endangered felids: developing linkage of in situ–ex situ populations. *Society of Reproduction and Fertility*, 65(suppl.), 417.

Świderska-Kiełbik, S., Krakowia, K. A., Wiszniewska, M., et al. (2009). Work-related respiratory symptoms in bird zoo keepers: questionnaire data. *International Journal of Occupational Medicine and Environmental Health*, 22, 393–399.

Swim, J. and Fraser, J. (2014). Zoo and aquarium professionals' concerns and confidence about climate change education. *Journal of Geoscience Education*, 62, 495–501.

Swinnerton, K. J., Groombridge, J. J., Jones, C. G., Burn, R. W. and Mungroo, Y. (2004). Inbreeding depression and founder diversity among captive and free-living populations of the endangered pink pigeon *Columba mayeri*. *Animal Conservation*, 7, 353–364.

Switzer, W. M., Bhullar, V., Shanmugam, V., et al. (2004). Frequent simian foamy virus infection in persons occupationally exposed to nonhuman primates. *Journal of Virology*, 78, 2780–2789.

Szleszkowski, L., Thannhäuser, A. and Jurek, T. (2017). Compound mechanism of fatal neck injury: a case report of a tiger attack in a zoo. *Forensic Science International*, 277, e16–e20.

Taber, A. B. (1987). The behavioral ecology of the mara, *Dolichotis patagonum*. Doctoral dissertation, Oxford University, Oxford.

Tafalla, M. (2017). The aesthetic appreciation of animals in zoological gardens. *Contemporary Aesthetics*, 15. https://ddd.uab.cat/pub/artpub/2017/178154/2017-zoos-contempa-publicado.pdf.

Tait, P. (2016). *Fighting Nature: Travelling Menageries, Animal Acts and War Shows*. Sydney University Press, Sydney.

Takasu, M., Morita, N., Tajima, S., et al. (2016). Cryopreservation of lar gibbon semen collected by manual stimulation. *Primates*, 57, 303–307.

Tanner, J. E. and Byrne, R. W. (1999). The development of spontaneous gestural communication in a group of zoo-living lowland gorillas. In: Parker, S. T., Mitchell, R. W. and Miles, H. L. (eds.), *The Mentalities of Gorillas and Orangutans: Comparative Perspectives*. Cambridge University Press, Cambridge, pp. 211–239.

Taronga Zoo (2017). Indigenous programs. https://taronga.org.au/education/education-sydney/indigenous-programs (accessed 17 August 2017).

Taylor, G., Canessa, S., Clarke, R. H., et al. (2017). Is reintroduction biology an effective applied science? *Trends in Ecology & Evolution*, 32, 873–880.

Taylor, S. M. (1986). Understanding processes of informal education: a naturalistic study of visitors to a public aquarium. Unpublished PhD dissertation, University of California at Berkeley, Berkeley.

Taylor, V. J. and Poole, T. B. (1998). Captive breeding and infant mortality in Asian elephants: a comparison between twenty western zoos and three eastern elephant centers. *Zoo Biology*, 17, 311–322.

Ten, D. C. Y., Edinur, H. A., Jani, R., Hashim, N. H. and Abdullah, M. T. (2021). Covid 19 and the Malaysian zoo preventative measure readiness. *Journal of Sustainability Science and Management*, 16, 46–54.

Tetley, C. L. and O'Hara, S. J. (2012). Ratings of animal personality as a tool for improving the breeding, management and welfare of zoo mammals. *Animal Welfare*, 21, 463–476.

Tetzlaff, S. J., Sperry, J. H. and DeGregorio, B. A. (2019a). Effects of antipredator training, environmental enrichment, and soft release on wildlife translocations: a review and meta-analysis. *Biological Conservation*, 236, 324–331.

Tetzlaff, S. J., Sperry, J. H., Kingsbury, B. A. and DeGregorio, B. A. (2019b). Captive-rearing duration may be more important than environmental enrichment for enhancing turtle head-starting success. *Global Ecology and Conservation*, 20. https://doi.org/10.1016/j.gecco .2019.e00797.

Thiangtum, K., Swanson, W. F., Howard, J., et al. (2006). Assessment of basic seminal characteristics, sperm cryopreservation and heterologous in vitro fertilisation in the fishing cat (*Prionailurus viverrinus*). *Reproduction, Fertility and Development*, 18, 373–382.

Thode-Arora, H. (2008). Hagenbeck's European tours: the development of the human zoo. In: Blanchard, P., Bancel, N., Boëtsch, G., et al. (eds). *Human Zoos: Science and Spectacle in the Age of Colonial Empires*. Liverpool University Press, Liverpool, pp. 165–173.

Thomas, J. A., Kastelein, R. A. and Awbrey, F. T. (1990). Behavior and blood catecholamines of captive belugas during plackbacks of noise from an oil drilling platform. *Zoo Biology*, 9, 393–402.

Thompson, K. A., Henderson, E., Fitzgerald, S. D., Walker, E. D. and Kiupel, M. (2021). Eastern equine encephalitis virus in Mexican wolf pups at zoo, Michigan, USA. *Emerging Infectious Diseases*, 27, 1173–1176.

Thorpe, S. K. S., Crompton, R. H. and Wang, W. J. (2004). Stresses exerted in the hindlimb muscles of common chimpanzees (*Pan troglodytes*) during bipedal locomotion. *Folia Primatologica*, 75, 253–265.

Tidière, M., Gaillard, J.-M., Berger, V., et al. (2016). Comparative analyses of longevity and senescence reveal variable survival benefits of living in zoos across mammals. *Science Reports*, 6, 36361. http://doi.org/10.1038/srep36361.

Tilson, R. L., Sweeny, K. A., Binczik, G. A. and Reindl, N. J. (1988). Buddies and bullies: social structure of a bachelor group of Przewalski horses. *Applied Animal Behaviour Science*, 21, 169–185.

The Times (2017). Danger: flying lemurs. *The Times*, 6 October, p. 6.

Tishler, C., Assaraf, O. B. Z. and Fried, M. N. (2020). How do visitors from different cultural backgrounds perceive the messages conveyed to them by their local zoo. *Interdisciplinary Journal of Environmental and Science Education,* 16(3), e2216. https://doi.org/10.29333/ ijese/8335.

Titus, S. E., Patterson, S., Prince-Wright, J., Dastjerdi, A. and Molenaar, F. M. (2022). Effects of between and within herd moves on elephant endotheliotropic herpesvirus (EEHV) recrudescence and shedding in captive Asian elephants (*Elephas maximus*). *Viruses*, 14, 229. https://doi.org/10.3390/v14020229.

Tlustry, M. F., Rhyne, A. L., Kaufman, L., et al. (2013). Opportunities for public aquariums to increase the sustainability of the aquatic animal trade. *Zoo Biology*, 32, 1–12.

Toledo, D., Agudelo, M. S. and Bentley, A. L. (2011). The shifting of ecological restoration benchmarks and their social impacts: digging deeper into Pleistocene re-wilding. *Restoration Ecology*, 19, 564–568.

Toni, P. (2017). Combat leads to intraspecific killing in Eastern grey kangaroos. *Australian Mammalogy*, 40, 109–111.

Toscano, G. (1997). Dangerous jobs. In *Fatal Workplace Injuries in 1995: A Collection of Data and Analysis*. US Department of Labor, Washington, pp. 38–41.

Toscano, M. J., Friend, T. H. and Nevill, C. H. (2001). Environmental conditions and body temperature of circus elephants transported during relatively high and low temperature conditions. *Journal of the Elephant Managers Association*, 12, 115–149.

Traylor-Holzer, K., Leus, K. and Bauman, K. (2019). Integrated collection assessment and planning (ICAP) workshop: helping zoos move towards the One Plan Approach. *Zoo Biology*, 38, 95–105.

Tribe, A. and Booth, R. (2003). Assessing the role of zoos in wildlife conservation. *Human Dimensions of Wildlife*, 8, 65–74.

Tuite, E. K., Moss, S. A., Phillips, C. J. and Ward, S. J. (2022). Why are enrichment practices in zoos difficult to implement effectively? *Animals*, 12, 554. https://doi.org/10.3390/ani12050554.

Turkowski, F. J. (1972). Education at zoos and aquariums in the United States. *Bioscience*, 22, 468–475.

Turley, S. K. (1999). Exploring the future of the traditional UK zoo. *Journal of Vacation Marketing*, 5, 340–355.

Turley, S. K. (2001). Children and the demand for recreational experiences: the case of zoos. *Leisure Studies*, 20, 1–18.

Turner, J. W., Nemeth, R. and Rogers, C. (2003). Measurement of fecal glucocorticoids in parrotfishes to assess stress. *General and Comparative Endocrinology*, 133, 341–352.

Turvey, S. T., Walsh, C., Hansford, J. P., et al. (2019). Complementarity, completeness and quality of long-term faunal archives in an Asian biodiversity hotspot. *Philosophical Transactions of the Royal Society B*, 374(1788), 20190217.

Tyson, E. (2021). *Licensing Laws and Animal Welfare: The Legal Protection of Wild Animals*. Palgrave Macmillan, London.

Úbeda, Y., Ortín, S., Robeck, T. R., Llorente, M. and Almunia, J. (2021). Personality of killer whales (*Orcinus orca*) is related to welfare and subjective well-being. *Applied Animal Behaviour Science*, 237, 105297.

Ueno, A. and Matsuzawa, T. (2005). Response to novel food in infant chimpanzees: do infants refer to mothers before ingesting food on their own? *Behavioural Processes*, 68, 85–90.

Uher, J. and Asendorpf, J. B. (2008). Personality assessment in the Great Apes: comparing ecologically valid behavior measures, behavior ratings, and adjective ratings. *Journal of Research in Personality*, 42, 821–838.

USFWS (2007). U.S. Fish and Wildlife Service reintroduces black-footed ferrets in Logan County, Kansas. News release, 20 December. www.fws.gov/mountain-prairie/PRESSREI/07-74.htm (accessed 25 April 2017).

Van De Bunte, W., Weerman, J. and Hof, A. R. (2021). Potential effects of GPS collars on the behaviour of two red pandas (*Ailurus fulgens*) in Rotterdam Zoo. *PLoS ONE*, 16, e0252456.

Van der Poel, W. H. M., Van der Heide, R., Van Amerongen, G., et al. (2000). Characterisation of a recently isolated lyssavirus in frugivorous zoo bats. *Archives of Virology*, 145, 1919–1931.

van Hoek, C. S. and Ten Cate, C. (1998). Abnormal behavior in caged birds kept as pets. *Journal of Applied Animal Welfare Science*, 1, 51–64.

van Hooff, J. A., Dienske, H., Jens, W. and Spijkerman, R. P. (1996). Differences in variability, interactivity and skills in social play of young chimpanzees living in peer groups and in a large family zoo group. *Behaviour*, 133, 717–739.

Van Houtan, K. S., Gagne, T., Jenkins, C. N. and Joppa, L. (2020). Sentiment analysis of conservation studies captures successes of species reintroductions. *Patterns*, 1(1), 100005.

Van Praag, H., Kempermann, G. and Gage, F. H. (2000). Neural consequences of enviromental enrichment. *Nature Reviews Neuroscience*, 1, 191–198.

Van Reybrouck, D. (2005). Archaeology and urbanism: railway stations and zoological gardens in the 19th-century cityscape. *Public Archaeology*, 4, 225–241.

Vandevoort, C. A., Neville, L. E., Tollener, T. L. and Field, P. P. (1993). Noninvasive semen collection from an adult orang-utan. *Zoo Biology*, 12, 257–265.

Vandiver, M. M. (1978). Keeper training at Santa Fe Community College: a progress report. *International Zoo Yearbook*, 18, 237–240.

Vargas, A. and Anderson, S. H. (1996). Effects of diet on captive black-footed ferret (*Mustela nigripes*) food preference. *Zoo Biology*, 15, 105–113.

Varner, G. (2008). Personhood, memory, and elephant management. In: Wemmer, C. and Christen C. A. (eds.), *Elephants and Ethics: Towards a Morality of Coexistence*. Johns Hopkins University Press, Baltimore, pp. 41–67.

Veado, B. V. (1997). Parental behaviour in maned wolf *Chrysocyon brachyurus* at Belo Horizonte Zoo. *International Zoo Yearbook*, 35, 279–286.

Veado, B. V. (2005). Paternal behaviour of maned wolf *Chrysocyon brachyurus* at Fundação Zoo-Botânica de Belo Horizonte. *International Zoo Yearbook*, 39, 198–205.

Veasey, J. S. (2020). Can zoos ever be big enough for large wild animals? A review using an expert panel assessment of the psychological priorities of the Amur tiger (*Panther tigris altaica*) as a model species. *Animals*, 10(9), 1536. http://doi.org/10.3390/ani10091536.

Veasey, J. S., Waran, N. K. and Young, R. J. (1996). On comparing the behaviour of zoo housed animals with wild conspecifics as a welfare indicator, using the giraffe (*Giraffa camelopardalis*) as a model. *Animal Welfare*, 5, 139–153.

Vergne, A. L., Aubin, T., Martin, S. and Mathevon, N. (2012). Acoustic communication in crocodilians: information encoding and species specificity of juvenile calls. *Animal Cognition*, 15, 1095–1109.

Vester, B. M., Burke, S. L., Dikeman, C. L., Simmons, L. G. and Swanson, K. S. (2008). Nutrient digestibility and fecal characteristics are different among captive exotic felids fed a beef-based raw diet. *Zoo Biology*, 27, 126–136.

Vick, S.-J., Waller, B. M., Parr, L. A., Smith Pasqualini, M. C. and Bard, K. (2007). A cross-species comparison of facial morphology and movement in humans and chimpanzees using the Facial Action Coding System (FACS). *Journal of Nonverbal Behavior*, 31, 1–20.

Victoria, J. K. and Collias, N. E. (1973). Social facilitation of egg-laying in experimental colonies of a weaverbird. *Ecology*, 54, 399–405.

Videan, E. N., Fritz, J. and Murphy, J. (2007). Development of guidelines for assessing obesity in captive chimpanzees (*Pan troglodytes*). *Zoo Biology*, 26, 93–104.

Vincze, O., Colchero, F., Lemaître, J.-F., et al. (2022). Cancer risk across mammals. *Nature*, 601, 263–267.

Wagner, W. M., Hartley, M. P., Duncan, N. M. and Barrows, M. G. (2005). Spinal spondylosis and acute intervertebral disc prolapse in a European brown bear (*Ursus arctos arctos*): clinical communication. *Journal of the South African Veterinary Association*, 76, 120–122.

Walker, B. N., Kim, J. and Pendse, A. (2007). Musical soundscapes for an accessible aquarium: bringing dynamic exhibits to the visually impaired. In *International Computer Music Conference*, Copenhagen.

Walker, E. H., Verschueren, S., Schmidt-Küntzel, A. and Marker, L. (2022). Recommendations for the rehabilitation and release of wild-born, captive-raised cheetahs: the importance of pre- and post-release management for optimizing survival. *Oryx*. https://doi.org/10.1017/S0030605321000235.

Wallace, C., Marinkovich, M., Morris, P. J., Rideout, B. and Pye, G. W. (2016). Lessons from a retrospective analysis of a 5-yr period of quarantine at San Diego Zoo: a risk-based approach to quarantine isolation and testing may benefit animal welfare. *Journal of Zoo and Wildlife Medicine*, 47, 291–296.

Wallace, E. K., Altschul, D., Körfer, K., et al. (2017). Is music enriching for group-housed captive chimpanzees (*Pan troglodytes*)? *PLoS ONE*, 12(3), e0172672.

Wallace, E. K., Herrelko, E. S., Koski, S. E., et al. (2019). Exploration of potential triggers for self-directed behaviours and regurgitation and reingestion in zoo-housed chimpanzees. *Applied Animal Behaviour Science*, 221, 104878.

Wallace, E. K., Kingston-Jones, M., Ford, M. and Semple, S. (2013). An investigation into the use of music as potential auditory enrichment for moloch gibbons (*Hylobates moloch*). *Zoo Biology*, 32, 423–426.

Wallace, M. P. (1994). Control of behavioral development in the context of reintroduction programs for birds. *Zoo Biology*, 13, 491–499.

Wallace, R. S. and Bush, M. (1987). Exertional myopathy complicated by a ruptured bladder in a dama gazelle (*Gazella dama*). *Journal of Zoo Animal Medicine*, 18, 111–114.

Wallace, R. S., Bush, M. and Montali, R. J. (1985). Deaths from exertional myopathy at the National Zoological Park from 1975 to 1985. *Journal of Wildlife Diseases*, 23, 454–462.

Walples, K. A. and Gales, N. J. (2002). Evaluating and minimising social stress in the care of captive bottlenose dolphins (*Tursiops aduncus*). *Zoo Biology*, 21, 5–26.

Walraven, V., Van Elsacker, L. and Verheyen, R. (1995). Reactions of a group of pygmy chimpanzees (*Pan paniscus*) to their mirror-images: evidence of self-recognition. *Primates*, 36, 145–150.

Walsh, B. (2017). Asian elephant (*Elephas maximus*) sleep study: long-term quantitative research at Dublin Zoo. *Journal of Zoo and Aquarium Research*, 5, 82–85.

Walzer, C., Pucher, H. and Schwarzenberger, F. (2000). A restraint chute for semen collection in white rhinoceros (*Ceratotherium simum simum*): preliminary results. In *Proceedings of the European Association of Zoo and Wildlife Veterinarians (EAZWV) Third Scientific Meeting*, Paris.

WAP (2022). World Animal Protection. www.api.worldanimalprotection.org (accessed 4 April 2022).

Ward, M. P., Ramer, J. C., Proudfoot, J., et al. (2003). Outbreak of salmonellosis in a zoological collection of lorikeets and lories (*Trichoglossus, Lorius*, and *Eos* spp.). *Avian Diseases*, 47, 493–498.

Ward, P. I., Mosberger, N., Kistler, C. and Fischer, O. (1998). The relationship between popularity and body size in zoo animals. *Conservation Biology*, 12, 1408–1411.

Ward, S. J. and Melfi, V. (2015). Keeper–animal interactions: differences between the behaviour of zoo animals affect stockmanship. *PLoS ONE*, 10(10), e0140237.

Warin, R. and Warin, A. (1985). *Portrait of a Zoo: Bristol Zoological Gardens 1835–1985*. Redcliffe Press Ltd, Bristol.

Wark, J. D., Kuhar, C. W. and Lukas, K. E. (2014). Behavioral thermoregulation in a group of zoo-housed colobus monkeys (*Colobus guereza*). *Zoo Biology*, 33, 257–266.

Wark, J. D., Wierzal, N. K. and Cronin, K. A. (2020). Mapping shade availability and use in zoo environments: a tool for evaluating thermal comfort. *Animals*, 10(7), 1189.

Waroff, A. J., Fanucchi, L., Robbins, C. T. and Nelson, O. L. (2017). Tool use, problem-solving, and the display of stereotypic behaviors in the brown bear (*Ursus arctos*). *Journal of Veterinary Behavior*, 17, 62–68.

Wascher, C. A. F., Baur, N., Hengl, M., et al. (2021). Behavioural responses of captive corvids to the presence of visitors. *Animal Behavior and Cognition*, 8, 481–492.

Watson, S. K., Townsend, S. W., Schel, A. M., et al. (2015). Vocal learning in the functionally referential food grunts of chimpanzees. *Current Biology*, 25, 495–499.

Watts, P. C., Buley, K. R., Sanderson, S., et al. (2006). Parthenogenesis in komodo dragons. *Nature*, 444, 1021–1022.

Waugh, D. (1988). Training in zoo biology, captive breeding, and conservation. *Zoo Biology*, 7, 269–280.

WAZA (2020). *Guidelines for Animal–Visitor Interactions*. World Association of Zoos and Aquariums, Barcelona.

Webster, K., Narayan, E. and De Vos, N. (2017). Fecal glucocorticoid metabolite response of captive koalas (*Phascolarctos cinereus*) to visitor encounters. *General and Comparative Endocrinology*, 244, 157–163.

Wechsler, B. (1995). Coping and coping strategies: a behavioural review. *Applied Animal Behaviour Science*, 43, 123–134.

Wei, W., Swaisgood, R. R., Dai, Q., et al. (2018). Giant panda distributional and habitat-use shifts in a changing landscape. *Conservation Letters*, 11(6), e12575.

Weinstein, T. A., Capitanio, J. P. and Gosling, S. D. (2008). Personality in animals. *Handbook of Personality: Theory and Research*, 3, 328–348.

Weiss, E. and Wilson, S. (2003). The use of classical and operant conditioning in training Aldabra tortoises (*Geochelone gigantea*) for venipuncture and other husbandry issues. *Journal of Applied Animal Welfare Science*, 6, 33–38.

Weissenbacher, A., Preininger, D., Ghosh, R., Morshed, A. G. J. and Praschag, P. (2015). Conservation breeding of the Northern river terrapin *Batagur baska* at the Vienna Zoo, Austria, and in Bangladesh. *International Zoo Yearbook*, 49, 31–41.

Welden, L. A., Stelvig, M., Nielsen, C. K., et al. (2020). The contributions of EAZA zoos and aquaria to peer-reviewed scientific research. *Journal of Zoo and Aquarium Research*, 8, 133–138.

Weldon, C., Crottini, A., Bollen, A., et al. (2013). Pre-emptive national monitoring plan for detecting the amphibian chytrid fungus in Madagascar. *EcoHealth*, 10, 234–240.

Wells, D. L. and Egli, J. M. (2004). The influence of olfactory enrichment on the behaviour of black-footed cats, *Felis nigripes*. *Applied Animal Behaviour Science*, 85, 107–119.

Wells, D. L. and Irwin, R. M. (2008). Auditory stimulation as enrichment for zoo-housed Asian elephants (*Elephas maximus*). *Animal Welfare*, 17, 335–340.

Wells, D. L. and Irwin, R. M. (2009). The effect of feeding enrichment on the moloch gibbon (*Hylobates moloch*). *Journal of Applied Animal Welfare Science*, 12, 21–29.

Wells, D. L., Coleman, D. and Challis, M. G. (2006). A note on the effect of auditory stimulation on the behaviour and welfare of zoo-housed gorillas. *Applied Animal Behaviour Science*, 100, 327–332.

Wemmer, C. and Christen, C. A. (eds.) (2008). *Elephants and Ethics: Towards a Morality of Coexistence*. Johns Hopkins University Press, Baltimore.

Wemmer, C. M. and Derrickson, S. R. (1987). Vertebrate reintroductions and translocations: an overview. *International Journal of Primatology*, 8, 424–424.

Wemmer, C., Pickett, C. and Teare, J. A. (1990). Training zoo biology in tropical countries: a report on a method and progress. *Zoo Biology*, 9, 461–470.

Wemmer, C., Rodden, M. and Pickett, C. (1997). Publication trends in zoo biology: a brief analysis of the first 15 years. *Zoo Biology*, 16, 3–8.

Wemmer, C., Krishnamurthy, V., Shrestha, S., et al. (2006). Assessment of body condition in Asian elephants (*Elephas maximus*). *Zoo Biology*, 25, 187–200.

Wenker, C. J., Stich, H., Müller, M. and Lussi, A. (1999). A retrospective study of dental conditions of captive brown bears (*Ursus arctos* spp.) compared with free-ranging Alaskan grizzles (*Ursus arctos horribilis*). *Journal of Zoo and Wildlife Medicine*, 30, 208–221.

West, P. C. and Virgene, H. (1990). Minorities and the Detroit Zoo. *Visitor Studies*, 2, 149–152.

Westlund, K. (2014). Training is enrichment: and beyond. *Applied Animal Behaviour Science*, 152, 1–6.

Wheeler, C. L. and Fa, J. E. (1995). Enclosure utilisation and activity in Round Island geckos (*Phelsuma guentheri*). *Zoo Biology*, 14, 361–369.

White, B. C., Beare, J., Fuller, J. A. and Houser, L. A. (2003). Social spacing in a bachelor group of related captive brown woolly monkey (*Lagothrix lagotricha*). *Neotropical Primates*, 11, 35–38.

White, J. and Barry, S. (1984). *Science Education for Families in Informal Learning Settings: An Evaluation of the Herp Lab Project*. National Zoological Park, Washington.

Whitehouse, J., Waller, B. M., Chanvin, M., et al. (2014). Evaluation of public engagement activities to promote science in a zoo environment. *PLoS ONE*, 9(11), e113395.

Whiten, A., Allan, G., Devlin, S., et al. (2016). Social learning in the real-world: 'over-imitation' occurs in both children and adults unaware of participation in an experiment and independently of social interaction. *PLoS ONE*, 11(7), e0159920.

Whitham, J. C. and Wielebnowski, N. (2009). Animal-based welfare monitoring: using keeper ratings as an assessment tool. *Zoo Biology*, 28, 545–560.

WHO (2022). Multi-country monkeypox outbreak: situation update 27 June 2022. www.who .int/emergencies/disease-outbreak-news/item/2022-DON396 (accessed 3 July 2022).

Wich, S. A. and Koh, L. P. (2018). *Conservation Drones: Mapping and Monitoring Biodiversity*. Oxford University Press, Oxford.

Wich, S. A., Shumaker, R. W., Perkins, L. and de Vries, H. (2009). Captive and wild orangutan (*Pongo* sp.) survivorship: a comparison and the influence of management. *American Journal of Primatology*, 71, 680–686.

Wiedenmayer, C. (1998). Food hiding and enrichment in captive Asian elephants. *Applied Animal Behaviour Science*, 56, 77–82.

Wielebnowski, N. (1996). Reassessing the relationship between juvenile mortality and genetic monomorphism in captive cheetahs. *Zoo Biology*, 15, 353–369.

Wielebnowski, N. C., Fletchall, N., Carlstead, K., Busso, J. M. and Brown, J. L. (2002). Noninvasive assessment of adrenal activity associated with husbandry and behavioral factors in the North American clouded leopard population. *Zoo Biology*, 21, 77–98

Wienker, W. R. (1986). Giraffe squeeze cage procedures. *Zoo Biology*, 5, 371–377.

Wiese, R. J. (2000). Asian elephants are not self-sustaining in North America. *Zoo Biology*, 19, 299–309.

Wiese, R. J., Hutchins, M., Willis, K. and Becker, S. (eds.) (1992). *AAZPA Annual Report on Conservation and Science*. American Association of Zoological Parks and Aquariums, Bethesda.

Wijeyamohan, S., Treiber, K., Schmitt, D. and Santiapillai, C. (2015). A visual system for scoring body condition of Asian elephants (*Elephas maximus*). *Zoo Biology*, 34, 53–59.

Wildwood Trust (2022). Wilder Blean. https://wildwoodtrust.org/Wilder-blean (accessed 17 December 2022).

Wiley, J. N., Leeds, A., Carpenter, K. D. and Kendall, C. J. (2018). Patterns of wounding in Hamadryas baboons (*Papio hamadryas*) in North American zoos. *Zoo Biology*, 37, 74–79.

Wiley, J. W., Snyder, N. F. R. and Gnam, N. F. R. (1992). Reintroduction as a conservation strategy for parrots. In: Beissinger, S. R. and Snyder, N. F. R. (eds.), *New World Parrots in Crisis*. Smithsonian Institution Press, Washington, pp. 165–200.

Williams, E., Cabana, F. and Nekaris, K. A. I. (2015). Improving diet and activity in insectivorous primates in captivity: naturalizing the diet of Northern Ceylon grey slender lorises (*Loris lydekkerianus nordicus*). *Zoo Biology*, 34, 473–482.

Williams, E., Carter, A., Rendle, J. and Ward, S. J. (2021a). Impacts of COVID-19 on animals in zoos: a longitudinal multi-species analysis. *Journal of Zoological and Botanical Gardens*, 2, 130–135.

Williams, E., Carter, A., Rendle, J. and Ward, S. J. (2021b). Understanding impacts of zoo visitors: quantifying behavioural changes of two popular zoo species during COVID-19 closures. *Applied Animal Behaviour Science*. https://doi.org/10.1016/j.applanim.2021.105253.

Williams, L. E., Abee, C. R., Barnes, S. R. and Ricker, R. B. (1988). Cage design and configuration for an arboreal species of primate. *Laboratory Animal Science*, 38, 289–291.

Williams, R. L., Porter, S. K., Hart, A. G. and Goodenough, A. E. (2012). The accuracy of behavioural data collected by visitors in a zoo environment: can visitors collect meaningful data? *International Journal of Zoology*. http://doi.org/10.1155/2012/724835.

Willis, G. P., Kapustin, N., Warrick, J. M., et al. (1999). Preventing dental calculus formation in lemurs (*Lemur catta, Eulemur fulvus collaris*) and baboons (*Papio cynocephalus*). *Journal of Zoo and Wildlife Medicine*, 30, 377–382.

Wilson, E. O. (1993). Biophilia and the conservation ethic. In: Kellert, S. and Wilson, E. O. (eds.), *The Biophilia Hypothesis*. Island Press, Washington, pp. 31–41.

Wilson, J. W., Bergl, R. A., Minter, L. J., Loomis, M. R. and Kendall, C. J. (2019). The African elephant *Loxodonta* spp. conservation programmes of North Carolina Zoo: two decades of using emerging technologies to advance in situ conservation efforts. *International Zoo Yearbook*, 53, 151–160.

Wilson, S. and Zimmerman, A. (2005). *Zoos and Conservation Bibliography 2005*. North of England Zoological Society. www.chesterzoo.org/conservation.asp (accessed 1 June 2006).

Wilson, S. C., Holder, H. W., Martin, J. M., et al. (2006). An indoor air quality study of an alligator (*Alligator mississippiensis*) holding facility. *Journal of Zoo and Wildlife Medicine*, 37, 108–115.

Wilson, S., Davies, T., Hazarika, N. and Zimmermann, A. (2015). Understanding spatial and temporal patterns of human–elephant conflict in Assam, India. *Oryx*, 49, 140–149.

Wilson, V., Lefevre, C. E., Morton, F. B., et al. (2014). Personality and facial morphology: links to assertiveness and neuroticism in capuchins (*Sapajus [Cebus] apella*). *Personality and Individual Differences*, 58, 89–94.

Wilting, A., Courtiol, A., Christiansen, P., et al. (2015). Planning tiger recovery: understanding intraspecific variation for effective conservation. *Science Advances*, 1(5), e1400175.

Wise, S. (2015). Update on the Sandra Orangutan Case in Argentina. Nonhuman Rights Project. www.nohumanrights.org/blog/update-on-the-sandra-orangutan-case-in-argentina (accessed 17 April 2022).

Wojtusik, J., Roth, T. L. and Curry, E. (2021). Evaluation of polar bear (*Ursus maritimus*) sperm collection and cryopreservation techniques. *Reproduction, Fertility and Development*, 34, 247.

Wojtusik, J., Roth, T. L. and Curry, E. (2022). Case studies in polar bear (*Ursus maritimus*) sperm collection and cryopreservation techniques. *Animals*, 12(4), 430.

Wood, J., Ballou, J. D., Callicrate, T., et al. (2020). Applying the zoo model to conservation of threatened exceptional plant species. *Conservation Biology*, 34, 1416–1425.

Woods, B. (1998). Animals on display: principles for interpreting captive wildlife. *Journal of Tourism Studies*, 9, 28–39.

Woods, B. (2002). Good zoo/bad zoo: visitor experiences in captive settings. *Anthrozoös*, 15, 343–360.

Woodside, D. P. and Kelly, J. D. (1995). The development of local, national and international zoo-based education programmes. *International Zoo Yearbook*, 34, 231–246.

Wright, D. W. M. (2018). Cloning animals for tourism in the year 2070. *Futures*, 95, 58–75.

Wünschmann, S., Wüst-Ackermann, P., Randler, C., Vollmer, C. and Itzek-Greulich, H. (2017). Learning achievement and motivation in an out-of-school setting: visiting amphibians and reptiles in a zoo is more effective than a lesson at school. *Research in Science Education*, 47, 497. https://doi.org/10.1007/s11165-016-9513-2.

WWF-UK (2021). *WWF-UK Annual Report Summary 2020–21*. World Wildlife Fund, Woking.

WWF-US (2021). *2021 WWF-US Annual Report*. World Wildlife Fund, Washington.

Wyatt, J. R. and Eltringham, S. K. (1974). The daily activity of the elephant in Rwenzori National Park, Uganda. *East African Wildlife Journal*, 12, 273–289.

Wyse, S. V., Dickie, J. B. and Willis, K. J. (2018). Seed banking not an option for many threatened plants. *Nature Plants*, 4, 848–850.

Xu, N., Yu, J., Zhang, F., et al. (2022). Colony composition and nutrient analysis of *Polyrhachis dives* ants, a natural prey of the Chinese pangolin (*Manis pentadactyla*). *Zoo Biology*, 41, 157–165.

Yaacob, M. N. (1994). Captive-breeding and reintroduction project for the milky stork *Mycteria cinerea*: at Zoo Negara, Malaysia. *International Zoo Yearbook*, 33, 39–48.

Yalcin, E. and Aytug, N. (2007). Use of fluoxetine to treat stereotypical pacing behaviour in a brown bear (*Ursus arctos*). *Journal of Veterinary Behavior*, 2, 73–76.

Yarrell, W. (1830). December 28, 1830. Letter from Mr J. C. Cox Esq. FLS. *Proceedings of the Zoological Society of London*, 1, 17–20.

Yasui, S., Konno, A., Tanaka, M., et al. (2012). Personality assessment and its association with genetic factors in captive Asian and African elephants. *Zoo Biology*, 32, 70–78.

Yasumuro, H. and Ikeda, Y. (2011). Effects of environmental enrichment on the behaviour of the tropical octopus *Callistoctopus aspilosomatis*. *Marine and Freshwater Behaviour and Physiology*, 44, 143–157.

Yasumuro, H. and Ikeda, Y. (2016). Environmental enrichment accelerates the ontogeny of cryptic behavior in pharaoh cuttlefish (*Sepia pharaonis*). *Zoological Science*, 33, 255–265.

Yasumuro, H. and Ikeda, Y. (2018). Environmental enrichment affects the ontogeny of learning, memory, and depth perception of the pharaoh cuttlefish *Sepia pharaonis*. *Zoology*, 128, 27–37.

Yerke, R. and Burns, A. (1991). Measuring the impact of animal shows on visitor attitudes. In *AAZPA Annual Conference Proceedings*, American Association of Zoological Parks and Aquariums, Wheeling, pp. 532–539.

Yew, W. (1991). *Noah's Art: Zoo, Aquarium, Aviary and Wildlife Park Graphics*. Quon Editions, Singapore.

Yoelinda, V. T., Arifiantini, R. I., Agil, M., et al. (2020). The use of transrectal massage combined with artificial vagina as semen collection in Javan banteng (*Bos javanicus*) bull. In *E3S Web of Conferences 151*, EDP Sciences.

Yorzinski, J. L., Penkunas, M. J., Platt, M. L. and Coss, R. G. (2014). Dangerous animals capture and maintain attention in humans. *Evolutionary Psychology*, 12, 534–48.

Yoshida, H. (1997). On the training of a female bonobo for artificial insemination in the Columbus Zoo. *Pan Africa News*, 4, 16–17.

Young, R. J. (2003). *Environmental Enrichment for Captive Animals*. Blackwell Publishing, Oxford.

Young, R. P., Hudson, M. A., Terry, A. M. R., et al. (2014). Accounting for conservation: using the IUCN Red List Index to evaluate the impact of a conservation organization. *Biological Conservation*, 180, 84–96.

Yu, J. H., Brown, J., Boisseau, N., Barthel, T. and Murray, S. (2021). Effects of Lupron and surgical castration on fecal androgen metabolite concentrations and intermale aggression in capybaras (*Hydrochoerus hydrochaeris*). *Zoo Biology*, 40, 135–141.

Yuan, Y., Yuzhong, Y. and Liu, Q. (2021). Inbreeding depression and population viability analysis of the South China tigers (*Panthera tigris amoyensis*) in captivity. *Mammalian Biology*, 101, 803–809.

Žagar, Ž., Šmalc, K., Primožič, P. K., Kvapil, P. and Nemec, A. (2021). Oral and dental examinations findings in 15 zoo Bolivian squirrel monkeys (*Saimiri boliviensis*) and black-tufted marmosets (*Callithrix penicillata*). *Journal of Veterinary Dentistry*, 38, 67–74.

Zahid, U. N., Dar, L. M., Amin, U., et al. (2018). Delayed peracute capture myopathy in a Himalayan Ibex *Capra sibirica* (Mammalia: Cetartiodactyla: Bovidae). *Journal of Threatened Taxa*, 26, 12363–12367.

Zeppel, H. and Muloin, S. (2007). Indigenous wildlife interpretation at Australian zoos and wildlife parks. *Tourism Review International*, 11, 265–277.

Zhang, G., Swaisgood, R. R. and Zhang, H. (2004). Evaluation of behavioral factors influencing reproductive success and failure in captive giant pandas. *Zoo Biology*, 23, 15–31.

Zhang, L., Jiang, W., Wang, Q. J. et al. (2016). Reintroduction and post-release survival of a living fossil: the Chinese giant salamander. *PLoS ONE*, 11(6), e0156715.

Zhao, D., Wang, Y., Han, K., Zhang, H. and Li, B. (2015). Does target animacy influence manual laterality of monkeys? First answer from Northern pig-tailed macaques (*Macaca leonina*). *Animal Cognition* 18, 931–936.

Zharkikh, T. L. and Andersen, L. (2009). Behaviour of bachelor males of the Przewalski horse (*Equus ferus przewalskii*) at the reserve Askania Nova. *Der Zoologische Garten*, 78, 282–299.

Zhiyun, O., Zhenxin, L., Jianguo, L., et al. (2002). The recovery processes of giant panda habitat in Wolong Nature Reserve, Sichuan China. *Acta Ecologica Sinica*, 22, 1840–1849.

Ziegler, T. (2015). In situ and ex situ reptile projects of the Cologne Zoo: implications for research and conservation of South East Asia's herpetodiversity. *International Zoo Yearbook*, 49, 8–21.

Zimbler-DeLorenzo, H. S. and Dobson, F. S. (2011). Demography of squirrel monkeys (*Saimiri sciureus*) in captive environments and its effect on population growth. *American Journal of Primatology*, 73, 1041–1050.

Zimmerman, C., Chen, Y., Hardt, D. and Vatrapu, R. (2014). Marius, the giraffe: a comparative informatics case study of linguistic features of the social media discourse. In *Proceedings of the 5th ACM International Conference on Collaboration Across Boundaries: Culture, Distance & Technology*, pp. 131–140.

Zimmermann, A., Davies, T. E., Hazarika, N., et al. (2009). Community-based human-elephant conflict management in Assam. *Gajah*, 30, 34–40.

ZIMS (2018a). Species holding report: *Elephas maximus*/Asian elephant. Zoological Information Management System. Species 360. www.species360.org (accessed 26 March 2018).

ZIMS (2018b). Species holding report: *Loxodonta africana*/African elephant. Zoological Information Management System. Species 360. www.species360.org (accessed 26 March 2018).

Zinner, D., Wygoda, C., Razafimanantsoa, L., et al. (2014). Analysis of deforestation patterns in the central Menabe, Madagascar, between 1973 and 2010. *Regional Environmental Change*, 14, 157–166.

Zoological Society of San Diego (2008). About the Zoological Society of San Diego. www .sandiegozoo.org/disclaimers/aboutus.html (accessed 18 July 2008).

ZSI (2019). *Annual Report 2019*. Zoological Society of Ireland, Phoenix Park, Ireland.

ZSL (1829). *The Charter, By-laws and Regulations of the Zoological Society of London, Incorporated 27 March, 1896*. Zoological Society of London. Waterlow and Sons. Ltd, Printers, London.

ZSL (2015). *ZSL's Annual Report and Accounts 2015*. Zoological Society of London, London. www.zsl.org/sites/default/files/media/2016-06/ZSL%20-%20Annual%20Report%20and% 20Accounts%202015.pdf

ZSL (2018). *Zoological Society of London's Annual Report and Accounts 2017–18*. Zoological Society of London, London.

ZSL (2019). *ZSL Annual Report and Accounts 2018–19*. Zoological Society of London, London.

ZSL (2021a). *ZSL Annual Report and Accounts 2020–21*. Zoological Society of London, London.

ZSL (2021b). HEAT-seeking. www.zsl.org/blogs/conservation/heat-seeking (accessed 25 February 2021).

ZSL (2022). *ZSL London Zoo Stocklist 2022*. Zoological Society of London, London. www.zsl .org/about-us/list-of-animals-and-animal-inventory (accessed 24 March 2022).

Zuolo, F. (2016). Dignity and animals: does it make sense to apply the concept of dignity to all sentient beings? *Ethical Theory and Moral Practice*, 19, 1117–1130.

Zwartepoorte, H. (2015). Captive breeding the Critically Endangered Egyptian tortoise *Testudo kleinmanni* Lortet, 1883, for an in situ recovery project in Egypt. *International Zoo Yearbook*, 49, 42–48.

Zwiefelhofer, E. M., Mastromonaco, G. F., Gonzalez-Marin, C., Zwiefelhofer, M. L. and Adams, G. P. (2021). 14 pregnancies produced after fixed-time artificial insemination using sex-sorted sperm in wood bison. *Reproduction, Fertility and Development*, 34, 241.

Subject Index

All zoos, aquariums and other animal facilities mentioned in the text are listed under 'zoos (list of)'. Page numbers in italics refer to figures or tables.

Animal Species Index

The scientific names of taxa listed here are those used by the authors of studies quoted in this work, and not necessarily the most recent names by which they were known at the time of its publication. In older publications authors may not have followed the modern convention of only capitalising the first letter of the name of a genus. These 'errors' have been retained wherever text has been directly quoted. Alternative scientific names are included in square brackets.